Biochemistry and Physiology of Plant Hormones

Second Edition

Thomas C. Moore

Biochemistry and Physiology
of Plant Hormones

Second Edition

With 177 Figures

Springer-Verlag
New York Heidelberg Berlin
London Paris Tokyo Hong Kong

Thomas C. Moore
Professor of Botany
Department of Botany and Plant Pathology
Oregon State University
Corvallis, Oregon   97331-2902
USA

Library of Congress Cataloging-in-Publication Data
Moore, Thomas C.
  Biochemistry and physiology of plant hormones.
  Includes bibliographical references.
  1. Plant hormones.   I. Title.
QK898.H67M66   1989   581.19′27   89-19737
ISBN 0-387-96984-5 (alk. paper)

Printed on acid-free paper

Typeset by Technical Typesetting Inc., Baltimore, Maryland.
Printed and bound by Edwards Brothers, Inc., Ann Arbor, Michigan.
Printed in the United States of America.

9 8 7 6 5 4 3 2 1

ISBN 0-387-96984-5 Springer-Verlag New York Berlin Heidelberg
ISBN 3-540-96984-5 Springer-Verlag Berlin Heidelberg New York

Preface to the First Edition

Biochemistry and Physiology of Plant Hormones is intended primarily as a textbook or major reference for a one-term intermediate-level or advanced course dealing with hormonal regulation of growth and development of seed plants for students majoring in biology, botany, and applied botany fields such as agronomy, forestry, and horticulture. Additionally, it should be useful to others who wish to become familiar with the topic in relation to their principal student or professional interests in related fields. It is assumed that readers will have a background in fundamental biology, plant physiology, and biochemistry.

The dominant objective of *Biochemistry and Physiology of Plant Hormones* is to summarize, in a reasonably balanced and comprehensive way, the current state of our fundamental knowledge regarding the major kinds of hormones and the phytochrome pigment system. Written primarily for students rather than researchers, the book is purposely brief. Biochemical aspects have been given priority intentionally, somewhat at the expense of physiological considerations. There are extensive citations of the literature—both old and recent—but, it is hoped, not so much documentation as to make the book difficult to read. The specific choices of publications to cite and illustrations to present were made for different reasons, often to illustrate historical development, sometimes to illustrate ideas that later proved invalid, occasionally to exemplify conflicting hypotheses, and most often to illustrate the current state of our knowledge about hormonal phenomena. The lists of references at the ends of the chapters, containing some references which are cited and others that are not, are not intended as comprehensive bibliographies of the most recent, or even exclusively the most important, publications on each subject. Each list is intended both to document the text and provide other examples of the extensive literature on each topic.

An explanation should be given for inclusion of the subject matter comprising Chapter 1, since it is acknowledged that many readers will regard Chapter 1 as quite elementary information with which they already are familiar. That is fully to be expected. But for those readers whose background may be deficient, as has been found to be true of a fair percentage of students, Chapter 1 will provide a reasonable overall introduction to and perspective about growth and development of whole plants throughout ontogeny and set the stage for consideration of hormonal regulation.

Books such as this invariably disappoint some readers, which is to say that they cannot be—perhaps should not even purport to be—all things that all readers might wish or expect. In my judgment, *Biochemistry and Physiology of Plant Hormones* most likely might disappoint some readers in each of two ways. First, the book does not contain as lengthy and integrated a discussion either of the physiological roles of the different kinds of hormones or of hormonal interactions as some readers will wish, although, of course, these topics definitely are covered. To the extent that this is true it is by design. For in my ten years of experience teaching a graduate course in hormonal regulation of plant growth and development, I personally have found that it is more effective to guide students from an information base such as this book provides to a more integrated understanding of regulation of growth and development than to undertake the converse approach. Another way this book might disappoint some prospective users is that it lacks detailed and comprehensive coverage of practical uses of synthetic plant growth regulators, except for synthetic auxins and auxin-type herbicides. Such information is largely beyond the scope of this small volume. Moreover, practical uses of plant growth regulators are covered in many specialized books in agronomy, forestry, and horticulture.

It seems to be a good time in some ways, and not so good a time in other ways, for a new book on the biochemistry and physiology of plant hormones. On the negative side, so far during the decade of the 1970s there seems to be a relative lull in the field as regards dramatic new developments—the "acid growth theory" and other important advancements notwithstanding—compared, let us say, to either of the previous two decades. In view of the relative scarcity of "big news," it could be argued that it is not a particularly good time. But, on the other hand, there is really good and highly significant research going on, and there is a steady output of important new knowledge. The literature—the state of the science—probably is in the best shape ever as far as unequivocal validation of facts and concepts is concerned. It is a time of separation of fact from fiction and devising new approaches to old problems, as well as asking new, important, exciting questions. For these reasons, it seems, therefore, timely for a new book to call attention to this healthy state of the science. In any case, it is an excellent time to be a student at any level of the fascinating subject discussed in *Biochemistry and Physiology of Plant Hormones*.

Acknowledgments

The real credit for *Biochemistry and Physiology of Plant Hormones* ultimately should go to the many Plant Physiologists whose research during the last half century disclosed information comprising the book. While too numerous to mention individually, the names of many of these scientists are contained in the literature lists at the ends of the chapters. Certain specific contributions by particular authors, of course, are acknowledged also in the forms of citations in the text and notations in legends to figures and tables.

The actual writing and production of the book naturally has involved several forms of assistance by many persons, to all of whom I express my sincere gratitude. For directly supplying or assisting to make available certain illustrations, I thank Douglas O. Adams, James D. Anderson, Gerard W. M. Barendse, Michael L. Evans, Peter Hedden, Hans Kende, Anton Lang, A. Carl Leopold, Morris Lieberman, Bernard O. Phinney, Folke Skoog, Nobutaka Takahashi and Jan A. D. Zeevaart. Donald J. Armstrong and Ralph S. Quatrano are thanked for the advice and technical assistance that they provided regarding various topics. I thank Ellen Witt and Leona Nicholson for typing and clerical assistance, and E. Kay Fernald for photographic service. Mark Licker, Science Editor, and Judi Allen, Production Editor, at Springer-Verlag Inc., New York, and their staff were very helpful throughout the review and production processes. Finally, I wish to acknowledge the financial support provided by the National Science Foundation for those of my own investigations during the past fifteen years which are cited in the book.

Corvallis, Oregon   Thomas C. Moore
January 1979

Preface to the Second Edition

Ten years ago, when writing the Preface to the first edition of *Biochemistry and Physiology of Plant Hormones*, I observed that, "The literature (on plant hormones)—the state of the science—probably is in the best shape ever as far as unequivocal validation of facts and concepts is concerned," and further that, "It is a time of separation of fact from fiction and devising new approaches to old problems, as well as asking new, important, exciting questions." Assuredly, those observations are as valid in 1989 as they were in 1979.

Progress during the last decade has not occurred by dramatic major discoveries, but rather by a careful, methodical, and definitive slow pace. The increasingly widespread application of a large number of techniques of plant molecular biology in hormone investigations has enabled many impressive advances, such as (1) elucidation of the nucleotide sequences of phytochrome cDNA clones and deduction of the amino acid sequences by Peter H. Quail and others; (2) isolation and partial characterization of hormone receptors by Michael A. Venis and others; and (3) identification of early gene products resulting from the action of hormones on chromatin by Gretchen Hagen, Tuan-hua David Ho, and others. We seem to be very close to understanding in precise molecular terms the early events in the mechanisms of action of the auxins, gibberellins, and abscisic acid in particular.

Because of the efforts of Anthony J. Trewavas and others, there is a growing appreciation of the concept that plant responses are not correlated only with the amounts or concentrations of endogenous growth substances but as well to the state of sensitivity of the tissues to those substances. The increasing use of mutants in hormone research by Bernard O. Phinney, James B. Reid, and others is enabling an ever closer connection between genes and the hormones they control. Through the efforts of N. Bhushan Mandava, Nobutaka Takahashi, Takao Yokota, and others we are gaining increased appreciation of the brassinosteroids, which, in the opinion of many plant physiologists, deserve to be treated as a new class of plant growth substances. And problems of long standing continue to receive attention. Kenneth V. Thimann and an associate only recently reported new evidence concerning the role of ethylene in apical dominance, for example.

As exciting and important as the emerging discoveries of plant molecular biology are, the challenge to understand hormonal regulation at the organismal level is ever present. We still seem a long way from understanding the role of hormones in phenomena such as photoperiodic induction of flowering, endogenous circadian and other rhythms, gravitropism, and senescence.

The changes made in the second edition of *Biochemistry and Physiology of Plant Hormones* are substantial. Part of the former Chapter 1 has been deleted, as that subject matter has become common knowledge; a new chapter on brassinosteroids has been added, and every other chapter has been revised and updated. The use of references continues to be as it was described in the Preface to the first edition, in that I have not attempted to provide comprehensive lists of references on each subject. The chapters are not comprehensive reviews but rather guides for students to what I believe to be the most relevant and important subject matter on each topic.

Acknowledgments

Many persons have contributed in many ways to the production of the second edition of *Biochemistry and Physiology of Plant Hormones*, and my sincere gratitude is extended to them all. In particular I wish to thank Donald J. Armstrong, Derek J. Baisted, Ronald C. Coolbaugh, Tuan-hua David Ho, Terri L. Lomax, N. Bhushan Mandava, Noboru Murofushi, B. W. Poovaiah, Peter H. Quail, Nobutaka Takahashi, and Takao Yokota. I thank Ellen Witt for her typing of the manuscript and clerical assistance.

Corvallis, Oregon   Thomas C. Moore
January 1989

Contents

CHAPTER 4   CYTOKININS   158

CHAPTER 5   ABSCISIC ACID AND RELATED COMPOUNDS   196

CHAPTER 6   ETHYLENE   228

CHAPTER 1

Introduction

Fundamental Terms and Concepts

The term "development," as it applies to whole seed plants arising by sexual reproduction, denotes the gradual and progressive changes in size, structure, and function that collectively comprise the transformation of a zygote into a mature, reproductive plant. It is also a correct and common practice to speak of the development of particular organs from initials or primordia, and to refer to development of a whole plant from any single cell. Whatever the specific case, development is a gradual process that takes time to be fully realized, generally is accompanied by increases in size and weight, involves the appearance of new structures and functions and the loss of former ones, is characterized by temporal and spatial discontinuities and changes in rate, and eventually slows down or ceases when mature dimensions are reached.

Unfortunately, but perhaps not surprisingly, there is no rigorously standardized terminology applied to the phenomena of plant growth and development. Some physiologists, the author included, consider that there are three interrelated processes that together comprise development, namely, "growth," "cellular differentiation," and "morphogenesis." "Growth" is defined as an irreversible increase in size that is commonly, but not necessarily (e.g., the growth of an etiolated seedling), accompanied by an increase in dry weight and in the amount of protoplasm. Alternatively, it may be viewed as an increase in volume or in length of a plant or plant part. In any case, it must be emphasized that growth can occur only by an increase in volume of the individual cells. Some authors consider cell division as a separate process that accompanies growth in meristems, but a more generally held view is that growth includes cell division as well as cell enlargement.

"Cellular differentiation" is the transformation of apparently genetically identical cells of common derivation from a zygote or other single cell into diversified cells with various biochemical, physiological, and structural specializations. It is the sum of the processes by which specific metabolic competences are acquired or lost and distinguish daughter cells from each other or from the progenitor cell.

"Morphogenesis" is the integration and coordination of growth and differentiative events occurring at the cellular level and is the process which accounts for the origin of morphological characters and gross form.

Other authors have used the terminology somewhat differently. For example, E. W. Sinnot (1960) in a book entitled *Plant Morphogenesis* wrote, "The process of organic *development*, in which are posed the chief problems for the science of morphogenesis, occurs in the great majority of cases as an accompaniment of the process of growth. The association between these two activities (growth and development) is not an invariable one, for there are a few organisms in which growth is completed before development and differentiation are finished, but far more commonly the form and structure of a living thing change while it grows.'' One example of an exception is the development of the female gametophyte from an 8-nucleate stage in embryo sac development in angiosperms.

Some have employed the term ''morphogenesis'' in a strictly descriptive sense, essentially as synonymous with classical developmental morphology. More generally and properly, however, it includes, besides descriptive facts as to the origin of form, a study of the results of experimentally controlled development and an analysis of the effects of factors, external and internal, that determine how the development of form proceeds. In other words, it attempts to get at the underlying formativeness in the development of organisms and especially to reach an understanding of the basic fact of which form is the most obvious manifestation, namely, biological organization itself. According to morphogeneticists like Sinnott, ''The organism may thus be said to make the cells rather than the cells to make the organism.''

F. B. Salisbury and Cleon Ross (1985) used the term ''development'' (or ''morphogenesis'') as an inclusive term and regard the phenomenon as consisting of two primary functions: growth and differentiation. They consider growth primarily as an increase in size, and differentiation as the process by which cells become specialized. P. F. Wareing and I. D. J. Phillips (1981) likewise adopted the view that development should be applied in its broadest sense to the whole series of changes that an organism goes through during its life cycle, while noting that it may also be applied to individual organs, to tissues, or even to cells. According to these authors, ''plant development'' involves both ''growth'' and ''differentiation.'' ''Growth'' is used to denote quantitative changes occurring during development and is defined as an irreversible change in the size of a cell, organ, or whole plant. ''Differentiation'' is applied to qualitative changes. Thus, in their view growth and differentiation are the two major developmental processes. They reserve the term ''morphogenesis'' as one used by experimental morphologists (morphogeneticists) to denote origin of form. F. C. Steward (1968) employed a quite inclusive connotation of growth. Essentially he used the term ''growth'' in a very general way to include what others consider more explicitly to be growth, differentiation, and morphogenesis.

With the variable usage of terminology, as has been illustrated, perhaps it is understandable why the coupled terms ''growth and development'' are used so prevalently. This couplet connotes the kind of concept that James Bonner and A. W. Galston (1952) had when they wrote. ''The changing shape, form, degree of differentiation, and state of complexity of the organization constitute the process of development.'' They viewed ''growth'' as a quantitative matter concerned with the increasing amount of the organism. On the other hand, ''development,'' in their view, refers to changes in the nature of the growth made by the organism.

Many biologists have emphasized the biological importance of organization and emphasize that the characteristics of life itself are characteristics of a system arising from, and associated with, the organization of materials and processes. It is of utmost importance to keep in mind the fact that there are unique emergent qualities associated with each successively higher level of biological organization—from molecular and subcellular to the levels of cells, tissues, organs, whole organisms, and beyond. In no instance is the cliché that the whole is more than the sum of its parts more vividly exemplified than when we observe the complicated changes that a seed plant manifests during the repeating cycle of development. We can arbitrarily conceive of this cycle as starting with the germination of a seed and continuing with the passage of a juvenile phase of growth and the graduation into maturity. With maturity the organism is capable of shifting from vegetative to reproductive development, with the development of flowers, the development of fruits, and the production again of a new generation of seed. Ultimately, the development of the individual plant ends with senescence and death. This book is concerned with the processes involved in and the mechanisms which control the growth and development of seed plants.

Patterns and Kinetics of Growth in Cells, Tissues, Organs, and Whole Plants

The curve which typically describes the changing size of a growing organism, organ, tissue, cell population, or individual cell is sigmoid in shape (Fig. 1.1a). The sigmoid growth curve can, for convenience, be considered in three parts. First, there is an accelerating phase in which growth starts slowly and gathers momentum. During this period of constantly accelerating growth rate, increase in size is exponential. When a plant is growing exponentially, it is said to be in the logarithmic phase of its growth. Second, there is either a point of inflection or a phase, more or less protracted, in which the course of growth is approximately linear with time. That is, equal increments of growth tend to occur in equal intervals of time. This is often termed the linear phase of growth. In some cases, however, there is no linear phase but merely a point of inflection in the curve, when the first rising rate of growth gives way to a decreasing rate of growth. Third, there is a phase of declining growth rate until, in fact, growth subsides and the organism may maintain only the size it has already achieved. In annual plants this is followed by senescence and death. In the case of woody perennials, each period of shoot growth is approximately sigmoidal and is followed by dormancy. Shoots of biennials, of course, exhibit two seasons of sigmoidal growth. Instead of plotting merely size versus time (Fig. 1.1a), it is often useful to plot the $\log_e$ of size versus time (Fig. 1.1b) or growth rate versus time (Fig. 1.1c) when analyzing growth. Examples of various specific growth curves and a brief discussion of each follow.

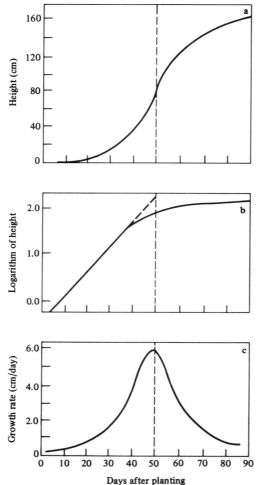

FIGURE 1.1. Generalized growth curves. **a** Size-versus-time sigmoid growth curve; **b** logarithmic growth curve; and **c** growth rate curve. (Based on data for corn, *Zea mays*, in Whaley, 1961.)

Even the growth of single cells has been demonstrated to be sigmoidal (Fig. 1.2). Organs such as leaves follow a sigmoidal pattern also on the basis of several parameters, including fresh weight, area of lamina, length of lamina, and cell number (Fig. 1.3).

Many fruits exhibit common sigmoid growth curves—e.g., apple (Fig. 1.4), pineapple, strawberry, pea, and tomato. However, fruits of many species exhibit more complicated growth curves, which are essentially double sigmoid growth curves. This type of curve is common to probably all the stone fruits such as cherry (Fig. 1.5), apricot, plum, and peach. Such curves are also typical of some nonstone fruits such as fig, grape, and currant. However, neither type of growth curve seems to be distinctive for a particular morphological type of fruit because there are berries, pomes, and simple and accessory fruits that manifest each.

FIGURE 1.2. Growth curves of single cells in the corn (*Zea mays*) root tip. **a** Fresh weight; **b** dry weight. (Redrawn, with permission, from Boss, 1964.)

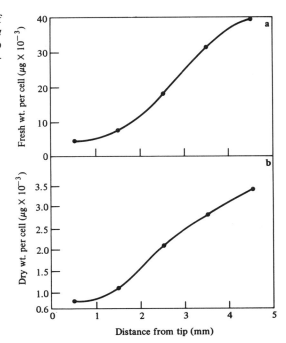

The double sigmoid growth curve can be explained on the basis of asynchrony of growth of the different parts of the fruit. Two general types of growth centers occur in fleshy fruits, the pericarp and ovule(s). The growth of the pericarp commonly accounts for the initial enlargement. Growth in later stages generally is associated with seed development.

There is wide variation in the extent to which cell division participates in the growth of fruits, ranging from cases in which cell division (except in the developing seeds) has been completed at the time of pollination (*Ribes, Rubus*), to cases in which there is a brief period of cell division just following pollination (tomato, *Citrus*, cucurbits, apples, *Prunus*), or rather extended periods of cell division (strawberry).

The curve describing the kinetics of growth of whole plants also is typically sigmoidal. Herbaceous annual plants, such as garden pea (*Pisum sativum* L.), exhibit a single sigmoid growth curve throughout ontogeny from germination to senescence and death (Fig. 1.6). Size versus time plots of the shoot growth of woody perennials are sigmoidal throughout ontogeny (Fig. 1.7), although there is a gradual progression toward the asymptote. Notably, however, the size versus time plot for shoot growth of woody plants is composed of increments of growth, each sigmodal, which alternate with the periods of dormancy. This is true of both primary shoot growth (Fig. 1.8) and secondary shoot growth (Fig. 1.9).

Growth and rate curves for different parameters in woody plants reveal other

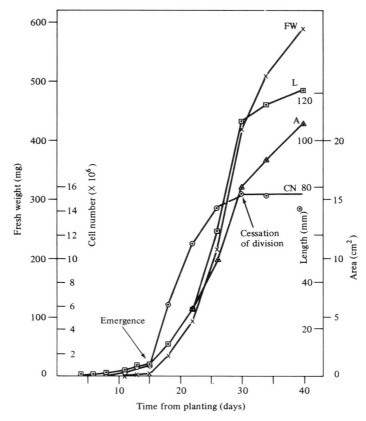

FIGURE 1.3. Growth curves for leaves of sunflower (*Helianthus annuus*). FW, fresh weight; A, leaf lamina area; L, length; CN, number of cells. (Redrawn, with permission, from Sunderland, 1960.)

noteworthy features. For one thing, growth is not synchronous for all parts of the plant (Fig. 1.10). Primary shoot growth, or shoot extension, is initiated first by bud break. Production of new leaves, a primary growth process, follows bud break closely. Secondary shoot growth, or increase in diameter, generally follows initiation of primary shoot growth and, in fact, is activated by a hormonal stimulus produced by the growing buds. Root growth of trees tends to be sporadic and to occur whenever physical and chemical environmental factors in the soil and the nutritional status of the shoot permit (Fig. 1.10), true root dormancy apparently being of rare occurrence.

Diurnal variations or periodicity in growth parameters are of common occurrence. Table 1.1 contains data revealing a marked diurnal periodicity in primary shoot growth in two species of *Pinus*.

Seasonal periodicity of shoot growth in trees is well known in temperate regions, indeed at all latitudes where there are distinct seasons. Shoot growth commonly is restricted to a single growing season, and growth alternates with

FIGURE 1.4. Growth curves for apple fruits. **a** Cell number versus fruit weight; **b** volume and weight versus time after pollination. (Redrawn from data of Bain and Robertson, 1951, in: *Plant Growth and Development*, 2nd ed., by A. C. Leopold and P. E. Kriedemann. Copyright © 1975 by McGraw-Hill Book Company. Used with permission of McGraw-Hill Book Company.)

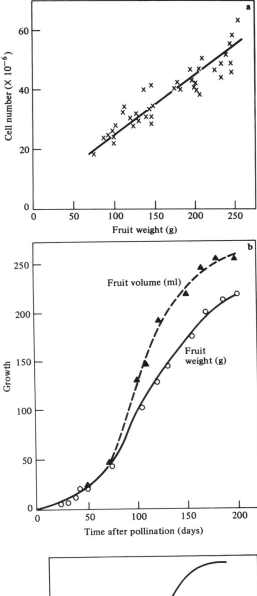

FIGURE 1.5. Generalized double sigmoid growth curve of cherry. (Redrawn from: *Plant Growth and Development*, 2nd ed., by A. C. Leopold and P. E. Kriedemann. Copyright © 1975 by McGraw-Hill Book Company. Used with permission of McGraw-Hill Book Company.)

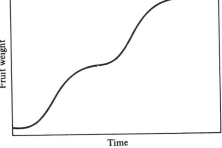

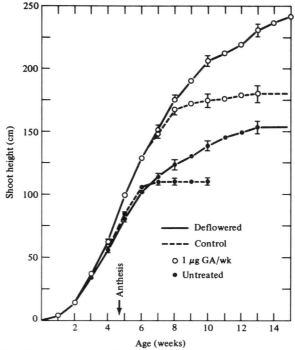

FIGURE 1.6. Growth curve of a whole hebaceous dicot, pea (*Pisum sativum*). Basic sigmoidicity of the curve is unaffected by removal of flowers or treatment with GA or both. (Redrawn, with permission, from Ecklund and Moore, 1968.)

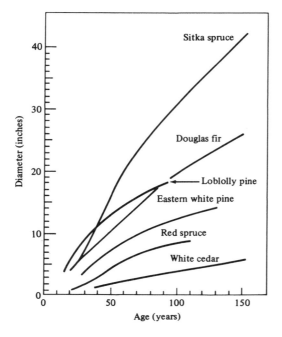

FIGURE 1.7. Cumulative secondary stem growth (diameter increase) in several conifers versus age. (Redrawn, with permission, from data of Baker, 1950, in Kramer and Kozlowski, 1960.)

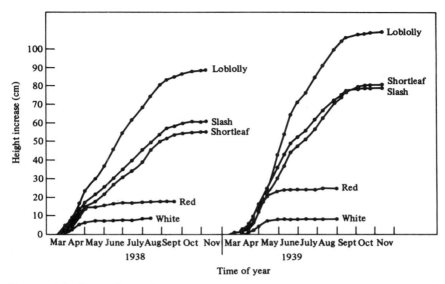

FIGURE 1.8. Seasonal growth curves (primary shoot or height growth) for several pines for each of two years. (Redrawn, with permission, from data of Kramer, 1943, in Kramer and Kozlowski, 1960.)

periods of dormancy or rest. In these plants, both the resumption and cessation of shoot growth generally can be correlated with particular physical environmental factors.

However, even in tropical environments trees generally exhibit striking and most intriguing periodicity of shoot growth. An individual shoot exhibits alternate phases of growth and dormancy. During each period of growth new

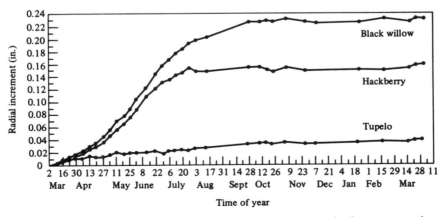

FIGURE 1.9. Seasonal secondary stem growth (radial expansion) in three tree species. (Redrawn, with permission, from data of Eggler, 1955, in Kramer and Kozlowski, 1960.)

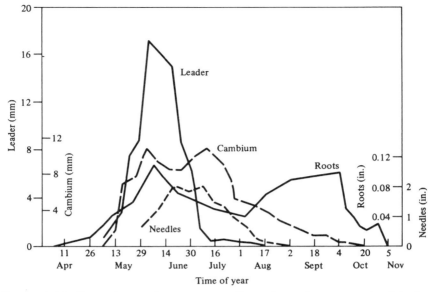

FIGURE 1.10. Seasonal growth rate curves, illustrating primary growth of leader, secondary growth of stem, primary growth of roots, and growth of new needles. Data are for 10-year-old white pine in southern New Hampshire. (Redrawn, with permission, from data of Kienholz, 1934, in Kramer and Kozlowski, 1960.)

leaves are formed and the young internodes elongate. The terminal bud then enters a state of dormancy until the next growth period begins. Both growing and dormant shoots occur on a single plant, and several "flushes" of shoot growth commonly are exhibited in the course of 1 year.

In tropical regions, where photoperiod is near constant and temperature and other physical environmental factors vary comparatively little, it appears that growth periodicity is controlled by internal factors that are independent of the physical environment. In other words, the controlling factors of the periodicity of shoot growth in the tropical species are endogenous.

A particularly interesting investigation of endogenous shoot-growth rhythm in tropical species was conducted by Greathouse et al. (1971) with shoot cuttings of

TABLE 1.1. Diurnal periodicity of primary shoot growth in two species in Pinus.[a]

	Average day growth (inches)	Average night growth (inches)
Loblolly pine	1.64	3.21
Shortleaf pine	1.27	2.13

[a] Data of Reed (1939) in Kramer and Kozlowski (1960) by permission.

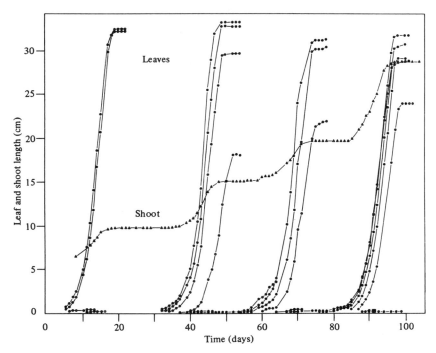

FIGURE 1.11. Endogenous growth rhythm in the shoot of a tropical tree, *Theobroma cacao*, grown under controlled environmental conditions, expressed as elongation of the shoot and its leaves. The period of shoot-growth rhythm was 26.2 ± 1.1 days measured from the beginning of one dormant period to the beginning of the next. (With permission, from Greathouse et al., 1971.)

Theobroma cacao. Shoot cuttings were placed in a growth chamber under constant conditions of a 12-hour photoperiod at a light intensity of 600 fc, a temperature of 29 ± 1°C, and a relative humidity of 70 to 100%. Growth measurements were initiated after the plants had been in the chamber for 4 months, and were repeated at 1- and 2-day intervals for about 100 days. The results were most interesting (Fig. 1.11). There was a remarkable rhythmicity to shoot elongation and leaf production, with a period of 26.6 ± 1.1 days measured from the beginning of one dormant period to the beginning of the next. Furthermore, the plants were not synchronous with regard to shoot growth. They were completely out of phase, as are field-grown *Theobroma cacao* trees in Trinidad. That is to say, the branches were out of phase with each other so far as growth and dormancy were concerned. Dissection of shoot tips revealed further that the total number of leaves and leaf primordia in the shoot apex remained constant during the dormant period and did not increase until the onset of growth. This indicated that activity of the apical meristem, as well as leaf growth and shoot elongation, was rhythmic.

Discontinuities in Growth, Growth Periodicities, and Problems of Relative Growth Rate

Growth does not normally proceed at a steady rate, even under uniform environmental conditions, but it involves certain discontinuities, which sometimes vary sporadically with environmental conditions but frequently are of a cyclical, rhythmic, or periodic nature. Some examples of discontinuities in growth and development will now be cited. The discontinuity in growth by day and by night, as shown by the effects of day length (or night length) and by the contrasted effects of night and day temperature, are evidence that any diurnal cycle is the summation of different effects of day and of night. These is no essential continuity between growth in length and in girth of a plant, and as the growth of a new lateral organ (a bud or a lateral branch, a lateral root, new tillers, adventitious roots) is initiated, new centers are established that become, in large measure, independent. The formation of floral organs, the occurrence of meiosis, and the onset of a sexual or gametophytic phase of growth, syngamy, and later seed development are all stages in development that effectively interrupt the smooth time course of growth and effectively destroy any illusion that it can be uniform in time. The occurrence of storage organs and the phenomena of quiescence and dormancy also show that growth does not bear a uniform, or consistent relation to time, and therefore, any analysis of growth that assumes this must be to this extent inadequate.

There are also problems of relative growth that should be considered when we analyze the growth of whole plants. The concept of relative growth recognizes the unequal growth rates of different parts in one plant body. Examples of different growth rates among the parts of single organs and among different organs of a single plant abound: (1) individual leaflets of a palmately lobed leaf such as that of *Aesculus* (horsechesnut) obviously grow in area and in length at different rates; (2) the shape of dicot leaf lamina such as those of tobacco is determined by the differential growth of the leaf in different parts of the lamina (blade); (3) in particular types of plants, the exponential growth of leaves is arrested with the advancing season, and by a photoperiodic or other stimulus, an underground tuber or storage root develops.

Examples of "growth periodicities" also are abundant. That is to say, development cycles are characterized not only by discontinuities but by numerous discontinuities which are cyclical, rhythmic, or periodic. Many plants exhibit a diurnal periodicity in shoot growth, with more shoot elongation occurring at night than during the day (e.g., Table 1.1). During unusually cold nights, however, growth at night may be less than during the day. This behavior was investigated for a number of kinds of plants, most notably tomato in which stem elongation occurs only at night, by F. W. Went some 35 years ago. Noting that many species of plants, particularly those native to temperate regions in which diurnal temperature fluctuations are a characteristic feature of the environment, grow best in an appropriately fluctuating diurnal temperature cycle, Went coined the term "thermoperiodicity" or "thermoperiodism." Tropical plants, which grow normally in regions of little diurnal temperature change, frequently do not exhibit thermoperiodicity but grow best at constant temperatures.

As we have already seen, in perennial plants quiesence and dormancy are conspicuous in temperate latitudes and these phenomena exemplify "seasonal periodicity" in growth and development. Even in tropical climates cycles exist, often in the absence of any apparent correlation with physical environmental stimuli (e.g., Fig. 1.11). The whole range of photoperiodic responses in plant development exemplify periodicities—onset of flowering, onset of dormancy, formation of such organs of perennation as tubers, bulbs, corms, etc. R. C. Friesner showed, as early as 1920, that frequency of mitoses in the roots of pea and corn seedlings showed a diurnal cycle, with more mitoses occurring at night than in the daytime. This rhythm in mitosis and cell division in a root tip is already induced in the seedlings at germination, and, once established, it becomes relatively independent of light and temperature, and tends to continue.

When time is an obvious parameter it is as though organisms have built-in biological clocks. The German physiologist E. Bünning was the first to extensively characterize what we term endogenous circadian rhythms in plants, and they are known for virtually all kinds of organisms (plant and animal) which have been investigated. Circadian rhythm means that the phenomenon recurs at intervals that comprise an approximate 24-hour cycle (e.g., Fig. 1.12), that is, it

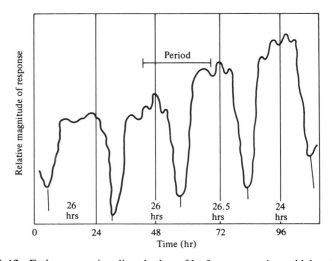

FIGURE 1.12. Endogenous circadian rhythm of leaf movement in cocklebur (*Xanthium strumarium*). A plant that was previously entrained to a day–night regimen was placed in continuous light, and the tip of one leaf was monitored by time-lapse photography. Measurements were made of the position of the leaf tip above and below a baseline on the photographs. The data then were plotted. Thus high points on the graph indicate "day" positions of the leaf tip, and low points indicate "night" positions. The endogenous circadian rhythm persisted even in continuous light throughout the length of the experiment. (Adapted from *Plant Physiology*, 3rd ed., by Frank B. Salisbury and Cleon W. Ross, © 1985 by Wadsworth Publishing Company, Inc., Belmont, California. By permission of the publisher.)

displays maxima and minima or presence and absence on a cycle of about a 24-hour period. While the nature of the timing mechanism is not known, endogenous circadian rhythms are fascinating and profoundly influence numerous plant responses to environmental stimuli and numerous developmental events, e.g., flowering. A remarkable feature about endogeneous circadian rhythms is that the length of the period is only slightly influenced by wide variations in temperature, as between 5 and 30°C, even though the rates of metabolic processes on which these rhythms presumably depend are very much slower at lower temperatures. This type of rhythm obviously has great biological value. Endogenous circadian rhythms enable plants to judge the lengths of the day and night and hence to time reproduction, dormancy, leaf fall, and other activities in relation to the march of the seasons despite erratic changes in temperature in the environment. More is said about endogeneous circadian rhythms in Chapter 8.

Mechanisms Controlling Cellular Differentiation

Genetic Basis of Development

That every organism is the product of the interaction of its genetic material and its environment is one of the most fundamental of all biological principles. The genetic material (DNA) determines the potentialities for development. The environment determines the degree and kind of the genetic potential that is actually manifested. Or, to put it a slightly different way, the form and function of plants are dependent on their genetic makeup and the ability of the genes to express themselves through regulation of protein synthesis. The environment influences the growth, differentiation, and metabolism of a plant by controlling, through effects on both protein synthesis and enzyme activity, expression of its genetic makeup.

Nature of Gene Action

Since the impressive, Nobel prize-winning research of G. W. Beadle and E. L. Tatum in the early 1950s on the mechanism of gene action in *Neurospora* that culminated in the one gene–one enzyme theory, we have understood that the fundamental action of a gene is to control the synthesis of an enzyme or structural protein. The structure of each polypeptide chain is governed by an independent gene. In the case of several enzymes, known to be made up of two or more polypeptide chains, it is apparent that the structure of each polypeptide chain is governed by a separate independent gene. Furthermore, we now know that there are functionally different types of genes, and that much of the DNA of a eukaryotic organism acts to control the ultimate expression of the structural genes, which contain the genetic code for specific proteins.

Evidence for Totipotency

There is abundant evidence that at least some and probably all of the kinds of nucleated specialized cells of the mature multicellular organism are in principle "totipotent." A cell that is totipotent contains all the genetic information that was present in the fertilized egg from which it was derived and possesses the capacity to regenerate the adult organism, that is, to behave like a zygote. Hence the term "totipotency" means "total genetic potential."

The intriguing enigma is that all the living cells of the higher plant body potentially have all the attributes of the organism even though, during development, they do not individually display them all. Hence our problem is to try and identify briefly some of the means by which variations from cell to cell of the same individual may have a genetic basis. Current working theory holds that the various types of differentiated cells differ from one another most basically and importantly in the kinds of enzymes and other proteins that they contain and that differentiation therefore is brought about by mechanisms that control the production of the kinds of proteins that each cell contains during its life.

Let us examine initially some of the evidence for totipotency of angiosperm cells. Professor F. C. Steward and associates of the Laboratory for Cell Physiology, Growth and Development at Cornell University at Ithaca, New York, did some particularly impressive research illustrating the totipotency of higher plant cells. In a series of papers dating from about 1958, Steward et al. first showed that cells derived from the secondary phloem region of mature tap roots of carrot (*Daucus carota*) grow and divide freely when suspended in a liquid medium containing coconut milk (liquid endosperm from coconut seeds or fruits) or liquid endosperm from other sources, and that they do this in various ways that are different from those that obtain in the intact plant body. The freely suspended cells, nourished by coconut milk and a basal medium, dedifferentiate (revert to a meristematic condition) and form loose, randomly growing cultures. It is important to note that if tissue is merely excised and started en masse in a stationary culture on semisolid medium, only callus would form. The loose, randomly growing cultures quickly organize, however, producing first roots, then shoots, to form complete plantlets (Fig.1.13).

In giving rise to plantlets, a multicellular nodule or growing center is often organized within the proliferated cell culture, which, in its behavior, replaces the zygote or proembryo, since it gives rise to a minute embryo-like structure that develops entirely from vegetative cells without the stimulus of fertilization. Stages in this growth from free cells strikingly resemble the normal growth of the carrot embryo (Fig.1.14).

The most important requirement for the cells to grow like embryos instead of merely proliferating to form callus is to start that development from free cells. These early investigations stressed the inherent totipotency of cells of the carrot root. They also revealed certain essential conditions for the expression of the inherent totipotency of the cells of the carrot plant. The essential conditions are basically two. First, the cells of an explant should first be set free from each other and thus be able to grow independently, so that their behavior is unrestricted by the controls that operate in a preformed tissue mass that had developed within the plant body. This is accomplished by growing tissue explants in rotating liquid cultures. Second, the free cells should be furnished with a medium that is fully competent to make them grow, and the best such

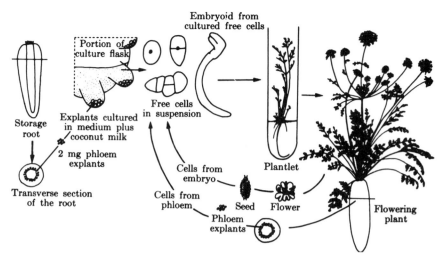

FIGURE 1.13. Diagram of cycle of growth of the carrot plant (*Daucus carota*), beginning with explants of the secondary phloem region of storage root. (From F. C. Steward, *Growth and Organization in Plants*, © 1968, Addison-Wesley, Reading, Massachusetts. Fig.10-7. Reprinted with permission.)

FIGURE 1.14. Embryo-like structures which develop from free cell suspension of carrot (*Daucus carota*). Structures **a–i** represent stages in normal embryogeny; structures **j–p** denote growth forms in the outgrowth of single cell suspension. (With permission, from Steward et al., in *Synthesis of Molecular and Cellular Structure*, edited by D. Rudnick, © 1961, The Ronald Press, New York.)

medium is still a basal medium (containing certain mineral salts, vitamins, and sugars) supplemented by reduced nitrogen compounds (casein hydrolysate) and the liquid endosperm from a developing ovule and embryo sac, such as coconut milk or the juice of immature corn grains or *Aesculus* (horsechestnut) fruits. Additional synergists are often useful, or necessary also, e.g., hexitols, auxins, and cytokinins.

Steward et al. developed two additional techniques that advanced their work. They found that they could pour a dilute free cell suspension onto the surface of a semisolid nutrient medium; in this way, the fate of single cell isolates could be followed. Second, they found that they could establish free cell suspensions from relatively young embryos. These free cell suspensions could be spread on agar plates. When this is done, these free cells of embryo origin have been shown to develop into surprisingly faithful replicas of zygotic embryos, and they do so with such frequency as to imply that virtually all these carrot cells are totipotent. The most outstanding results were obtained by F. C. Steward and associates with a cultivar of carrot (*Daucus carota*) called Queen Anne's Lace. However, this type of work with carrot cells and the demonstration of the ability to rear plants from free cells was rapidly succeeded by other examples, including endive, water parsnip (*Sium*), *Arabidopsis thaliana, Nicotiana suaveolens, Cymbidium* sp., and others.

Marked variations were witnessed in the behavior of carrot cells isolated from different tissues and organs. Early in the development of zygotes, each cell if freed would readily become an embryo and polyembryony would result. Parenthetically it may be noted that plants known to develop embryos apomictically carry this propensity. The most potent source of somatic cells that develop freely and in large numbers into plantlets is still a free cell culture propagated from cells originating from young embryos removed from ovules (immature seeds). As organogensis proceeds and tissues and organs become mature, the control or restrictions over the behavior of the totipotent cells may tighten. Therefore, it is not surprising that more drastic devices often are needed to evoke totipotent growth. Evidence of this is the needed interplay which has been observed of the factors present in coconut milk, on the one hand, and synergistically active substances on the other. Thus it is often the case that combinations of growth factors (such as coconut milk + naphthaleneacetic acid or 2,4-dichlorophenoxyacetic acid) will cause mature tissue to proliferate actively and pave the way for later growth on a basal medium.

The chromosome complement does not seem to be significant to a demonstration of totipotency. Polyploidy is commonplace in cell and tissue cultures such as those derived from diploid tissue explants by F.C. Steward and associates. In a fascinating report J.P. Nitsch and C. Nitsch (1969) of the Laboratoire de Physiologie Pluricellulaire at Gif-sur-Yvette, France, reported on the production of haploid plants from pollen grains (actually uninucleate microspores). These authors reported that they grew pollen from numerous varieties of 14 *Nicotiana* species *in vitro* on a relatively simple medium and that some of the pollen grains proliferated into embryo-like structures that developed in stages similar to those of zygotic embryos. The plantlets matured and flowered profusely but, since they were haploid, did not set seed.

The methods employed by the Nitsches really were very simple. They excised anthers from normal flowers and planted them on a nutrient medium. Plantlets emerged from the anthers and could be transplanted to individual tubes on a simplified medium. Once they had formed an adequate root system, they could be transplanted and reared to mature, flowering plants. Formation of plantlets occurred when the pollen grains (really microspores) were fully individualized, uninucleate and devoid of starch. Mature pollen grains simply produce pollen tubes but no plantlets. Auxins [indoleacetic acid (IAA) or 2,4-dichlorophenoxyacetic acid (2,4-D)] generally enhanced formation of plantlets but were not essential.

The practical implications of the types of work done by F.C. Steward, J.P. Nitsch, C. Nitsch, and others are very considerable indeed. By means of such techniques, it is theoretically possible to produce large clones of plants with identical genotypes, for example. Breeding programs can be expedited also by methods such as those employed by the Nitsches in that observation of phenotypic expression of a genotype of interest can be observed independent of crossing and the awaiting of seed set, germination, and growth of a particular generation.

Evidence for Differences in Protein Content and for Changes in Protein Content Correlated with Cellular Differentiation

Let us begin with the theory, which has been introduced previously, that cells differ from one another most basically and importantly in the types of proteins (including, of course, enzymes) that they contain and that differentiation is therefore brought about by mechanisms that control the production of proteins. We shall examine first some direct evidence for the first part of the theory, that is, that different kinds of differentiated cells differ qualitatively and quantitatively in protein content. Steward et al. (1965) performed one of the first investigations of this nature. What they sought to do was to determine whether there are detectable differences in the protein complements of different organs of seedling plants, and to ascertain also whether detectable changes occur in the protein complements of the same organs as they develop. They extracted the soluble proteins from 3-day-old pea seedlings (Fig. 1.15) with Tris-glycine buffer (0.1 M, pH 8.3), separated proteins in each extract by electrophoresis on acrylamide gels, stained the proteins, and compared the banding patterns on gels. The studies by Steward et al. revealed quite distinctive differences, as well as obvious similarities, in the protein patterns of different organs of the pea seedling and in root segments at different stages of development (Fig. 1.16). Their study corroborated the results of other investigators, some of whom have monitored changes in activities of various enzymes in the different parts of single organs, e.g., oat coleoptiles and pea roots, during their development.

Primary attention was devoted by Steward et al. to the analysis of soluble

FIGURE 1.15. Diagram of 3-day-old pea seedling showing parts from which protein extracts were prepared. 1–6, Successive millimeter sections from the root tips; 7, hypocotyl; 8, epicotyl; 9, plumule; and 10, cotyledons. Compare with Fig. 1.16. (Redrawn, with permission, from Steward et al., 1965.)

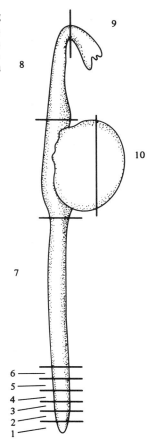

proteins in roots. As they pointed out, the growing root is a convenient experimental object for the study of the development of cells from their meristematic to mature state. Successive transverse segments, in a basipetal sequence from the root tip, comprise a series of cell populations that represent, in space, a developmental sequence in time. Cells pass from the meristematic, undifferentiated state in the root tip to specialized cell types in the more basal parts of the root. As the cells mature in this way, their protein patterns change, both qualitatively and quantitatively.

Thus, we may conclude that different types of specialized cells in a multicellular organism exhibit different protein complements in spite of the fact that all must be presumed, on the basis of other good evidence, to have the same genetic constitution and therefore the same DNA-encoded information in their chromosomes. Lest the reader be misled, however, it should be noted that, while we are now emphasizing the dissimilarities in protein complements among different kinds of cells, we should appreciate the probability that the qualitative similarities in protein content greatly exceed qualitative dissimilarities.

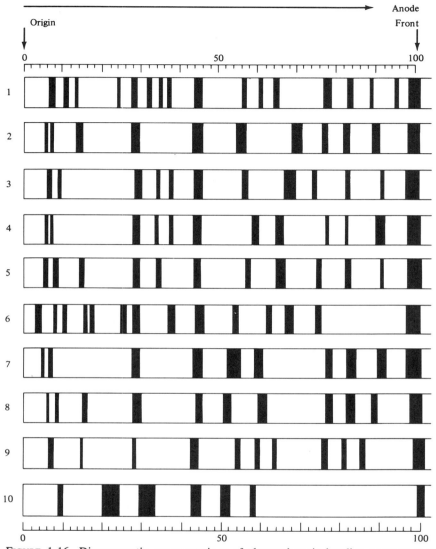

FIGURE 1.16. Diagrammatic representations of electrophoretic banding patterns seen after electrophoresis of soluble proteins from different parts of 3-day-old pea seedlings. 1–6, Successive millimeter sections from the root tip; 7, hypocotyl; 8, epicotyl; 9, plumule; and 10, cotyledons. Compare with Fig. 1.15. (Redrawn, with permission, from Steward et al., 1965.)

Viewing the evidence just presented, one could postulate, as many biologists have, that the observed differences in protein (enzyme) content are the cause of differentiation rather than its result. The problem we turn to now is: If all cells in a multicellular organism contain the genetic information necessary for the manufacture of all the different types of proteins found in individuals of that species, what is it that controls the expression of this information? Some subsidiary questions are: (1) How does gene expression result in one protein complement in one part of the plant and an entirely different complement in another part? (2) What is it that controls gene expression such that the same organ can exhibit different protein complements at different developmental stages? (3) Is there direct experimental evidence for selective switching on and off of genes during the course of cellular differentiation and morphogenesis? and (4) Is there any direct evidence for a specific differentiative event being correlated (in a cause and effect manner) with the synthesis of one or more unique proteins?

Control of Gene Expression

Without going into detail, the salient features of protein synthesis or gene action are:

$$\text{DNA} \xrightarrow{\text{Transcription}} \text{RNA} \xrightarrow{\text{Translation}} \text{Protein} \rightarrow \begin{array}{l}\text{Structure or}\\\text{Catalysis}\end{array}$$

There is abundant evidence that all genes are not active at all times, that each kind of specialized cell contains its characteristic enzymes, and that each kind of specialized cell produces only a portion of all the enzymes for which its DNA contains information. Major problems now incompletely resolved, however, are (1) the levels of control of gene expression (of protein synthesis), whether at the transcriptional level or translational level, or both, or others; and (2) the molecular mechanism(s) of control of gene action.

We shall now identify sites of control of gene expression in higher plant systems and describe briefly specific investigations which have provided examples of each.

The role of the collective genes of a eukaryotic organism (one whose cells contain definite nuclei, as opposed to prokaryotes such as the bacteria and blue-green algae) is twofold: to control other genes and to control the synthesis of specific structural proteins and enzymes. As previously noted, in the case of enzymes and other proteins that are composed of two or more polypeptide units, the structure of each polypeptide is governed by a separate structural gene.

Consideration of the complex series of events between transcription of a segment of DNA in the formation of a molecule of messenger RNA (mRNA) and the final production of a polypeptide destined to serve as all or part of a structural protein or active enzyme readily reveals that there are many potential sites of control of gene expression. Not surprisingly, it is becoming increasingly apparent that gene expression in eukaryotes does indeed occur at multiple sites, or, to put it another way, control of gene expression occurs by multiple mechanisms in eukaryotes.

Some mechanisms of control of gene expression that have been demonstrated in higher plants are

1. selective *de novo* mRNA synthesis or selective gene transcription (transcriptional control);
2. processing and translocation of mRNA (posttranscriptional control);

3. synthesis of new proteins using preformed mRNA (translational control); and
4. activation of preexisting enzymes (posttranslational control).

Another mode of control that has been demonstrated in some animal and primitive plant systems but that has not yet been shown in higher plants is selective gene replication. The clearest, best documented example of this mode of control of gene expression (i.e., synthesis of unique DNA molecules from a specific portion of the total genome) is the amplification of ribosomal RNA genes (rDNA) in oöcytes of amphibians. The extra copies of the rDNA genes are contained in nucleoli and serve as templates for the extraordinary demand for ribosomal RNA synthesis in immature oöcytes. A similar example of rRNA amplification has been reported for the slime mold *Physarum*. Whereas selective gene replication is an important mechanism, genes coding for specialized proteins characteristic of differentiated cells do not appear to be amplified, but rather are present in one or a few copies. Accumulation of gene products in the case of proteins typical of differentiated cells occurs, rather, by reiterated expression of genes (i.e., selective gene transcription).

Selective gene transcription is one major mechanism of control of gene expression. Two examples from higher plants have been particularly well documented. One is the GA-stimulated *de novo* synthesis of α-amylase and particular other hydrolases in the aleurone layers of cereal grains during early germination. The evidence concerning stimulation of *de novo* synthesis of α-amylase is the strongest support so far achieved for transcriptional control of a specific enzyme during plant development. This important phenomenon is discussed in detail in Chapter 3. Another example is the stimulation of *de novo* synthesis of cellulase in pea epicotyls by 2,4-D, a synthetic auxin. In the latter case, putative mRNA was isolated from ribosomes of pea epicotyls after treatment of the epicotyls with 2,4-D. Then the putative mRNA was translated in a cell-free enzyme system from wheat germ, and cellulase was assayed by immunoprecipitation. It was shown that 2,4-D treatment was essential for the synthesis of the mRNA for cellulase. Since the mRNA for cellulase was readily translated in wheat germ, specific factors influencing translation, which might vary among species, appeared not to be involved. Yet, since no delay occurred in mRNA production after 2,4-D treatment but a 24-hour lag period intervened between time of treatment and appearance of cellulase, some form of translational control also appeared to be implicated.

The evidence for other levels of control of gene expression in higher plants comes largely from studies conducted with germinating seeds. Indirect evidence for posttranscriptional control is derived from investigations of enzymic activities that increase during germination, and that are sensitive to inhibitors of protein synthesis but insensitive to inhibitors of RNA synthesis. For many years it has been suggested that some of the protein (enzyme) synthesis that occurs during early germination depends on mRNA already present in the dry seed and activated for translation during hydration of the seed. Conversion of monoribosomes to polyribosomes and initiation of protein synthesis in the apparent absence of transcription were early observations on which that suggestion was based. More recently there have been efforts to isolate and characterize a specific mRNA fraction from ungerminated seeds and to demonstrate its role in protein synthesis.

There is some evidence for posttranscriptional control from investigations conducted by L. Dure and associates with cotton seed. Major reserves stored in the cotyledons of a cotton seed are proteins and fats. Two enzymes which play key roles in the breakdown of these reserves are carboxypeptidase C and isocitrate lyase. The former degrades proteins, and the latter enzyme is needed in the glyoxylate cycle for utilization of the stored fat. Carboxypeptidase C and isocitrate lyase activities are not detectable in the dry cotton seed but activity appears in the cotyledons during the second day of germination. The appearance of both enzymic activities is inhibited by cycloheximide (inhibitor of cytoplasmic protein synthesis) but not by actinomycin D (inhibitor of RNA synthesis). Moreover, it has been shown that carboxypeptidase C is synthesized *de novo* from constituent amino acids. This and other evidence may indicate that the *de novo* synthesis of carboxypeptidase C depends on translation of preformed mRNA that is already present in the dry cotton seed prior to the initiation of seed germination, although this point at present is uncertain.

This brief discussion should serve to illustrate that the appearance of enzyme activities (hence gene expression) is regulated at several different levels, concomitantly in part and sequentially in part, during the development of higher plant systems. For an excellent early review on the regulation of RNA metabolism by auxins, gibberellins, and cytokinins, the article by John V. Jacobsen (1977) should be consulted. See Davies (1987) for recent treatments.

Introduction to Plant Hormones

Development obviously is a rigidly regulated process. One has but to contemplate the remarkable structural and functional unity of the plant body throughout ontogeny and the chronology of developmental events that occur in individuals of each species to recognize this fact.

The plant hormones are extremely important agents in the integration of developmental activities, and they also are concerned importantly in the response of plants to the external physical environment. Enviromental factors often exert inductive effects by evoking changes in hormone metabolism and distribution within the plant. Aside from participation in responses to inductive environmental effects, hormones also are the principal agents which otherwise regulate expression of the intrinsic genetic potential of plants.

The literature on plant hormones is replete with variable attempts to define or describe these important substances. Most so-called definitions are of the operational type; that is, they define hormones in terms of their biological activity, often to the virtual exclusion of any mention of their chemical nature. An example is the definition which the author has adopted for plant hormones or phytohormones generally.

A plant hormone may be defined—perhaps described would be more accurate—as an organic substance other than a nutrient (a substance which supplies either carbon and energy or essential mineral elements), active in very minute amounts (e.g., < 1 mM, often < 1 μM), which is formed in certain parts of the plant and which usually is translocated to other sites, where it evokes specific biochemical, physiological, and/or morphological responses. Plant hormones are active in tissues where they are produced as well as at a distance. A growth regulator ($=$ growth substance or plant growth regulator) is an organic compound (other than a nutrient) which in small amounts (< 1 mM) (or low concentrations) promotes, inhibits, or qualitatively modifies growth and development. Growth inhibitors are organic compounds that retard growth generally and that have no stimulatory range of concentrations. Hence, all hormones (natural plant products) are plant growth substances, but the converse is not true. There are literally hundreds of purely synthetic compounds which qualify as growth regulators but which are not hormones.

There is a controversy at present as to what extent control of growth and developmental phenomena by a hormone is correlated with (1) changes in the amount or concentration of the hormone, and (2) changes in sensitivity of the tissue(s) to the hormone. The latter concept has been emphasized especially by Trewavas (1981; Trewavas and Cleland, 1983). Trewavas has pointed out that there is frequently a lack of correlation between hormone concentrations measured in tissues and the response of the tissues. Additionally, in most plants growth is proportional to the logarithm of the concentration of applied hormone, such that there may be an increasing response over three orders of magnitude in concentration, whereas observed changes in the endogenous concentration in tissues usually are of much smaller magnitude. Thus Trewavas has argued that since concentration changes cannot account for the differences in physiological response, something else does, and he has asserted that it must be tissue sensitivity. Little is known about ''sensitivity'' of tissues to hormones. Firn (1986) pointed out that a change in sensitivity simply refers to an observation that the response to a given amount of a particular hormone has changed. This conceivably could be caused by a change in the concentration of hormone receptors, a change in receptor affinity, or a change in a subsequent chain of events that might involve other hormones, or other factors. Undoubtedly, in the author's view *both* changes in concentration *and* changes in sensitivity are important, although one factor probably outweighs or even occurs at the exclusion of the other in particular cases.

The commonly recognized classes of plant hormones are the auxins, gibberellins, cytokinins, abscisic acid and other growth inhibitors, ethylene, brassinosteroids, and the hypothetical florigens or anthesins. In the following chapters we shall devote attention to the following topics, where we can, about each class of hormones: definitions or descriptions; history of discovery; chemical characterization; occurrence; biosynthesis, enzymic degradation, and other features of metabolism; transport; physiological effects; and mechanisms of action. In a final chapter we shall devote extensive attention to phytochrome and its role in numerous photoresponses and the roles of hormones in flowering.

References

Avery, G.S., Jr. 1933. Structure and development of the tobacco leaf. *Am. J. Bot.* **20**: 565–592.

Bergmann, L. 1960. Growth and division of single cells of higher plants *in vitro*. *J. Gen. Physiol.* **43**: 841–851.

Black, M. and J. Shuttleworth. 1976. Inter-organ effects in the photocontrol of growth. In: Smith. H., ed. *Light and Plant Development*. Butterworth & Company, Limited, London. Pp. 317–331.

Bonner, J. 1965. *The Molecular Biology of Development*. Oxford University Press, New York.

Bonner, J. and A. W. Galston. 1952. *Principles of Plant Physiology*. W. H. Freeman and Company, San Francisco.

Boss, M. L. 1964. Equational relations between nutrient quantities and cellular growth. *New Phytol.* **63**: 47–54.

Britten, R. J. and E. H. Davidson. 1969. Gene regulation for higher cells: A theory. *Science* **165**: 349–357.

Bünning, E. 1977. Fifty years of research in the wake of Wilhelm Pfeffer. *Annu. Rev. Plant Physiol.* **28**: 1–22.

Carr, D. J. and K. G. M. Skene. 1961. Diauxic growth curves of seeds with special reference to French beans (*Phaseolus vulgaris* L.). *Aust. J. Biol. Sci.* **14**: 1–12.

Davies, P. J., ed. 1987. *Plant Hormones and Their Role in Plant Growth and Development*. Martinus Nijhoff Publishers, Dordrecht, The Netherlands.

Dure, L. S., III. 1975. Seed formation. *Annu. Rev. Plant Physiol.* **26**: 259–278.

Ecklund, P. R. and T. C. Moore. 1968. Quantitative changes in gibberellin and RNA correlated with senescence of the shoot apex in the 'Alaska' pea. *Am. J. Bot.* **55**: 494–503.

Eggler, W. A. 1955. Radial growth in nine species of trees in southern Louisiana. *Ecology* **36**: 130–136.

Erickson, R. O. 1976. Modeling of plant growth. *Annu. Rev. Plant Physiol.* **27**: 407–434.

Evans, G. C. 1972. *The Quantitative Analysis of Plant Growth*. University of California Press, Berkeley.

Firn, R. D. 1986. Growth substance sensitivity: The need for clearer ideas, precise terms and purposeful experiments. *Physiol. Plantarum* **67**: 267–272.

Friesner, R. C. 1920. Daily rhythms of elongation and cell division in certain roots. *Am. J. Bot.* **7**: 380–407.

Galston, A. W. and P. J. Davies. 1970. *Control Mechanisms in Plant Development*. Prentice-Hall, Englewood Cliffs. New Jersey.

Gibbs, J. L. and D. K. Dougall. 1963. Growth of single plant cells. *Science* **141**: 1059.

Greathouse, D. C., W. M. Laetsch, and B. O. Phinney. 1971. The shoot-growth rhythm of a tropical tree, *Theobroma cacao. Am. J. Bot.* **58**: 281–286.

Gross, P. R. 1968. Biochemistry of differentiation. *Annu. Rev. Biochem.* **37**: 631–660.

Haber, A. H., T. J. Long, and D. E. Foard. 1964. Is final size determined by rate and duration of growth? *Nature (London)* **201**: 479–480.

Heslop-Harrison, J. 1967. Differentiation. *Annu. Rev. Plant Physiol.* **18**: 325–348.

Heslop-Harrison, J. 1972. Genetics and the development of higher plants: a summary of current concepts. In: Steward. F. C., ed. *Plant Physiology. Vol. VIC. Physiology of Development: From Seeds to Sexuality*. Academic Press. New York. Pp. 341–366.

Higgins, T. J. V., J. A. Zwar, and J. V. Jacobsen. 1976. Gibberellic acid enhances the level of translatable mRNA for α-amylase in barley aleurone layers. *Nature (London)* **260**: 166–168.

Horgan, R. 1987. Instrumental methods of plant hormone analysis. In: Davies, P. J., ed. *Plant Hormones and Their Role in Plant Growth and Development.* Martinus Nijhoff Publishers, Dordrecht, The Netherlands. Pp. 222–239.

Jacobsen, J. V. 1977. Regulation of ribonucleic acid metabolism by plant hormones. *Annu. Rev. Plant Physiol.* **28**: 537–564.

Kienholz, R. 1934. Leader, needle, cambial and root growth of certain conifers and their relationships. *Bot. Gaz.* **96**: 73–92.

Kramer, P. J. 1943. Amount and duration of growth of various species of tree seedlings. *Plant Physiol.* **18**: 239–251.

Kramer, P. J. and T. T. Kozlowski. 1960. *Physiology of Trees.* McGraw-Hill Book Company, New York.

Leopold, A. C. and P. E. Kriedemann. 1975. *Plant Growth and Development.* 2nd ed. McGraw-Hill Book Company, New York.

Lockhart, J. A. 1965. An analysis of irreversible plant cell elongation. *J. Theoret. Biol.* **8**: 264–275.

Loomis, W. E. 1934. Daily growth of maize. *Am. J. Bot.* **21**: 1–6.

Medawar, P. B. 1945. Size, shape, and age. In: Clark, W. E. L. and P. B. Medawar, eds. *Essays on Growth and Development.* Oxford University Press, Oxford. Pp. 157–187.

Meins, F., Jr. 1975. Cell division and the determination phase of cytodifferentiation in plants. In: *Results and Problems in Cell Differentiation. Vol. 7. Cell Cycle and Differentiation.* Springer-Verlag, Berlin. Pp. 151–175.

Millerd, A. 1975. Biochemistry of legume seed proteins. *Annu. Rev. Plant Physiol.* **26**: 53–72.

Nelson, O. E. and B. Burr. 1973. Biochemical genetics of higher plants. *Annu. Rev. Plant Physiol.* **24**: 493–518.

Nitsch, J. P. and C. Nitsch. 1969. Haploid plants from pollen grains. *Science* **163**: 85–87.

Pence, V. C. and J. L. Caruso. 1987. Immunoassay methods of plant hormone analysis. In: Davies, P. J., ed. *Plant Hormones and Their Role in Plant Growth and Development.* Martinus Nijhoff Publishers, Dordrecht, The Netherlands. Pp. 240–256.

Plaisted, P. H. 1957. Growth of the potato tuber. *Plant Physiol.* **32**: 445–453.

Richards, F. J. 1969. The quantitative analysis of growth. In: Steward, F. C., ed. *Plant Physiology. Vol. VA. Analysis of Growth: Behavior of Plants and Their Organs.* Academic Press, New York, Pp. 3–76.

Salisbury, F. B. and C. W. Ross. 1985. *Plant Physiology.* 3rd ed. Wadsworth Publishing Company, Belmont, California.

Sangwan, R. S. and B. Norreel. 1975. Induction of plants from pollen grains of *Petunia* cultured *in vitro. Nature (London)* **257**: 222–224.

Sinnott, E. W. 1960. *Plant Morphogenesis.* McGraw-Hill Book Company, New York.

Steeves, T. A. and I. M. Sussex. 1972. *Patterns in Plant Development.* Prentice-Hall, Englewood Cliffs, New Jersey.

Steward, F. C. 1968. *Growth and Organization in Plants.* Addison-Wesley Publishing Company, Reading, Massachusetts.

Steward, F. C. 1970. Totipotency, variation and clonal development of cultured cells. *Endeavour* **29**: 117–124.

Steward, F. C. and A. D. Krikorian. 1971. *Plants, Chemicals and Growth*. Academic Press, New York.

Steward, F. C., R. F. Lyndon, and J. T. Barber, 1965. Acrylamide gel electrophoresis of soluble, plant proteins: a study on pea seedlings in relation to development. *Am. J.Bot.* **52**: 155–164.

Steward, F. C., E. M. Shantz, J. K. Pollard, M. O. Mapes, and J. Mitra. 1961. Growth induction in explanted cells and tissues: metabolic and morphogenetic manifestations. In: Rudnick, D., ed. *Synthesis of Molecular and Cellular Structure*. The Ronald Press, New York. Pp. 193–246.

Sunderland, N. 1960. Cell division and expansion in the growth of the leaf. *J. Exp. Bot.* **11**: 68–80.

Sweeney, B. M. 1969. *Rhythmic Phenomena in Plants*. Academic Press, New York.

Takahashi, N., ed. 1986. *Chemistry of Plant Hormones*. CRC Press, Boca Raton, Florida.

Thimann, K. V. 1977. *Hormone Action in the Whole Life of Plants*. University of Massachusetts Press, Amherst.

Torrey, J. G. 1967. *Development in Flowering Plants*. Macmillan Company, New York.

Trewavas, A. 1976. Post-translational modification of proteins by phosphorylation. *Annu. Rev. Plant Physiol.* **27**: 349–374.

Trewavas, A. 1981. How do plant growth substances work? *Plant, Cell Environ.* **4**: 203–228.

Trewavas, A. and R. E. Cleland. 1983. Is plant development regulated by changes in the concentration of growth substances or by changes in the sensitivity to growth substances? *Trends Biochem. Sci.* **8**: 354–357.

Tukey, H. B. 1934. Growth of the embryo, seed, and pericarp of the sour cherry *(Prunus cerasus)* in relation to season of fruit ripening. *Proc. Am. Soc. Hort. Sci.* **31**: 125–144.

Turrell, F. M., M. J. Garber, W. W. Jones, W. C. Cooper, and R. H. Young. 1969. Growth equations and curves for citrus trees. *Hilgardia* **39**: 429–445.

Verma, D. P. S., G. A. Maclachlan, H. Byrne, and D. Ewings. 1975. Regulation and *in vitro* translation of messenger ribonucleic acid for cellulase from auxin-treated pea epicotyls. *J. Biol. Chem.* **250**: 1019–1026.

Vince-Prue, D. 1975. *Photoperiodism in Plants*. McGraw-Hill Book Company, New York.

Wareing, P. F. and I. D. J. Phillips. 1981. *The Control of Growth and Differentiation in Plants*. 3rd ed. Pergamon Press, New York.

Went, F. W. 1957. *The Experimental Control of Plant Growth*. Ronald Press, New York.

Whaley, W. G. 1961. Growth as a general process. In: Ruhland, W., ed. *Handbuch der Pflanzenphysiologie*. Vol. XIV. Springer-Verlag, Berlin, Pp. 71–112.

Wilkins, M. B. 1969. Circadian rhythms in plants. In: Wilkins, M. B., ed. *Physiology of Plant Growth and Development*. McGraw-Hill Publishing Company Limited, London. Pp. 647–671.

Wilson, C. C. 1948. Diurnal fluctuations of growth in length of tomato stem. *Plant Physiol.* **23**: 156–157.

Ycas, M., M. Sugita, and A. Bensam. 1965. A model of cell size regulation. *J. Theoret. Biol.* **9**: 444–470.

Zimmermann, M. H. and C. L. Brown. 1971. *Trees—Structure and Function*. Springer-Verlag, Berlin.

CHAPTER 2

Auxins

Brief History of Discovery

The first type of plant hormone to be discovered was the auxins. The term "auxin" is derived from the Greek "auxein" which means "to grow," and was proposed originally by Kögl and Haagen-Smit and by F. A. F. C. Went to designate a particular substance that had the property of promoting curvature in the *Avena* Coleoptile Curvature Test, which will be described later.

Devising an all inclusive and exclusive definition of auxins is difficult, as is the case of all other plant hormones. A committee of plant physiologists in 1954 defined "auxin" as a generic term for compounds characterized by their capacity to induce elongation in shoot cells. It was emphasized that auxins resemble indole-3-acetic acid, the only known naturally occurring indole auxin, in physiological action and that they affect other processes besides elongation, even though elongation is considered critical. Years later K. V. Thimann (1969) defined "auxins" as "organic substances which at low concentrations (<0.001 M) promote growth (cell enlargement) along the longitudinal axis, when applied to shoots of plants freed as far as practical from their own inherent growth-promoting substance, and inhibit the elongation of roots." Some authors have restricted the term to substances active in the *Avena* Coleoptile Curvature Test.

A very extensive discussion would be required to recount in detail the discovery, isolation, and eventual identification of the well-established natural indole auxin, indoleacetic acid. We can only describe the highlights here. One could say that the development of the hormone concept in general had its beginning as long ago as 1758 when Duhamel du Monceau, a Frenchman, conducted experiments that led him to conceive of growth correlations in plants brought about by two saps, one moving downward, the other upward. But, actually, the German botanist Julius von Sachs, known as the Father of Plant Physiology, is credited in 1880 with having launched the first theory of substances, which we now call hormones, controlling plant growth. Sachs envisaged the existence of organ-forming substances moving in various polar patterns and controlling growth and development. His thesis was that morphological differences among the different organs are due to corresponding differences in their material composition, which must be already present at the time of initiation, even though in his time chemical reactions and other crude methods failed to show any difference. Sach's theory found little support among his contemporaries, and it was nearly half a century before the existence of a substance corresponding to his organ-forming substances was finally proved. And, in fact, to suggest that any kind of plant hormone is a specific organ-forming substance is misleading in

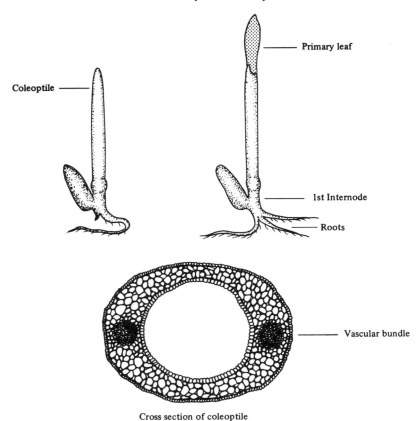

Cross section of coleoptile

FIGURE 2.1. The *Avena* coleoptile. The coleoptile is a hollow cylinder enclosing the epicotyl and is attached to the axis of the seedling at the first node. It is a leaf-like structure and the first part of the plant to emerge from the soil. Coleoptiles are characteristic of the grasses. The first leaf, as it expands, eventually pierces the coleoptile at its apex after which growth of the coleoptile ceases. Cell division takes place in an etiolated *Avena* coleoptile until it reaches about 1 cm in length, after which cell division ceases and it elongates for 4 days at a rate of approximately 1 mm/hour. (Redrawn, with permission, from Bonner and Galston, 1952.)

view of our present knowledge that each kind of hormone influences multiple developmental processes.

The discovery of auxins was the outcome of experiments on phototropism and, to a lesser extent, gravitropism. Charles Darwin, better known generally for his theory of evolution, was the first investigator in the field of auxinology. Darwin was interested in plant movements, especially phototropism, and his observations, published in 1880 in a delightful book entitled *The Power of Movement in Plants,* led ultimately to the definitive discovery of auxin. Darwin performed very simple experiments on the phototropism of coleoptiles (Fig. 2.1) of canary grass (*Phalaris*) seedlings. When coleoptiles of etiolated canary grass seedlings were illuminated from one side, a strong positive curvature (positive phototropism, growth toward the light source) resulted. If the tip of the coleoptile was darkened by a tin foil cap and only the lower part was unilaterally illuminated, curvature usually did not result. However, when he reversed this procedure, that is, when only the lower part was darkened and the coleoptile tip was exposed, curvature resulted. Darwin also demonstrated that a coleoptile does not react phototropically when 2.5 to 4 mm of the tip is removed. From these empirical experiments, Darwin concluded ''that when seedlings are freely exposed to a lateral light, some influence is transmitted from the upper to the lower part, causing the latter to bend.''

Several other investigators extended Darwin's experiments, but it remained for a young Dutch botany student in Utrecht named Frits W. Went to make the definitive discovery of auxin in 1926–1928. Went was, at the time, serving in the Dutch army, but at night he would go to his father's (F. A. F. C. Went) plant physiology laboratory at the university to keep up his graduate studies in botany. Went's favorite object of study was the coleoptile of oat (*Avena sativa*) (Fig. 2.1). The coleoptile of *Avena sativa* starts its development as a dome-shaped structure that covers the shoot apex. Cell divisions cease early in the life of the coleoptile, and it enlarges principally by cell elongation for the first 70 to 100 hours of seedling growth. During most of its growth period, elongation of the coleoptile takes place primarily in the center of the organ, almost to the exclusion of the extreme apical end and basal portion. Under environmental conditions usually employed, oat coleoptiles reach a height of 3 cm in light and 6 cm in darkness and at maturity are about 1.5 mm in diameter.

Went's basic experimental design really was quite simple. Tips were excised from etiolated oat coleoptiles and placed uniformly on a thin layer of 3% agar. After 1 to 4 hours, the tips were removed and the agar was sliced into small blocks. The small blocks of agar were then placed on the stumps of decapitated coleoptiles. The result was that the rate of elongation of the coleoptiles was promoted just as if the stumps had been capped with fresh coleoptile tips (Fig. 2.2). By this simple, if tedious, procedure the definitive discovery of auxin was

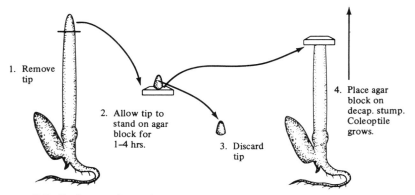

1. Remove tip

2. Allow tip to stand on agar block for 1–4 hrs.

3. Discard tip

4. Place agar block on decap. stump. Coleoptile grows.

FIGURE 2.2. Basic experiment by F. W. Went that led to the definitive discovery of auxin. (Redrawn, with permission, from Bonner and Galston, 1952.)

made and was first published in 1928. Reportedly it was 3:00 AM on the morning of April 17, 1926 that F. W. Went saw the successful result of his experiment. He dashed home and awakened his father to report the news. His father bawled him out for awakening him, and told Frits he would look at his results the next morning!

Went's *Avena* Coleoptile Curvature Test

The growth-promoting property of coleoptile tips, or of agar blocks that have been in physical contact with them, may be demonstrated in a striking manner by applying the tip or agar block asymmetrically (one-sidedly) to the cut surface of a decapitated coleoptile stump. Under these conditions the growth-promoting substance passes down the coleoptile in an asymmetric flux, causing more growth to occur on the side of the stump subtending the source of auxin than on the other, resulting in a curvature of the coleoptile. Since curvature is away from the applied block or tip, it is referred to as a negative curvature. The angle of curvature is proportional, over a wide range, 0 to 20°, to the concentration of auxin in the agar block.

F. W. Went proceeded to devise a bioassay for auxin, the *Avena* Coleoptile Curvature Test, often called simply the *Avena* test, which remains the most sensitive and specific bioassay for auxins. A bioassay is, of course, a biological testing procedure according to which the response of a plant or plant part is used to detect the presence of an active substance and to measure the amount present. The essential steps of the *Avena* Coleoptile Curvature Test are outlined in Fig. 2.3.

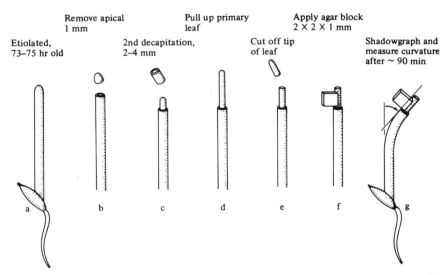

FIGURE 2.3. Procedures in the *Avena* Coleoptile Curvature Test. (Redrawn, with permission, from Leopold, 1955.)

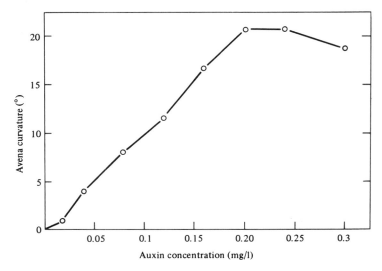

FIGURE 2.4. Typical dose–response curve for the *Avena* Coleoptile Curvature Test, showing direct proportionality of degrees of negative curvature to arithmetic auxin concentration, up to about 0.2 mg/liter. (Redrawn, with permission, from Leopold, 1955.)

In the *Avena* Coleoptile Curvature Test, response (negature curvature) is linearly proportional to arithmetic concentration of auxin (up to about 0.20 ppm) (Fig. 2.4). However, in other bioassays (e.g., *Avena* straight growth test), responses are linearly proportional to logarithm of concentration, over specific ranges of concentration.

Early Isolations of IAA

Although Went had succeeded in isolating auxin by a diffusion technique, he was not able to isolate it in pure form or to identify the substance chemically. The first attempt to isolate auxin, in sufficient quantity to permit a chemical analysis of its structure, was not carried out on oat coleoptiles or indeed even on plant tissue, as it occurs there in much too small quantities for the analytical methods available at that time. Biochemists in their searches for sources of substances active in the *Avena* test found other biological materials which were far richer in substances active in the *Avena* test. Kögl and Haagen-Smit (1931) at Utrecht, Holland, obtained an active substance called "auxin A" (auxentriolic acid) from 33 gal of human urine. Kögl, Haagen-Smit, and Erxleben (1934) analyzed urine again and found substances called "auxins A and B" + "heteroauxin." On subsequent analysis heteroauxin was found to be identical with indole-3-acetic acid (IAA) (Fig. 2.5). Kögl and Kostermans (1934) isolated IAA from yeast plasmolysate. K. V. Thimann isolated IAA from large-scale cultures of the

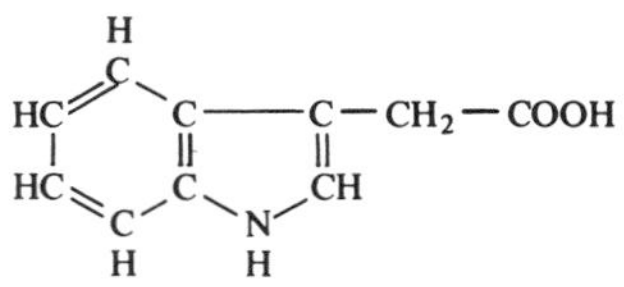

FIGURE 2.5. Structural formula of indole-3-acetic acid.

$$HC \overset{\overset{\displaystyle H}{\underset{\displaystyle C}{\|}}}{\underset{\displaystyle HC}{|}} \begin{matrix} C \\ \| \\ C \end{matrix} \begin{matrix} --- C-CH_2-COOH \\ \| \\ CH \end{matrix}$$

fungus *Rhizopus suinus* in 1935. The first generally accepted report of the occurrence of IAA in a higher plant was published by A. J. Haagen-Smit et al. in 1946. Since then there have been reports of the occurrence of IAA in dozens of species of seed plants. It is now generally agreed that indoleacetic acid is the major, perhaps the only, native indole auxin of higher plants. Interestingly, IAA had long been known to chemists since it was first synthesized in Germany in 1904 by Ellinger, but never before had it been suspected of having biological activity.

Synthetic Auxins

There are many purely synthetic compounds which may correctly be called auxins (Fig. 2.6), since they exhibit physiological action similar to that of IAA. While these compounds are auxins, of course they technically may not be called hormones.

Synthetic auxins are chemically diverse, but may for convenience be categorized in five major groups: indole acids, naphthalene acids, chlorophenoxy acids, benzoic acids, and picolinic acid derivatives. Examples of the first group are indolepropionic acid and indolebutyric acid. However, these two compounds apparently are not exclusively synthetic, since each has been reported to occur naturally in a few species. The second group is typified by naphthaleneacetic acid and β-naphthoxyacetic acid, both of which are synthetic auxins of long standing. Among the synthetic auxins in the chlorophenoxy acid series, the best known are 2,4-dichlorophenoxyacetic acid (2,4-D), 2,4,5-trichlorophenoxyacetic acid (2,4,5-T), and 2-methyl-4-chlorophenoxyacetic acid (MCPA). Several true auxins that promote growth in a manner like IAA at physiological concentrations are best known as selective herbicides when used at sufficiently high concentrations. More will be said about the chlorophenoxy auxins in the succeeding section.

Common synthetic auxins in the benzoic acid group are 2,3,6- and 2,4,6-trichlorobenzoic acids and dicamba, the latter being a powerful herbicide. It is effective against some species of deep-rooted perennials, e.g., wild morning glory (*Convolvulus arvenis*) and Canada thistle (*Cirsium arvense*), which are not readily killed by 2,4-D.

Among the best known of the picolinic acid series is the auxin-type herbicide known as picloram or Tordon (4-amino-3,5,6-trichloropicolinic acid), which is one of the most powerful selective herbicides known.

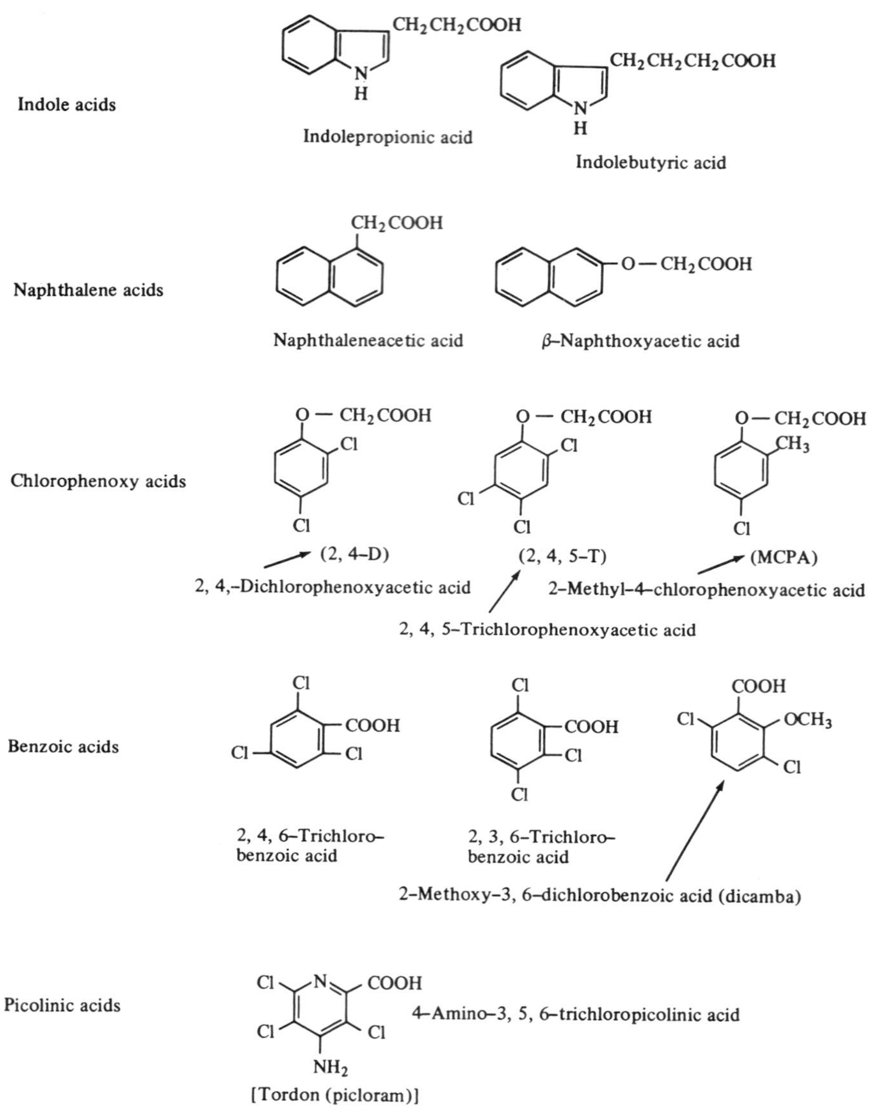

Indole acids

CH_2CH_2COOH

Indolepropionic acid

$CH_2CH_2CH_2COOH$

Indolebutyric acid

Naphthalene acids

CH_2COOH

Naphthaleneacetic acid

$O—CH_2COOH$

β–Naphthoxyacetic acid

Chlorophenoxy acids

$O—CH_2COOH$
Cl
Cl
(2, 4–D)
2, 4,–Dichlorophenoxyacetic acid

$O—CH_2COOH$
Cl
Cl
Cl
(2, 4, 5–T)
2, 4, 5–Trichlorophenoxyacetic acid

$O—CH_2COOH$
CH_3
Cl
(MCPA)
2-Methyl–4-chlorophenoxyacetic acid

Benzoic acids

Cl
COOH
Cl
Cl
2, 4, 6–Trichloro–
benzoic acid

Cl
COOH
Cl
Cl
2, 3, 6–Trichloro–
benzoic acid

COOH
Cl
OCH$_3$
Cl
2-Methoxy–3, 6–dichlorobenzoic acid (dicamba)

Picolinic acids

Cl
N
COOH
Cl
Cl
NH$_2$
[Tordon (picloram)]
4-Amino–3, 5, 6–trichloropicolinic acid

FIGURE 2.6. Examples of synthetic auxins and auxin-type herbicides.

Controversy Surrounding the Use of Certain Chlorophenoxy Acids as Herbicides and Defoliants

Background

Among the auxin-type herbicides, certain chlorophenoxy acids are among the oldest known, most selective as to species affected, and most effective. Best known among compounds of this type are 2,4-D, 2,4,5-T, and MCPA.

Research on these compounds was conducted in secrecy during World War II because of their potential use as agents of biological warfare, and they became available for commercial use in the late 1940s.

2,4-D, the first of the organic selective herbicides to be discovered and applied widely, and 2,4,5-T, the two compounds to which further discussion will be restricted, are both very active auxins. At concentrations equivalent to physiological concentrations of IAA, they are active in most auxin bioassays and are widely used instead of the native auxin IAA in experimentation. This is because they are less rapidly destroyed by the IAA oxidase system than IAA and less readily form conjugates, e.g., with amino acids. Incidentally, it should also be noted that 2,4-D and 2,4,5-T exhibit only weak polar transport, as compared to IAA.

2,4-D and 2,4,5-T are very potent and highly selective herbicides, being lethal in adequate concentration to most dicotyledonous (broad-leaved) plants and comparatively ineffective on most monocotyledonous plants. However, it is to be stressed that selectivity in herbicidal action is relative, as regards species specificity, as well as of course a function of herbicide concentration. Hence while 2,4-D will kill many dicotyledonous species at concentrations that are harmless to some monocots, at sufficiently high concentrations it will also kill monocot species.

2,4,5-T is used in a variety of commercial formulations and in recent years this herbicide has received major attention. The formulations used in commercial sprays include the free acid, salts, and a variety of esters and amine salts. The free acid form of 2,4,5-T is practically insoluble in water; the sodium salt is only soluble to a limited extent ($<3\%$). The amine salts are considerably more soluble in water, but they are somewhat difficult to prepare. Ester formulations are most commonly used as oil–water emulsions. Esters and amine salts are most commonly used. Because of the problem of drift through air and damage to nontarget species, highly volatile esters are no longer commercially available. 2,4,5-T is more expensive than 2,4-D, hence it has been used primarily to control certain woody plants (e.g., sagebrush and mesquite in the western United States) and a few herbaceous species on which it is more effective than 2,4-D. Because of the cost difference, commercial formulations containing 2,4,5-T usually are mixtures of the two compounds. The n-butyl ester of 2,4,5-T in a 1:1 mixture with 2,4-D is known as "Agent Orange," which the United States military used extensively as a defoliant in Vietnam during the United States' involvement in the conflict between North and South Vietnam. The spray concentrations usually vary between 0.1 and 2.5% and the rates of application are usually between 0.5 and 8 lb per acre. Higher rates and concentrations were used in Vietnam for military purposes.

Chlorinated dibenzo-*para*-dioxins have been recognized as possible by-products in manufacturing certain chlorinated phenols since 1950. As many as 75 chlorodioxins are theoretically possible, but only a few have been isolated and investigated. One very prominent exception is 2,3,7,8-tetrachlorodibenzo-*para*-dioxin (TCDD) (Fig. 2.7), which is produced as a by-product in the manufacture of 2,4,5-T. TCDD has been called the most toxic synthetic

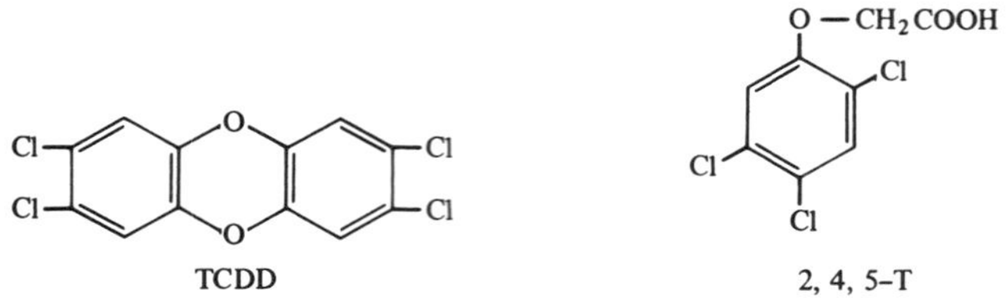

FIGURE 2.7. Structural formulas of 2,3,7,8-tetrachlorodibenzo-*para*-dioxin (TCDD) and 2,4,5-trichlorophenoxyacetic acid (2,4,5-T).

chemical known. Perhaps it will be of interest to examine the procedures for the chemical synthesis of 2,4,5-T and to show where and how TCDD is formed.

Chemical Synthesis of 2,4,5-T[1]

The usual starting material for the chemical synthesis of 2,4,5-T is 1,2,4,5-tetrachlorobenzene that can be reacted with methanol and sodium hydroxide under high temperature and pressure conditions for several hours to produce the sodium salt of 2,4,5-trichlorophenol [Reaction (1)]:

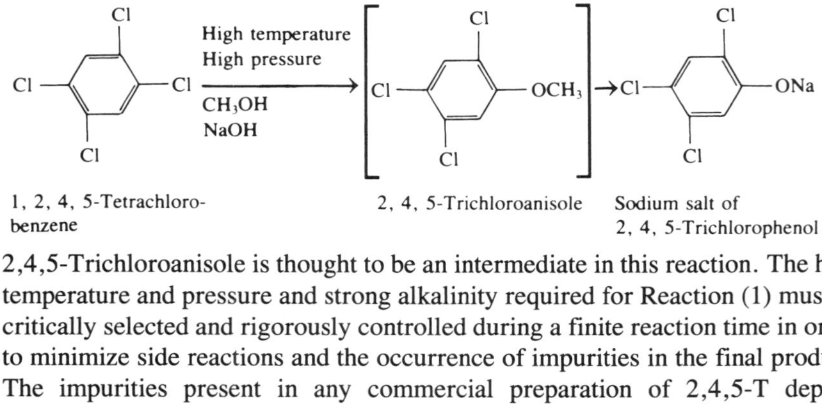

(1)

1, 2, 4, 5-Tetrachlorobenzene 2, 4, 5-Trichloroanisole Sodium salt of 2, 4, 5-Trichlorophenol

2,4,5-Trichloroanisole is thought to be an intermediate in this reaction. The high temperature and pressure and strong alkalinity required for Reaction (1) must be critically selected and rigorously controlled during a finite reaction time in order to minimize side reactions and the occurrence of impurities in the final product. The impurities present in any commercial preparation of 2,4,5-T depend strongly on the purity of the starting materials and the reaction conditions.

The aqueous trichlorosodium phenoxide is next reacted with chloroacetic acid under mildly alkaline conditions [Reaction (2)]:

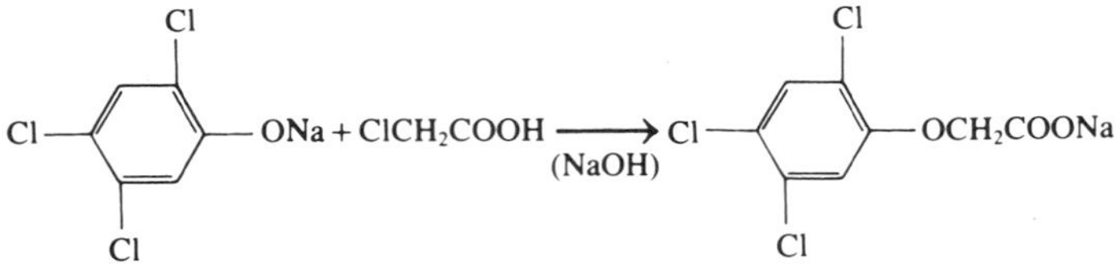

(2)

[1] From Report on 2,4,5-T. A report of the Panel on Herbicides of the President's Science Advisory Committee, Executive Office of The President, Office of Science and Technology, March 1971. 68 pp.

This product is then acidified with H_2SO_4 to produce 2,4,5-T [Reaction (3)]:

$$\text{(3)}$$

The conditions for Reactions (2) and (3) are mild compared with those required for the hydrolysis in Reaction (1). TCDD results primarily as a side product of Reaction (1), but also apparently from impurities in the starting materials.

Unfortunately, formulations of 2,4,5-T that were commercially available prior to about 1968 were contaminated with from 2 to 50 parts per million (ppm) of TCDD. Commercial preparations of 2,4,5-T available in the United States are now legally required to contain 0.1 ppm or less of TCDD.

Toxicities of 2,4,5-T and TCDD

Initial concern for the possible hazard to humans and other animals exposed to 2,4,5-T came from research conducted by the Bionetics Research Institute under contract with the National Cancer Institute of the U.S. Public Health Service, circa 1964 to 1968. In those studies it was found that mice and rats treated during early pregnancy with large doses of 2,4,5-T gave birth to defective offspring. Announcement of those results, together with reports of alleged increased occurrence of human birth defects by South Vietnamese newspapers in June and July 1969, elicited prompt and far-reaching reactions from governmental agencies, segments of the scientific community, various citizen's groups, and public mass communication media.

The concern was so great that in the United States, by joint order of the Secretaries of Agriculture and Health, Education and Welfare, the use of 2,4,5-T in the United States was curtailed for a time to the point that, practically speaking, it could not be used. Two major manufacturers of 2,4,5-T in the United States, Dow Chemical Company and Hercules Incorporated, exercised their legal rights to petition for referral of the matter to an Advisory Committee. Such an Advisory Committee subsequently was appointed from members of the National Academy of Sciences, and a U.S. government bulletin entitled *Report on 2,4,5-T—A Report of the Panel on Herbicides of the President's Science Advisory Committee* was published in March 1971.

What, if anything, can be scientifically concluded about the deleterious effects of 2,4,5-T vis-à-vis TCDD on humans and other animals? As this question is addressed, it must be noted first of all that it is impossible to prove scientifically that a given chemical is without hazard. The small number of known cases in which human ingestion of 2,4,5-T led to clinical illness offer no information on the minimal dosage of the compound that is toxic to man. In animals, however, the toxicity of 2,4,5-T is similar to that of 2,4-D. Thus some information on 2,4-D is of interest.

In an extreme case, ingestion of a single dose of more than 6500 mg (in excess of 90 mg/kg) by a human in an attempted suicide was indeed fatal. But in contrast, a patient was given 18 intravenous doses of 2,4-D during 33 days without any apparent side-effects. Each of the last 12 doses in the series was 800 mg (approximately 15 mg/kg) or more, and the last was 2000 mg (about 37 mg/kg). In a study reported in 1973, five human male volunteers ingested a single dose of 5 mg/kg without incurring detectable clinical effects. The 2,4,5-T was rapidly excreted by the subjects. $LD_{50}s$[2] for highly purified 2,4,5-T and 2,4-D, following oral ingestion by laboratory animals, are usually on the order of hundreds of milligrams of chemical per kilogram of body weight. In marked contrast, the $LD_{50}s$ for TCDD ingested orally by laboratory animals vary from 0.0006 mg/kg for male guinea pigs to 0.115 mg/kg for rabbits (mixed sexes). As far as human hazard is concerned it seems certain that any danger of 2,4,5-T formulations resides in their TCDD content. While data are too limited for a firm conclusion, there is no evidence presently available that TCDD as a contaminant in 2,4,5-T is likely to be encountered by animal or man in sufficient dosage to cause toxic reaction. In its report, the U.S. President's Advisory Committee concluded that, "No evidence has been found of adverse effects on human reproduction in three separate locations, namely Vietnam; Globe, Arizona; and Sweden where pregnant women have allegedly been exposed to high levels of 2,4,5-T."

In concluding this subject two hypothetical examples may be of interest. First, a normal application (e.g., 2 lb/acre) of 2,4,5-T, for example, usually would constitute dispersal of no more than 450 μg of the dioxin on each treated acre, assuming a dioxin content of <0.5 ppm. Second, data from experimental animals indicate that immediately after spraying 2,4,5-T at accepted rates on forests, a 100-lb human would have to absorb all of the TCDD on 5 acres of forest to suffer a minimum detectable toxic effect. That would be the amount of TCDD suspected to be sufficient to cause birth defects in humans, if humans are as sensitive as laboratory animals.

Consideration of 2,4-D and other phenoxy-acid-type auxins calls to mind another important feature of those compounds, namely, the relationship between positions of halogen atoms on the ring and biological activity. For example, 2,4,5-T is a very potent auxin, whereas 2,4,6-T is inactive. In fact the latter compound competitively inhibits the action of the former and therefore is logically termed an "antiauxin." There are numerous other structure–activity relationships that could be mentioned, but the number of generalizations that hold true for all cases is few.

Natural Occurrence of Auxins

IAA is also synthesized in very many species of nonseed plants, including many bacteria, fungi, and algae, and may be ubiquitous in the plant kingdom. For many years, IAA was the one critically identified auxin, and it came to be

---

[2] LD_{50} = dose that kills 50% of a population of test animals; "lethal dose for 50%."

FIGURE 2.8. Structure of phenylacetic acid (PAA).

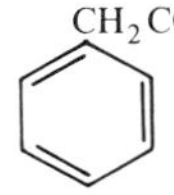

thought of as the only natural auxin. However, there have been numerous reports of other substances in plants that show similar capacity to stimulate growth in auxin bioassays. Some of these substances are very obviously related to IAA and, in the opinion of many, they simply represent alternative sources or precursors for IAA and are really active only on conversion to IAA. At this time it must be said that some uncertainty still exists as to whether a number of indolic compounds may be active compounds, biosynthetic intermediates, degradation products, storage or transport forms, detoxification products, or artifacts. But, as K. V. Thimann wrote in 1969, "Thus, in spite of frequent suggestions to the contrary, IAA remains the only naturally-occurring (indole) auxin whose existence is definitely established, and the wide occurrence as well of such a number of its derivatives makes it evident that IAA is very widely—perhaps universally—distributed throughout the plant kingdom."

One nonindolic compound that is widely recognized as a naturally occurring auxin is phenylacetic acid (PAA) (Fig. 2.8), which has been studied extensively by Frank Wightman and his associates at Carleton University at Ottawa. PAA occurs with IAA as native auxins in tomato and sunflower shoots, for example, and experiments by Wightman and Rauthan (1974) with excised shoots of tomato, pea, sunflower, tobacco, barley, and maize showed that all these shoots formed [^{14}C] PAA from exogenously supplied [3-^{14}C]phenylalanine. Also, appreciable DL-[3-^{14}C]phenylalanine was converted to [^{14}C] PAA in cell-free enzyme extracts prepared from tomato shoot tips. The enzymatic conversion of phenylalanine to PAA evidently occurs via two pathways that involve the intermediate formation of phenylpyruvic acid or phenylethylamine. If this is the case, then PAA is formed in plants by a sequence of reactions that is parallel to that involved in the biosynthesis of IAA from tryptophan.

According to Thimann (1977), PAA has much lower activity than IAA, varying in different tests from 2 to 10% of that of IAA. Yet, Wightman and associates found much higher amounts of PAA than of IAA in several plants. Thus, it is not surprising that Wightman and Rauthan (1974), using bioassays, found that the PAA fraction of extracts of tomato and sunflower shoots showed growth-promoting activity comparable to that of the extracted IAA. Thus, in some plants PAA occurs with IAA and appears to be a native auxin that contributes to observed auxin phenomena in them.

Auxin Biosynthesis

For a compound to fulfill the role of a hormone, it should be apparent that its own concentration must be rigidly controlled. The control of concentration of auxin in the various parts of the plant is complex, involving multiple processes.

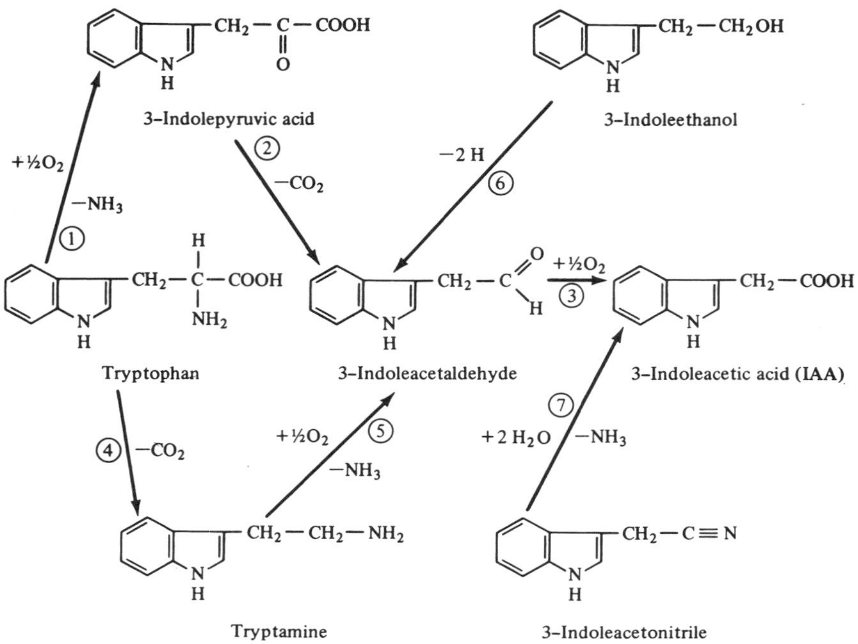

FIGURE 2.9. Pathways of auxin biosynthesis. Enzymes: 1, tryptophan transaminase; 2, indolepyruvate decarboxylase; 3, indoleacetaldehyde oxidase or dehydrogenase; 4, tryptophan decarboxylase; 5, amine oxidase; 6, indoleethanol oxidase; and 7, nitrilase.

One means of regulation of auxin levels is control of biosynthesis *in situ*. We shall now turn to a discussion of auxin biosynthesis and consider later other processes that contribute to regulation of auxin levels.

The amino acid tryptophan is commonly regarded as the precursor for the biosynthesis of auxin (IAA) in plants, and the biosynthetic pathways that have been worked out are illustrated in Fig. 2.9. By one pathway, tryptophan is converted to indolepyruvic acid via a transaminase reaction, which requires an α-keto acid and pyridoxal phosphate in addition to the enzyme. Indolepyruvic acid is next decarboxylated to indoleacetaldehyde in a reaction requiring a decarboxylase and thiamine pyrophosphate. Either an oxidase or a dehydrogenase then oxidizes indoleacetaldehyde to IAA. For the latter, nicotinamide adenine dinucleotide reportedly is the most effective coenzyme.

The second major pathway involves an initial decarboxylation of tryptophan to form tryptamine. Catalysis by an amine oxidase next converts tryptamine to indoleacetaldehyde, which is, in turn, oxidized to IAA.

In some plant systems, one or the other of these pathways apparently occurs to the exclusion of the other. In other systems both pathways are operative.

Several other indolic compounds besides those already mentioned can function as precursors of IAA in particular higher plant systems. For example, cucumber (*Cucurbita pepo*) seedlings contain appreciable quantities of indole-

3-ethanol (or tryptophol), which is readily converted to indoleacetaldehyde in that system. Members of the Cruciferae or Brassicaceae contain indole-3-acetonitrile and a nitrilase that converts the nitrile directly to IAA. Indoleacetamide occurs in some plants (e.g., tomato), and can reportedly go to IAA via a hydrolase reaction. Still other indolic compounds have been implicated as auxin precursors in particular higher plant systems.

An intriguing observation which has been reported many times is that D-tryptophan is often as active or even more active than L-tryptophan as an auxin precursor, even though the latter is the only isomer which occurs naturally in the proteins of higher plants. This enigma was answered by Miura and Mills (1971) for a system using cell cultures of tobacco (*Nicotiana tabacum L.*) They demonstrated the presence in the tissues of a tryptophan racemase that converts D-tryptophan to L-tryptophan in a reversible reaction. Since the pool size of free tryptophan in most higher plants is comparatively very small, it appears that the synthesis and utilization of tryptophan is subject to rigid control. Since D-tryptophan apparently can undergo transamination to form indolepyruvic acid and ultimately IAA, it seems possible that conversion of L-tryptophan to D-tryptophan prior to transamination may be important in controlling the levels of free tryptophan and of IAA. In other words, conversion of L-tryptophan to D-tryptophan may be a means of ensuring a pool of the amino acid for particular pathways including, but not restricted to, auxin biosynthesis.

There is another reaction for which D-tryptophan is a substrate in at least some plant tissues, and that is the formation of malonyl-D-tryptophan. M. H. Zenk suggested many years ago (1964) that the transformation of some unspecified indole auxin precursor to D-tryptophan, which is immobilized and accumulated as malonyl-D-tryptophan, might be significant in regulating the biosynthesis of IAA. This reinforces the notion that D- and L-tryptophan racemization may be important in the regulation of IAA biosynthesis.

It has been suggested by several investigators that the indole ring need not necessarily be derived from tryptophan, but might be synthesized directly in the course of auxin biosynthesis. In other words, the idea has been expressed that auxin biosynthesis may occur by a tryptophan-independent pathway. For example, Winter (1966) suggested, on the basis of very limited studies with *Avena* coleoptile sections, that tryptophan is not a precursor of IAA in that system. He proposed a hypothetical pathway according to which anthranilic acid is converted to indole, and that indole by unknown reactions could, in turn, be converted to tryptamine and on to IAA. However, Winter's scheme lacks proof. In the absence of convincing experimental evidence to the contrary, we shall conclude that tryptophan is the precursor of IAA.

A serious potential problem confronting investigations of auxin biosynthesis is microbial contamination, since numerous species of microorganisms are known to convert tryptophan to IAA. The influence of epiphytic bacteria on auxin metabolism in preparations from pea plants has been documented extensively by Libbert et al. in Germany. Hence several investigators have performed experiments to determine the possible influence of microbial metabolism in the system they have employed. Utilizing a variety of techniques—e.g., involving aseptic culture of seedlings and antibiotics—it has been concluded that in enzyme extracts of pea shoot tips and other higher plant tissues, tryptophan is indeed converted to IAA by enzymes present in the tissues.

At the present state of our knowledge, the possibilities must be considered that (1) there may be, and probably are in particular cases, different pathways of IAA biosynthesis from tryptophan in different species, (2) that more than one pathway may be operative in a single species, and (3) that these pathways may differ in different parts of the same plant or at different stages of growth in the same plant part.

"Free" and "Bound" Auxin

The natural auxin indoleacetic acid (IAA) exists in a variety of chemical states in plant tissues. Free auxin is considered to be that auxin that is readily extractable (e.g., 2 hours in diethyl ether at $0°C$ in the dark) while, traditionally, bound auxin is considered to be that auxin liberated from tissues when subjected to enzymolysis, hydrolysis, or autolysis. Free auxin, IAA per se, apparently is the form immediately utilizable in growth. Bound auxin apparently can exist in various forms, and the forms vary among different species (Fig. 2.10). Bound auxins generally are considered storage forms from which IAA can be released or detoxification products that form in the presence of relatively high levels of IAA. Auxin glycosyl esters are especially abundant in seeds and storage organs and apparently represent inactive, storage forms of hormone from which IAA can be enzymically released. Ascorbigen and glucobrassicin may also be considered bound forms of auxin in some species of the Cruciferae or Brassicaceae. Auxin peptides, e.g., indoleacetylaspartic acid, are generally considered detoxification products that are irreversibly formed and protect tissues from the accumulation of excessive auxin. Auxin–protein complexes are hypothetical associations, presumably by weak hydrogen bonds, between IAA and specific proteins.

Robert S. Bandurski and his associates at Michigan State University analyzed extensively the conjugated forms of IAA in corn (*Zea mays*) kernels. They found more than 16 different conjugates of IAA, including the isomeric esters of IAA and *myo*-inositol, IAA esters of *myo*-inositol glycosides, and IAA esters of high-molecular-weight glucans (Fig. 2.10). Free IAA accounted for only 0.8% of the total indolic compounds found in corn kernels. The indoleacetylinositols (one illustrated) accounted for 15.2%, the indoleacetylinositol-arabinosides (one illustrated), 23.2%, 5-*O*-β-L-galactopyranosyl-2-*O*-(indoleacetyl)-*myo*-inositol (illustrated), 8.1%, and IAA β(1-4) glucan (not illustrated) accounted for 52.5%; all other conjugates (two illustrated, IAA *myo*-inositols and IAA-D-glucopyranose) were present in relatively trace quantities and together accounted for only 0.3% of the total indolic compounds. The abundant IAA glucans (not illustrated) are β(1-4) cellulosic glucans with 7 to 50 glucose units per molecule of IAA.

Glycosyl esters

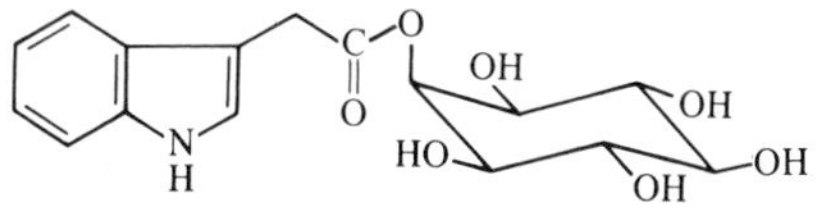

2-*O*-(Indole-3-acetyl)-*myo*-inositol

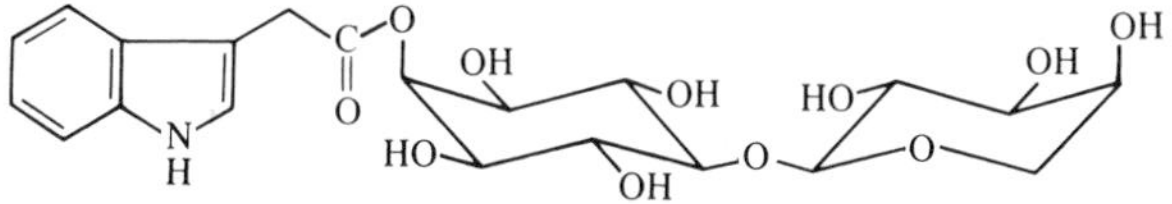

5-*O*-β-L-Arabinopyranosyl-2-*O*-(indole-3-acetyl)-*myo*-inositol

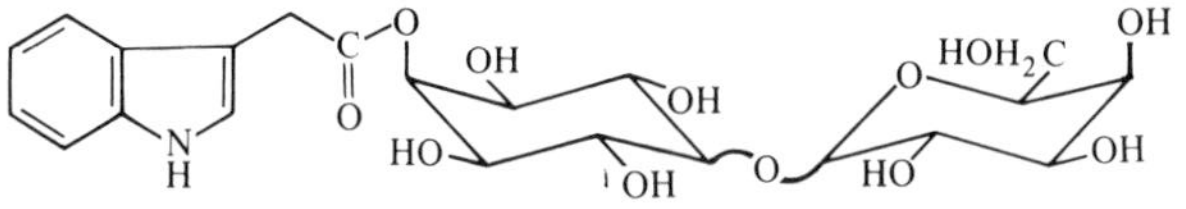

5-*O*-L-Galactopyranosyl-2-*O*-(indole-3-acetyl)-*myo*-inositol

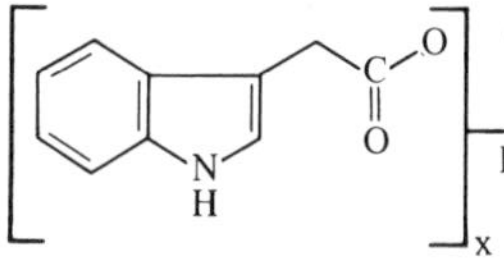

General structure of indole-3-acetyl-*myo*-inositols

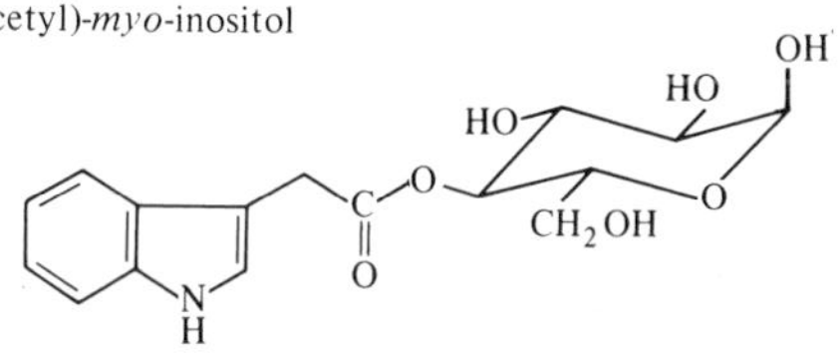

4-*O*-(Indole-3-acetyl)-D-glucopyranose

Thioglucosides

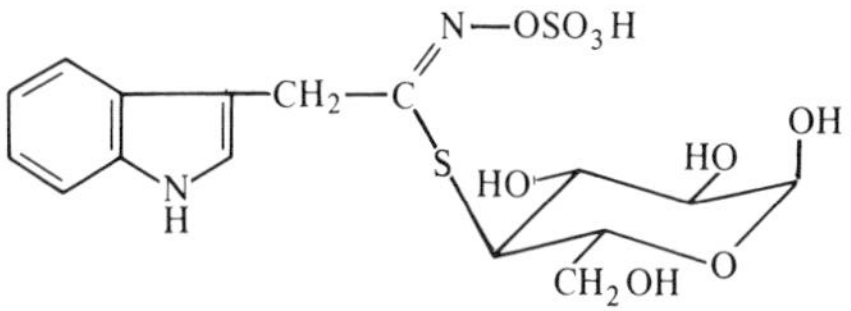

Glucobrassicin

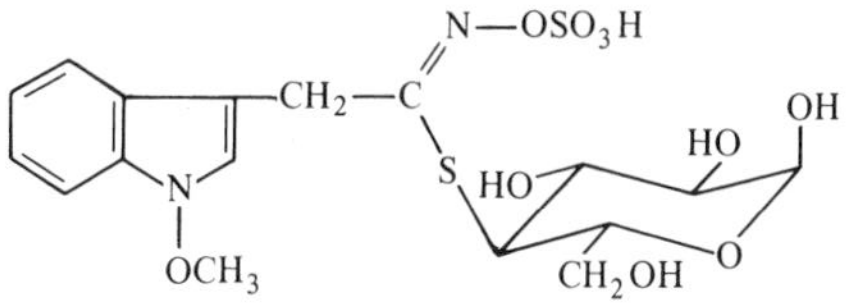

Neoglucobrassicin

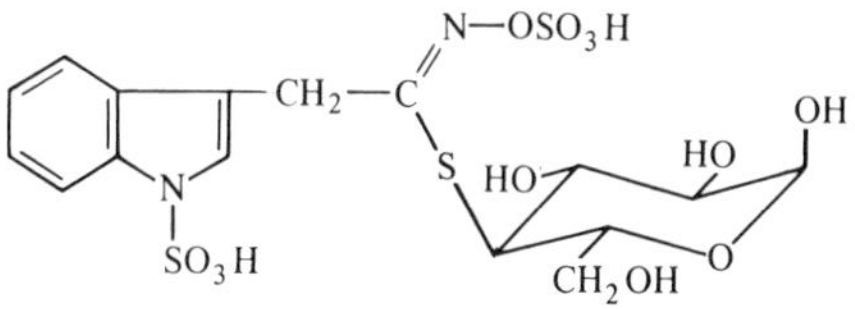

Sulfoglucobrassicin

IAA peptides

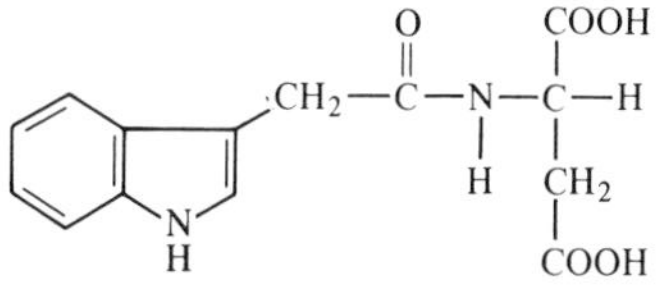

Indoleacetylaspartate

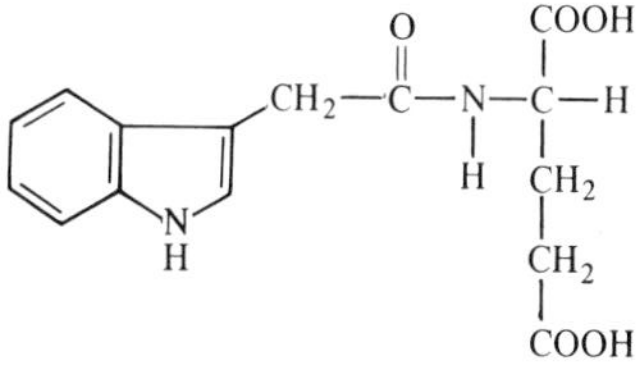

Indoleacetylglutamate

FIGURE 2.10. Examples of conjugated (''bound'') auxins from various plant sources.

There appear to be three metabolic roles of these conjugates of IAA: (1) IAA conjugates, and tryptophan, are sources for IAA during seed germination; thus an IAA conjugate is the so-called seed auxin precursor; (2) conjugation of IAA protects it against peroxidative attack; and (3) reversible synthesis and hydrolysis of IAA conjugates constitute a hormonal system that is responsive to environmental influence.

Some synthetic auxins also form bound auxin. For example, the following auxins (besides IAA) reportedly can conjugate with aspartic acid in pea epicotyl sections: indolepropionic acid, indolebutyric acid, benzoic acid, and 2,4-D. Actually, very little 2,4-D was destroyed or conjugated. This auxin is relatively very persistent and quite resistant to enzymic degradation and conjugation. Naphthaleneacetic acid (NAA) was reported to form a glycosylester in plants from several families, as well as to form a peptide with aspartate.

Destruction of IAA

As already emphasized, the level of free auxin in tissues must be rigidly controlled. A third process by which auxin levels are regulated, besides regulated synthesis *in situ* and reversible and irreversible formation of bound auxin, is degradation to inactive compounds. Auxin destruction is by two basic oxidative processes: (1) enzymatic reactions and (2) photooxidation.

A very important oxidative process physiologically is enzymatic destruction, catalyzed by an enzyme known as IAA oxidase. An accurate description of IAA oxidase is that it is a peroxidase acting as an oxidase (Fig. 2.11). The enzyme is at least partially independent of exogenous H_2O_2, when assayed *in vitro,* apparently because it carries a peroxide-producing system along with it either as an impurity or as a bound flavin moiety. Generally, in assays *in vitro,* only about 1/10 mole of exogenous H_2O_2 is needed per mole of IAA oxidized, and peroxidation in the strict sense, that is, by H_2O_2 in the absence of O_2, does not occur. One mole of O_2 is taken up, and 1 mole of CO_2 evolved from the carboxyl group. One of the dominant products is 3-methylene-oxindole. Other reported products of peroxidase oxidation of IAA are 3-hydroxymethyl oxindole, indole-3-methanol, and indole-3-carboxylic acid (Fig. 2.11). Indole-3-methanol and indole-3-aldehyde may be precursors to indole-3-carboxylic acid, and 3-hydroxymethyl oxindole is a precursor of 3-methylene oxindole. The ratio of the various products depends on several factors, including the enzyme:substrate ratio, cofactors, and pH of the reaction (Bandurski et al., 1986.) The enzyme system from higher plants requires Mn^{2+}. The IAA oxidase reaction is accelerated by monophenols (e.g., 2,4-dichlorophenol); it is inhibited by *ortho*-diphenols such as catechol and pyrogallol. Flavanols act in the same way; that is, kaempferol promotes and quercetin inhibits IAA oxidase.

The photooxidation of IAA is a very different reaction from the enzymatic ones and appears to have little physiological significance. Riboflavin, eosin, and other fluorescent dyes are effective catalysts in assays *in vitro.* One mole of O_2 is

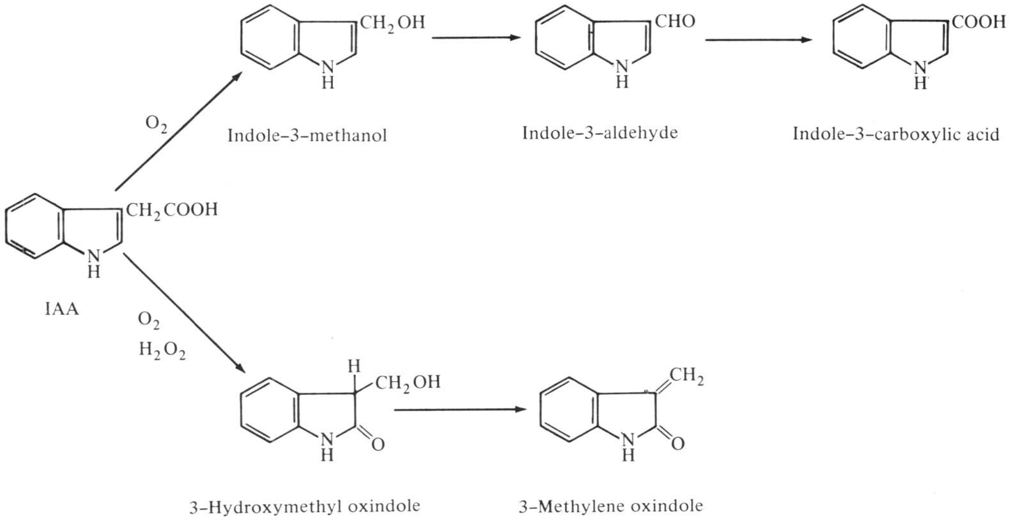

FIGURE 2.11. Oxidative decarboxylation pathway for IAA. (Redrawn, with permission, from Reinecke and Bandurski et al., 1987.)

taken up per mole of IAA destroyed, as in the enzymic, IAA oxidase reaction, and the products apparently are identical, including 3-methylene oxindole and 3-indolealdehyde. Relatively very large light dosages are required, and the process seems to have no bearing on phototropism or other physiological phenomena.

V. Tuli and H. S. Moyed (1976) showed directly that extracts of pea seedlings (*Pisum sativum* var. Alaska) oxidized IAA to 3-hydroxymethyloxindole, and at physiological pH this compound was readily dehydrated to 3-methylene oxindole. The extracts of pea seedlings also contained a reduced triphosphopyridine nucleotide-linked enzyme that reduced 3-methylene oxindole to 3-methyloxindole. 3-Methylene oxindole is known to be a potent bacteriostatic agent. It is a potent sulfhydryl reagent, which reacts rapidly with substances such as glutathione and CoA, and is a potent inhibitor of SH-containing enzymes. 3-Methyloxindole is nontoxic. The observed enzymic reactions were also seen to take place in intact pea seedlings.

It has been reported that the peroxidase of peas exists as a number of isozymes. Isozyme patterns apparently change during ontogeny and may vary among the organs of a single plant.

Addition of IAA to green tissues enhances peroxidase activity, but this does not appear to be true enzyme induction of IAA oxidase. While treatment of plant tissues with auxins (IAA or other) often increases the peroxidase activity, auxins that are not oxidized act as well as IAA. Hence the change is evidently not a true induction enzyme activity but a result of generally enhanced synthesis.

It was noted previously that monophenols and *ortho*-diphenols modify IAA oxidase activity, and there is evidence that some endogenous phenols affect enzymic activity in a significant way. For example, pea plants contain the

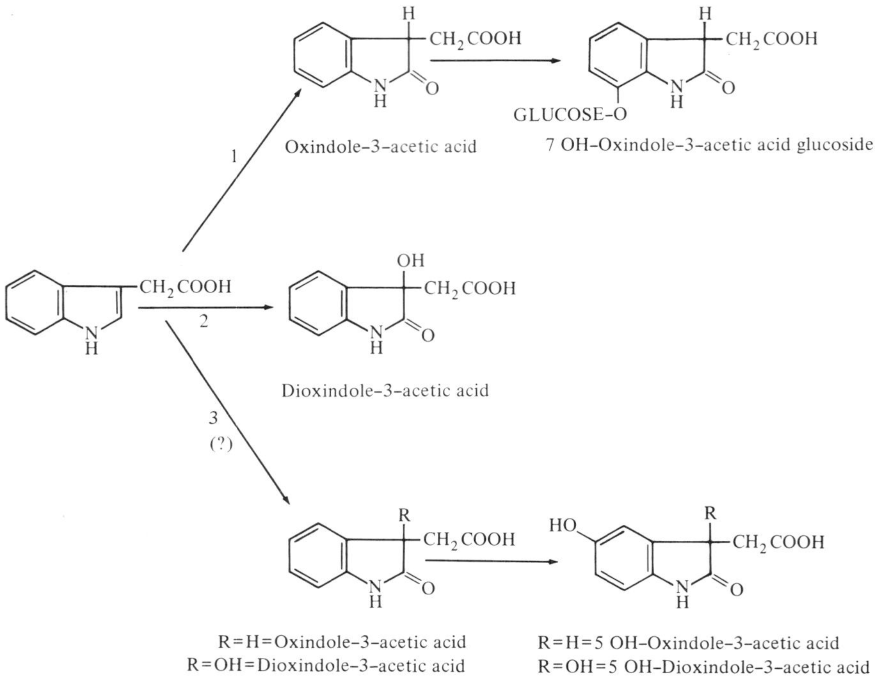

2. Auxins

FIGURE 2.12. Nondecarboxylation oxidative pathway for IAA in *Zea mays* (1), *Vicia faba* (2), and *Oryza sativa* (3). (Redrawn, with permission, from Reinecke and, Bandurski, 1987.)

flavonols kaempferol and quercetin as the triglucosides and coumaroyl (coumaric acid) triglucosides. When an etiolated pea plant is irradiated with white or red light, interesting correlative effects of red light occur with respect to the flavonols. Red light promotes apical bud growth and inhibits subapical stem elongation. In a reportedly correlative manner, red light stimulates the synthesis of kaempferol, a cofactor of IAA oxidase, in stems, and stimulates synthesis of quercetin, an inhibitor of the enzyme, in leaves. Since phenolic substances occur widely in plants, they must be assumed to have some importance as modifiers of auxin destruction.

There is also a nondecarboxylation oxidation pathway for IAA, which is separate from the peroxidase-catalyzed decarboxylation. This other pathway involves oxidation of IAA at the 2-position (Fig. 2.12) and yields oxindole-3-acetic acid (OxIAA) and dioxindole-3-acetic acid (DiOxIAA). Pathway 1 is the major oxidative pathway for IAA in maize or corn (*Zea mays*) (Bandurski et al., 1986), and pathways 2 and 3 have been reported to occur in broad bean (*Vicia faba*) and rice (*Oryza sativa*), respectively. OxIAA and DiOxIAA are produced from labeled IAA in maize and broad bean, respectively. Both OxIAA and DiOxIAA occur naturally in rice, and DiOxIAA has been shown to be produced from IAA in broad bean. It is not known whether the DiOxIAA pathway is separate from the OxIAA pathway, as illustrated in Fig. 2.12, or whether OxIAA can be oxidized to DiOxIAA. In maize OxIAA is further metabolized by hydroxylation and glucose addition at the 7-position of the indole nucleus forming 7-OH-OxIAA-glucoside.

Rice is interesting in that it is the only plant known to have both OxIAA and DiOxIAA. Rice also has the 5-OH analogs of OxIAA and DiOxIAA. However, the oxindole-3-acetic acids have not been isolated from vegetative tissues of rice, nor have radiolabel-feeding studies been undertaken to show the precursor–product relationship with IAA, so pathway 3 is less certainly known than pathways 1 and 2.

Thus, in summary, there are two pathways of enzymic IAA catabolism: the peroxidase-catalyzed decarboxylation of IAA and the oxidation without decarboxylation to OxIAA and DiOxIAA.

The enzymology of the OxIAA/DiOxIAA pathway is much less understood than the peroxidase pathway at the present time. As in the case of the decarboxylation pathway, the physiological significance of the OxIAA/DiOxIAA pathway is not well understood.

Auxin Transport

For many years, actually since the classical experiments of F. W. Went with *Avena* coleoptile sections in 1932, polarity of auxin transport has been recognized. Apical dominance and other growth correlations (e.g., cambial activation) in the plant are correlated with polar movement of endogenous auxin. Tropisms are correlated with asymmetrical movement of endogenous auxin. Indeed the influences of auxin on growth and development are, in general, reflections of the polar pattern of its distribution.

The movement of endogenous auxin in shoots is predominantly basipetally polar from morphological apex to morphological base (Fig. 2.13). This basipetal polar movement occurs not only in segments of coleoptiles or Coleus stems lacking vascular tissue but also in veins of leaves. Thus, both parenchymatous tissue and vascular tissue appear to be capable of polar transport. Evidently, in fact, endogenous IAA generally is not translocated through the phloem sieve tubes nor through xylem vessels and tracheids. Rather, movement occurs through living tissues such as phloem parenchyma and other parenchyma cells that surround vascular bundles in organs that have differentiated vascular tissues. There is no impediment to cell-to-cell transport in younger, relatively undifferentiated tissues. There actually is remarkably little direct evidence on the question of how endogenous auxin moves in the intact whole plant. Indirect evidence suggests that polar basipetal movement may occur in the cambia of trees. Polar basipetal movement diminishes in older parts of stems, coleoptiles, and petioles. Acropetal transport in at least two systems from shoots, oat

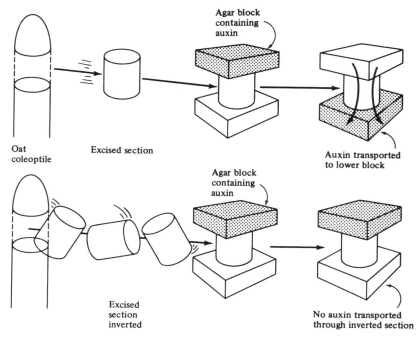

FIGURE 2.13. Early experiment of F. W. Went, demonstrating basipetal polar transport of endogenous auxin in excised sections of *Avena* coleoptile. (Redrawn, with permission, from Bonner and Galston, 1952.)

coleoptiles and bean petioles, apparently is nonmetabolic and appears to be diffusion.

In excised coleoptile and stem sections, IAA is not the only substance which can move in the polar transport system. Polarity of movement has also been demonstrated for several synthetic auxins, including indolebutyric acid, naphthaleneacetic acid, and 2,4-D. In general, growth regulators (including IAA, 2,4-D, etc.) applied to intact roots or shoots move in both xylem and phloem, but this movement is nonpolar, and velocities of 50 mm/hour or more are not uncommon.

Auxin transport in roots is less completely understood than movement in shoots. The most convincing data, from experiments in which movement of [^{14}C] IAA in root segments has been studied, indicates that movement is acropetal, that is, from the base of the stem toward the root tip. Polarity of movement in roots is an order of magnitude lower than in shoots.

There are several general features of polar movement of auxin that are based almost entirely on experiments in which measurements of transport have been made by applying auxin to one or the other end of a segment excised from a plant organ, and subsequently observing either the distribution of auxin along the segment or the amount appearing in a block of agar applied as a receiver to the

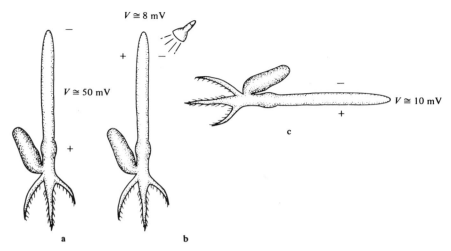

$V \cong 8$ mV

$V \cong 50$ mV

$V \cong 10$ mV

a   b

c

FIGURE 2.14. Differences in electrical potential in *Avena* coleoptiles. **a** Between tip and base; **b** between illuminated and dark sides; and **c** between upper and lower sides of a horizontally positioned coleoptile.

opposite end of the segment. For one thing, polar movement of auxin apparently is an active (energy-dependent) process. The rate (12-20 mm/hour in *Avena* coleoptile sections) is too rapid to be accounted for by diffusion. The process is temperature-sensitive. Polar transport is independent of O_2 concentrations up to about 5% in *Avena* coleoptile sections, but is abolished anaerobically. Additionally, movement through *Avena* coleoptile sections can be against a concentration gradient. Finally, there is a decline in the polarity of auxin transport with age of stems, leaf petioles, roots, and *Avena* coleoptiles.

Obviously an understanding of the mechanism of polar transport of auxin is essential to an understanding of the ways in which the hormone serves to assist in integration of growth and development of the whole plant. It must be admitted at the outset that the mechanism of polar auxin movement is quite incompletely understood. However, it will be worthwhile to examine some of the hypotheses that have been proposed.

The oldest idea is the electrical polarity theory. In 1932, F. W. Went suggested that lateral redistribution of auxin, to cause tropistic responses, might be mediated by an electrical field, and this idea was considered by others for many years. For example, A. R. Schrank in the years 1945–1951 studied electrical polarity and auxin movement extensively. He showed differences in electrical potential from apex to base of *Avena* coleoptiles (Fig. 2.14). When illuminated unilaterally, the shaded side became electropositive to the lighted side. When laid horizontally, the bottom side became electropositive to the other. Schrank reasoned that at physiological pH IAA would exist as an anion ($pK \cong 4.8$) and that auxin movement was along an electrical gradient; that is, that the observed electrical gradient caused the polar auxin movement.

Interestingly, the observed electric changes were found to occur even when bending in response to the stimuli was suppressed by a lack of auxin.

More recent experiments have tended to discount potential gradients as the cause of polar auxin movement. Recent evidence indicates that the longitudinal electrical potential gradient of coleoptiles and the transverse electric potential that develops after tropic stimulation are more a result rather than a cause of the polar transport of auxin. I. A. Newman, physicist at the University of Tasmania, described a wave of electrical disturbance which apparently accompanies the movement of an auxin front down decapitated coleoptiles of *Avena*. The electric wave traveled down the coleoptile at a speed of ~ 14 mm/hour, which is about equal to the speed of auxin transport. Furthermore, he showed that an electric wave is produced by the application of IAA in physiological concentration to the top of a decapitated coleoptile (Fig. 2.15). Newman's hypothesis is that IAA moves down the coleoptile not as a stream of constant concentration but that it moves down with partial or total bunching, in the form of a wave. He stated in 1963 that, ''The close relationship between the electric wave and a postulated wave in the downward-moving auxin stream suggests that an electric field is directly involved in the mechanism of auxin translocation.'' Newman agreed with B. I. H. Scott's concept of a feedback system of some sort as the cause of auxin translocation. Scott proposed a feedback system involving three elements (electric field, IAA concentration, and membrane permeability) to account for auxin translocation in coleoptiles. According to him, as auxin moves down a coleoptile it alters membrane permeability, which sets up an electric field that in turn may ''push the hormone farther down the coleoptile.''

The polar secretion hypothesis advanced by Leopold and Hall (1966) embodies ideas similar to some of those of Newman and Scott. Leopold and Hall proposed to account for polar auxin movement by preferential secretion of more auxin from one end of a cell than from the other end. They suggested that a secretion mechanism, which causes individual cells to drive auxin across a cell membrane by active transport that is preferential for the exit of auxin, accounts for polar transport of IAA. They noted that a difference of only 3%, for example, between acropetal and basipetal movement through an individual cell of corn coleoptile would be sufficient, after passing through 31 cells (4 mm), to produce 54 times as much auxin in basal as apical receivers. If 52.5% of the auxin secreted by a file of 100 cells were to come out of the basal end of each cell, over 10,000 times as much auxin would be found at the basal end of the tissue as at the acropetal end.

The modern view of the mechanism of polar auxin transport is represented by one or more chemiosmotic[3] models (Fig. 2.16). As at least a first approximation this scheme is well supported by experiments with tissue segments and with

---

[3] The basic chemiosmotic mechanism for auxin transport was originally proposed by Rubery and Sheldrake (1974), although the designation ''chemiosmotic polar diffusion hypothesis'' was adopted first by Goldsmith in a 1977 review.

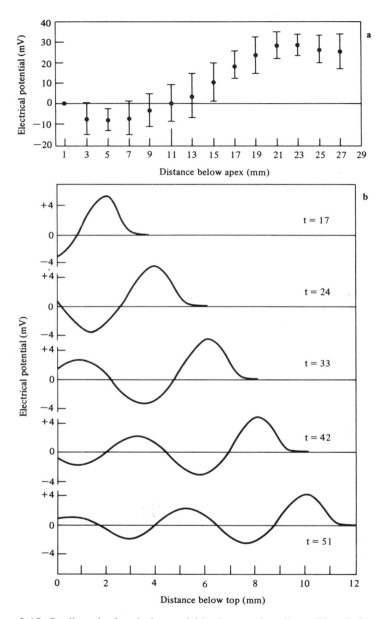

FIGURE 2.15. Gradients in electrical potential in *Avena* coleoptiles. **a** Electrical potential plotted against distance from the apex of an intact coleoptile; **b** wave of electrical disturbance that accompanies polar movement of exogenous IAA down a decapitated coleoptile. In **b** the potential plotted is the departure of the potential from its average value. The average potential distribution is illustrated at each of five times (*t*, in minutes) after the application of IAA. (Redrawn, with permission, from Newman, 1963.)

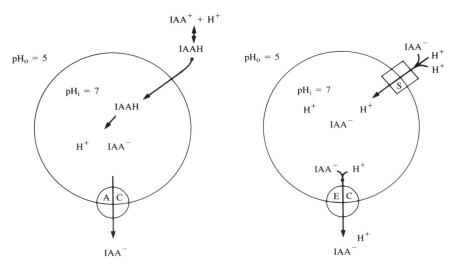

FIGURE 2.16. Two chemiosmotic models of polar auxin transport through a cell or an outside-out plasma membrane vesicle. S, electrogenic symport cotranslocating 2 H$^+$ with 1 IAA$^-$; AC, anion carrier; EC, efflux carrier (or a symport) for IAA$^-$ and H$^+$. (Redrawn, with permission, from Hertel, 1986.)

isolated plasma membrane vesicles. According to the chemiosmotic theory, the source of energy for auxin accumulation and polar auxin transport is the catalysis of ATP hydrolysis by an H$^+$-ATPase in the plasma membrane, which pumps H$^+$ from the cytosol into the cell wall. The lower pH of the cell wall (approximately 5) keeps the carboxyl group of IAA less dissociated (pK$_a$ for IAA equals about 4.76) than in the cytosol, where the pH is higher (approximately 7). This pH (H$^+$) gradient acts then as the thermodynamically driving force for uptake of IAA. The IAA moves into the cell either as undissociated molecules (IAAH, Fig. 2.16, left) or via an electrogenic symport cotranslocating 2 H$^+$ with 1 IAA$^-$ (Fig. 2.16, right). Uptake is increased by an outside-positive membrane potential. Lomax (1986) presented evidence that the accumulation of IAA in zucchini membrane vesicles can be facilitated by auxin influx via a saturable, specific, electrogenic plasma membrane carrier, rather than simple, nonmediated accumulation of a weak acid across a pH gradient. As the concentration of IAA$^-$ (dissociated auxin) increases internally (in the cytosol) either the IAA$^-$ (left in Fig. 2.16) or IAA$^-$ and H$^+$ (right in Fig. 2.16) leave the cell via a polarly distributed exit carrier. Thus, as Hertel (1986) stated, "A 'simple' transport scheme of auxin accumulation in the cytoplasm and auxin 'secretion' from each cell consists of two specific molecular transmembrane processes at the plasmalemma: (1) an accumulation of IAA into the cytoplasm driven by a pH gradient and by the electrical membrane potential; and (2) an efflux of IAA through polarly distributed exit carriers."

Current research utilizes preparations of membrane vesicles as model "cells" for polar auxin transport. Terri L. Lomax and associates have devised such an experimental *in vitro* system. Tightly sealed outside-out membrane vesicles that are primarily of plasma membrane origin are prepared from etiolated zucchini (*Cucurbita pepo*) hypocotyls. They have a pH gradient of the same direction as found in intact plants. Thus, these vesicles provide an apparently excellent system for studying the transport components. Extensive use is made of auxin transport inhibitors ("phytotropins") such as 2,3,5-triiodobenzoic acid (TIBA) and naphthylphthalamic acid (NPA).

In other important research, Lomax and associates utilized photoaffinity labeling of plasma membrane proteins by an azido-IAA (^{3}H-5N$_3$-IAA) as a preliminary step toward purification of the IAA/2H$^+$ uptake symport. Using sealed "right-side out" plasma membrane vesicles prepared from *Cucurbita pepo,* they found that about five polypeptides became labeled with ^{3}H-5N$_3$-IAA. Addition of unlabeled IAA or 2-NAA to the vesicle preparations blocked labeling, indicating specificity of the polypeptides for auxin.

As we have seen, IAA is transported directionally via a system that requires a pH gradient, an uptake carrier, and a basally located carrier. N-1-naphthylphthalamic acid (NPA), an inhibitor of IAA transport, binds to a basally located, membrane-bound receptor protein. Judith G. Voet and Jay R. Desai at Swarthmore College in Pennsylvania recently synthesized a photolabile 5′-azido analog of NPA to help isolate and characterize this NPA receptor protein. In *Cucurbito pepo* membrane vesicles Az-NPA competed with NPA for membrane binding sites. They went on to show that Az-NPA stimulates [^{14}C]IAA uptake in *Cucurbita pepo* hypocotyl stem segments. Stimulation of [^{14}C]IAA uptake in stem segments by NPA occurs by inhibition of IAA efflux, and is thus reportedly a good measure for inhibition of IAA transport. The fact that Az-NPA both competes with NPA for binding sites and inhibits IAA transport makes it a promising analog for covalently labeling the NPA receptor protein.

Until 1988, no endogenous ligand for the NPA receptor, capable of affecting polar auxin transport, had been reported. However, in that year Philip H. Rubery and Mark Jacobs of the University of Cambridge reported that a group of flavonoids—including quercetin, apigenin, and kaemferol—can specifically compete with [^{3}H]NPA for binding to its receptor and can perturb auxin transport in a variety of plant tissues and transport systems in a manner paralleling the synthetic inhibitors of polar auxin transport. The evidence indicated that certain naturally occurring flavonoids act as natural auxin transport regulators in plants.

A very intriguing question is: What is the source of acropetally moving auxin in the roots of intact plants? It is unlikely that such auxin is merely from the downward moving stream of auxin that the shoot produces, because—judging by results with sections—there is a graded decrease in the amount of auxin transported down the bean hypocotyl to the point where little or none moves through sections from the stem–root transition region. There is evidence that auxin is synthesized in root tips, but it is difficult to envisage how auxin from the root tip comes to be transported acropetally in subapical sections of the root. The point needs investigating, particularly with a critical appraisal of how accurately the results with sections apply to the intact plants.

Relationships between Auxin Content and Growth

Multiple factors determine the amounts of endogenous auxin in a particular part of a seed plant at a given time. Gain by import from primary centers of auxin synthesis and loss by export via polar transport are important factors, which have just been discussed. Auxin is synthesized in relatively large amounts in only a few localized centers, but it is transported through all the living tissues of the plant. Shoot tips, including particularly the young leaves, are the center of most abundant auxin synthesis in the vegetative seed plant. Other rich sources are enlarging leaves, flowers, fruits, and seeds. The enzymes for the conversion of tryptophan to IAA occur generally throughout the plant and are especially active in regions of intense metabolic activity, such as meristems, expanding leaves, fruits, and root tips. Other factors, already discussed, are synthesis *in situ,* enzymic destruction, and reversible and irreversible formation of bound auxin and release of free auxin.

In general, a reasonably good correlation is found between relative auxin content and relative growth in the various organs of seed plants, particularly taking into account the marked differences in sensitivity of the different parts of a plant to auxin (Fig. 2.17). And, there are available data to show that auxin

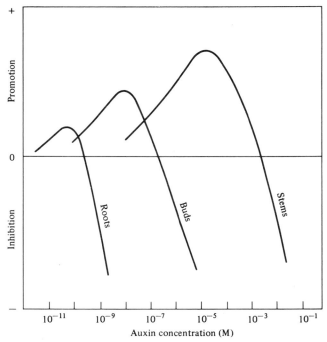

FIGURE 2.17. Biphasic dose–response curves for different plant organs and IAA. (Redrawn, with permission, from Leopold, 1955.)

content is indeed influenced by the various factors that have been previously discussed. For example, Figs. 2.18, 2.19, and 2.20 illustrate the amounts of auxin (presumably free auxin, specifically, as judged by the methods used) in different parts of a monocot and two species of dicot seedling, respectively.

IAA oxidase preparations from different parts of an angiosperm seedling tend to reveal, by *in vitro* assay, an inverse correlation between enzyme activity and auxin content. That is, IAA oxidase activity tends to be comparatively low in regions of relatively high auxin content (Fig. 2.21).

In 1969 investigations were conducted on the comparative net biosynthesis of IAA from tryptophan in cell-free enzyme extracts of different parts of pea (*Pisum sativum* L. cv. Alaska) seedlings and immature pea seeds (Table 2.1). Tryptophan was converted to IAA in enzyme extracts of all parts of the light-grown pea seedlings and in extracts of immature pea seeds. Auxin yields are expressed as the amount of radioactivity from DL-[2-^{14}C]tryptophan incorporated into [^{14}C]IAA per gram fresh weight of tissue. The unit fresh weight is considered a physiologically more meaningful parameter than unit protein or nitrogen for comparing the auxin-synthesizing capacities of the different parts of seedlings, because of the marked differences in protein and nitrogen concentration of the plant materials. On this basis, the regions of most active auxin production apparently were the terminal bud, young stem below the terminal bud, and young leaves. Lesser amounts of net auxin biosynthesis occurred in enzyme extracts prepared from older stems and leaves and from root tips. Thus, in general, the data agree well, to the extent that they are comparable, with the amounts of "free" auxin obtained by diffusion and extraction from young light-grown pea seedlings. Developing pea seeds exhibited an auxin-synthesizing capacity that was comparable to or higher than that of terminal buds of seedlings.

Two considerations ought to be borne in mind in interpreting the data in Table 2.1, however. In the first place, it is not known to what extent the auxin-synthesizing capacities in cell-free enzyme extracts reflect the actual *in vivo* capacities of the various plant parts to produce IAA from tryptophan. Second, in these experiments net IAA production was measured, and there was no accounting for concomitant auxin degradation by the IAA oxidase system in the preparations. Previously it was shown that [1-^{14}C]IAA (carboxyl-labeled), when added to cell-free enzyme extracts of shoot tips in estimated substrate quantities, was enzymically decarboxylated. Hence it is reasonable to assume that IAA oxidase activity was present, and at variable levels, in the enzyme extracts of the various plant parts.

In summary there appears to be a good correlation between capacity for net auxin biosynthesis from tryptophan and endogenous free auxin content of the various parts of young green pea seedlings. There still remains the question of whether there are differences in the capacities of the various parts of the seedling plant for gross auxin production, or whether the differences in apparent auxin-producing capacity noted in the present investigation are correlated primarily with differential IAA oxidase activity. Scott and Briggs, while also considering other possible contributing factors, felt in 1963 that auxin destruction seemed to

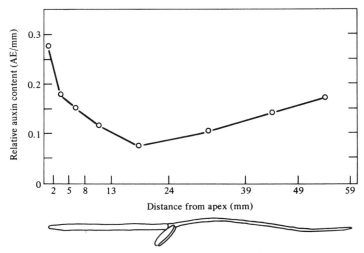

FIGURE 2.18. Distribution of endogenous auxin in an etiolated *Avena* seedling. The auxin content is expressed in *Avena* Einheits (AE), units no longer in common usage. (Redrawn, with permission, from data of Thimann, 1934, in Leopold, 1955.)

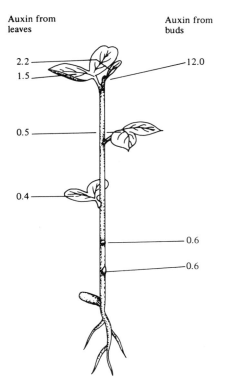

FIGURE 2.19. Comparative diffusible auxin content of different parts of a seedling of broad bean (*Vicia fava*). The numbers refer to *Avena* Einheits of diffusible auxin per hour. (Redrawn, with permission, from data of Thimann and Skoog, 1934, in Leopold, 1955.)

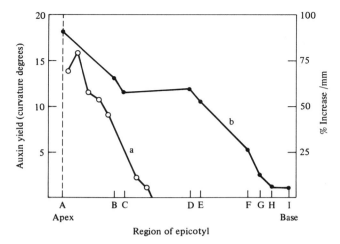

FIGURE 2.20. Relative growth rate, **a,** and distribution of diffusible endogenous auxin, **b,** in the epicotyl of a 9-day-old light-grown pea seedling. (Redrawn, with permission, from Scott and Briggs, 1960.)

Specific activities
(μg IAA destroyed $\cdot$ mg protein^{-1} $\cdot$ hour^{-1})

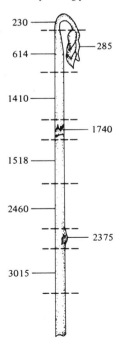

FIGURE 2.21. Relative magnitude of IAA ox-idase activity (specific activities) in different parts of an etiolated pea epicotyl. (Redrawn, with permission, from Galston, 1967.)

TABLE 2.1. Comparative net biosynthesis of [^{14}C]IAA in cell-free extracts of different parts of pea plants.[a,b]

Plant part(s)	dpm incorporated into [^{14}C]IAA/g fresh wt/2 hours	Mean fresh wt of parts/plant (mg)	mg N/g fresh wt
Terminal bud	61,166	24	6.31
Stem above 5th node	40,181	98	1.59
Leaves above 5th node[c]	51,776	77	4.96
Shoot tip (all above 5th node)	38,933	305	3.55
Stem below 5th node	19,200	478	0.68
Leaves below 5th node[c]	10,240	545	1.62
Apical 2 cm of primary root	7,786	34	0.98
Developing seeds[d]	56,000	171	2.22
Developing seeds[e]	89,266	—	—

[a] From Moore (1969) with permission.
[b] Collective data from three separate experiments.
[c] Including stipules, rachis, leaflets, and tendrils.
[d] Approximately half-grown "Alaska" pea seeds.
[e] Approximately half-grown "Telephone" pea seeds.

be the best alternative to account for the variations in "free" auxin content in different parts of green pea epicotyls. Based on all available data it appears probable that the different parts of the pea seedling do differ in potential auxin-synthesizing capacity as well as in IAA oxidase activity, and that differential auxin biosynthesis *in situ* is one factor among many that accounts for variations in auxin content in the various parts of pea seedlings.

Correlative Differences in Auxin Relations between Etiolated and Light-Grown and Dwarf and Normal Plants

There is considerable interest in the possible correlative differences in auxin relations between etiolated and light-grown seedlings of the same species and cultivar and between plants of the dwarf genotype and those of the normal or tall genotype of different cultivars of the same species. *Pisum sativum,* garden pea, is an example of an herbaceous annual species that has been used quite extensively in research in this area, and the hormone field generally. Both genetic dwarf and normal (tall) varieties of pea are available, and the genetics of several varieties is fairly well known. Dwarf varieties mature at heights of 30 cm or less, whereas some tall varieties exceed a meter in shoot height at maturity. Because of their self-fertilization and high degree of homozygosity, each variety is phenotypically very uniform.

Regarding etiolated and light-grown seedlings of the same cultivar of *Pisum sativum,* there is a marked difference in sensitivity of excised stem sections, with the former being about two orders of magnitude more sensitive than the latter (see Fig. 6.9). Other interesting physiological differences between the etiolated and light-grown specimens are known, and many of these were described by T. K. Scott and W. R. Briggs in 1963, using seedlings of the Alaska variety, a normal, early-flowering type (Table 2.2).

TABLE 2.2. Comparison of light-grown and dark-grown pea seedlings with respect to structure, growth patterns, and growth substance relationships.[a]

	Light-grown	Dark-grown
Morphology and anatomy	Five internodes, top two, true stem; basal three, transition zone	Three internodes, all transition zone
Growing region	Top internode, evenly distributed	Top internode, primarily apical region
Extractable auxin	High, no change from apex to base of top two internodes, decrease basally through transition zone	Decrease, apex to base
Diffusible auxin	Growing region, decrease, apex to base; no change through 2nd internode; decrease, apex to base of transition zone	Not obtainable
Auxin transport	Polar, 10–12 mm/hour, saturated with endogenous auxin only in transition zone	Possibly incompletely polar, apparently 6–7 mm/hour
IAA oxidase activity	At least partially masked by inhibitor	High, increasing, apex to base
IAA oxidase inhibitor activity	High, decreasing, apex to base	Low
Response, sections from growing zone, to IAA	Optimal at high concentrations	Optimal at low concentrations
Response, sections from growing zone, to gibberellic acid	Optimal at high concentrations	Optimal at low concentrations

[a] Adapted from Scott and Briggs (1963) with permission.

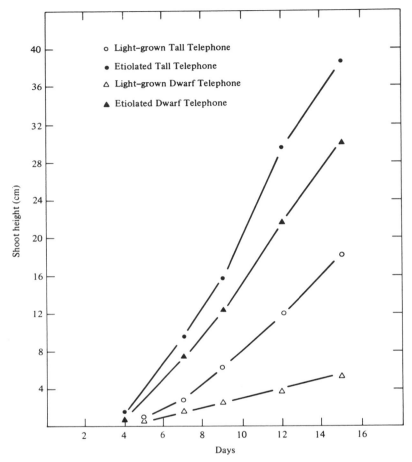

FIGURE 2.22. Growth curves of etiolated (closed symbols) and light-grown seedlings (open symbols) of Tall (circles) and Dwarf (triangles) Telephone peas grown in growth chambers. Each point represents the mean value for 20 plants. (Unpublished data of Moore, 1964.)

The hormonal basis for dwarfism presents some very interesting questions, which have been answered for only a few species. In the case of peas, etiolated specimens of both dwarf and normal genotype exhibit nearly identical rates of stem elongation (Fig. 2.22). That is, dwarfism becomes readily apparent only in the light-grown condition.

T. C. Moore in 1967 compared the net biosynthesis of IAA from tryptophan in cell-free enzyme extracts prepared from the shoot tips of etiolated and light-grown seedlings of the Dwarf and Tall Telephone cultivars. This comparison of maximum net IAA biosynthesis revealed an order, on a unit protein basis, of light-grown tall > light-grown dwarf > etiolated tall ≃ etiolated dwarf.

It was concluded in that study that the different rates of stem elongation among etiolated and light-grown dwarf and tall pea seedlings are correlated with small differences in auxin relations. However, it was also concluded that correlative differences in endogenous gibberellin relations probably are of much greater importance. The hormonal basis of dwarfism is discussed more comprehensively in Chapter 3.

Mechanism of Auxin Action

Some Fundamental Terms and Concepts

As L. G. Paleg stated in 1965, there is merit in applying different connotations to two terms, "mechanism" and "mode" of action, which often are used synonymously. When a hormone acts upon a responsive plant system, it of course enters into some direct and specific molecular interaction that results, eventually, in a manifestation of measurable effect (biochemical or physiological response, e.g., cell elongation). But there really are two aspects of the hormone action involved: (1) the direct and specific molecular interaction; and (2) the succeeding series of steps (events, processes) that result in the measurable biochemical or physiological response. The former is the "mechanism" of action; the latter is the "mode" of action.

By definition, therefore, every hormone has its own distinctive mechanism(s) of action, even though the manifestation of the hormonal mechanism may depend upon the prior, concomitant, or subsequent role of another factor(s). It follows that in one system a hormonal mechanism may lead to one physiological manifestation, while in a second system the same mechanism may lead to a completely different manifestation. This difference can be brought about because the second system has more (or less) of another hormone, is from a different tissue or organ, or other causes. Thus, it is to be emphasized that auxins have their own distinctive mechanisms of action, though their modes of action frequently may be responsive to other factors.

Preliminary Considerations Concerning the Mechanism of Auxin Action

Ever since the discovery of each type of plant hormone, great effort has been made to attain a biochemical understanding of the manner by which each type exerts its diverse effects. A major difficulty in elucidating the mechanism of auxin (and other hormone) action is the multitude of different kinds of physiological processes auxin controls. One might naturally wonder whether there is only one fundamental site and mechanism of auxin action, which leads to diverse physiological responses, or whether there are multiple sites and mechanisms of action. The truth is that we do not yet know. However, very significant progress toward ultimate elucidation of the mechanism of auxin action has been made. Let us tabulate some of the most important known facts about auxin action that any theory (model) of auxin action must accommodate:

1. All auxin-induced physiological responses require constant presence of the hormone. In the case of auxin-induced (IAA) cell elongation in coleoptiles, an increased growth rate is evident within 10 minutes or less after presentation of the auxin; on withdrawal of IAA a decline in growth rate is evident within 10 minutes and growth rate is back to control level within about 40 minutes after withdrawal of the auxin.
2. Auxin (like hormones generally) is active in extremely small concentrations (e.g., 10^{-6} M). Therefore, some form of amplification of an initial triggering response must occur; e.g., amplification might occur by the hormone (a) acting as an allosteric effector activating certain enzymes, (b) stimulating the synthesis of certain enzymes, or (c) evoking a change in membrane permeability.
3. It is clear that continued RNA (specifically mRNA) and protein synthesis are essential for cell elongation responses to be sustained for as long as several hours. The ability of auxin to enhance the rate of cell elongation is dependent upon new RNA (mRNA) and protein synthesis.
4. It is equally clear that some auxin responses may occur too rapidly to involve gene activation as the primary action of auxin. For examples: elongation in coleoptile and stem sections occurs in < 10 minutes, increase in protoplasmic streaming can often be observed within 20 seconds to a few minutes, and auxin alteration of membrane potentials occurs in 10 to 15 minutes or less.

Water Relations in Auxin-Induced Cell Enlargement

The auxin function that has been investigated most extensively is auxin-induced cell enlargement. Auxin is required specifically for cell wall loosening but not for cell enlargement itself. Reviewing the equation relating the major parameters of cell water potential, we see that:

$$\Psi \text{ (water potential)} = \Psi_\pi \text{ (osmotic potential)} + \Psi_p \text{ (pressure potential)}$$

For water uptake to be promoted by auxin, the water potential inside the cell must become more negative than outside. Conceivably, Ψ could become more negative either because

1. the osmotic potential (Ψ_π) becomes more negative, or
2. the pressure potential (Ψ_p) ($\equiv$ turgor pressure) decreases.

What actually happens in the auxin response are the following sequential events:

1. Auxin increases wall loosening, thus decreasing the resistance of the wall to stretching and allowing the wall to yield to pressure.
2. The pressure potential (Ψ_p) thus decreases.
3. More water osmoses in because of the more negative internal Ψ.
4. The cell volume is thereby increased and the wall is irreversibly extended.

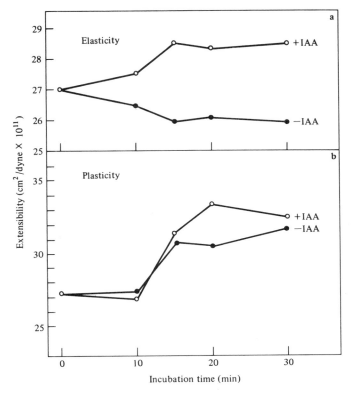

FIGURE 2.23. Effects of auxin on cell wall elasticity, **a,** and plasticity, **b,** in oat coleoptiles. (Redrawn, with permission, from Masuda, 1969).

The cell wall loosening action of auxin really is a dual one—an initial effect on elasticity and a succeeding effect on plasticity (Fig. 2.23). Of these two effects, the latter is most important, for cell enlargement is most directly correlated with the effect on cell wall plasticity (Fig 2.24). On completing an increment of enlargement due to one exposure to a growth-promoting concentration of auxin and withdrawal of the hormone, the rigidity of the primary cell wall reverts to its original state.

The Nature of Auxin-Induced Cell Wall Loosening

The primary cell wall consists of a framework of cellulose microfibrils that is cross-linked by other polysaccharides. Peter Albersheim and associates then of the University of Colorado worked out the most extensive analysis of the macromolecular components of primary cell walls, using cultured sycamore cells (Table 2.3) and also created a model of the molecular architecture of the wall (Fig. 2.25).

Albersheim and associates also worked out a model for auxin-induced cell

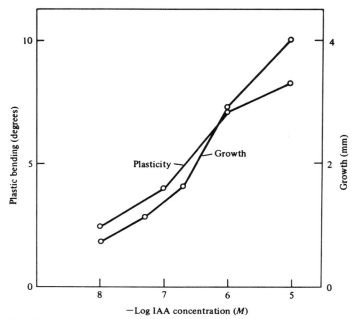

FIGURE 2.24. Direct correlation between auxin effects on growth and cell wall plasticity. (Redrawn with modifications, by permission, from Bonner, 1961.)

expansion, which includes the following features. Since walls extend throughout their length, for a wall to be extended, it is necessary that the cellulose fibrils be able to slide along their length relative to each other. Also, walls that have extended are essentially as strong as those that have not, indicating that the number of cross-links and the amount of material in an extended cell wall are insignificantly different from those in an equivalent area of cell wall prior to expansion. In other words, if a wall extends as a result of cleavage of cross-

TABLE 2.3. Macromolecular components of the walls of suspension-cultured sycamore cells. [a]

Wall component	Cell walls (%)
Arabinan	10
3,6-Linked arabinogalactan	2
4-Linked galactan	8
Cellulose	23
Protein	10
Rhamnogalacturonan	16
Tetraarabinosides (attached to hydroxyproline)	9
Xyloglucan	21
Total	99

[a] With permission, from Talmadge et al. (1973).

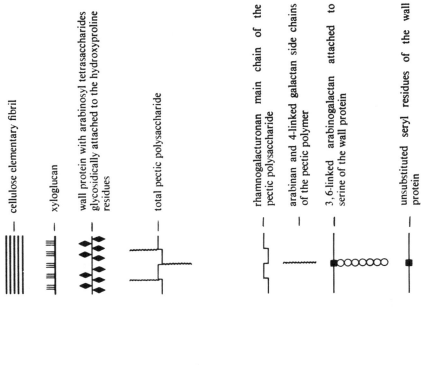

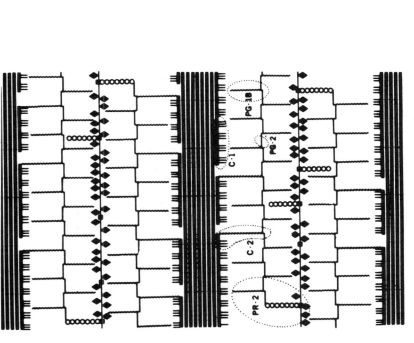

FIGURE 2.25. Tentative model of the molecular architecture of sycamore cell walls. (With permission, from Keegstra et al., 1973.)

's between polymers, a mechanism must exist for the formation of
ks at the termination of an increment of cell expansion.

..un of their model of the cell wall led Albersheim and associates to
...mulate a specific hypothesis regarding the breaking and reforming of cell wall
cross-linkages. They found that the only noncovalent cross-links among the
structural polymers of the cell wall evidently are the hydrogen bonds between
the xyloglucan polymers and the cellulose microfibrils. Movement of the
xyloglucan chains along the cellulose microfibrils could be accomplished by a
mechanism—enzymic or nonenzymic—that catalyzes breakage and reformation
of the hydrogen bonds, allowing the xyloglucan to creep inchworm-fashion
along the cellulose microfibrils. The rate by which the xyloglucan polymers
move along the cellulose microfibrils should be increased by conditions, such as
high hydrogen-ion concentrations, which weaken hydrogen bonds. An enhanced
rate of creep would be completely and immediately reversed by raising the pH or
lowering the temperature. Furthermore, one can readily conceive of unidirec-
tional creep in walls under tension since the xyloglucan polymers are covalently
linked at their reducing ends to noncellulosic wall components. Therefore, when
neighboring cellulose fibrils are pushed in an opposite direction by hydrostatic
pressure, the xyloglucan polymers on each fibril would creep in only one
direction relative to the cellulose fibril to which they are attached.

The further intriguing suggestion was made by Albersheim and associates that
auxin may activate a hydrogen-ion pump in the plasmalemma, hence lower the
pH of the wall, enhance xyloglucan creep, and thereby increase cell wall
loosening. We will return to this idea later in this chapter.

Sites of Auxin Action

Currently there are two major concepts as to the site of action of auxin. One of
these centers on the cell wall as the site of action, a concept already introduced in
considering the investigations by Peter Albersheim and associates. The other
concept focuses on nucleic acid metabolism. Many cell-wall loosening phenomena may be related to auxin treatment in a manner independent directly of
nuclear activities. Yet an enhancement of RNA synthesis is consistently
associated with sustained growth responses to the hormone. Hence, we have two
concepts that at first seem to be mutually exclusive and to indicate that there are
multiple sites (and perhaps mechanisms) of auxin action. However, after
examining the evidence on which each concept is based, we shall describe
hypothetical models that attempt to reconcile the two concepts into one. For now
we shall first examine the evidence that auxin exerts its hormonal action by an
effect on nucleic acid and protein metabolism. Then we shall discuss so-called
rapid responses to auxin that do not depend on a direct effect of the hormone on
nucleic acid or protein metabolism.

Effects of Auxins on Nucleic Acid and Protein Metabolism

Abundant evidence indicates quite conclusively, and not at all surprisingly of
course, that continued synthesis of RNA and protein is required for continued

cell enlargement. Most investigations have utilized excised plant parts (sections of coleoptiles, stems, tubers), have measured rates of growth and of incorporation of radioactive precursors into RNA and protein, and have relied heavily on the use of various inhibitors of RNA and protein synthesis.

Before getting into a detailed description of those types of investigations it will be well to provide a general characterization of effects of auxins on intact plants and excised tissues. For one thing, very few species of plants exhibit positive growth responses to applied auxin; cucumber (*Cucurbita pepo* cv. National Pickling) is one exception. Generally, intact plants respond to the auxin-type growth regulators by an inhibition of growth (both cell division and cell enlargement) and inhibition of nucleic acid synthesis in the normal growing points; the more mature stem tissues are activated relative to nucleic acid synthesis and massive cell proliferation. Treatment of intact seedlings with 2,4-D or some other auxin, while causing an increase in DNA and protein as well as in RNA, causes an apparent overproduction of RNA, resulting in increased RNA/DNA and RNA/protein ratios. The RNA accumulates mainly as ribosomal RNA. Thus, incidentally, the implications are that the herbicidal action of relatively high dosages of 2,4-D-type regulators relates to their hormonal effects, that is, evocation of massive RNA and protein synthesis rather than to some direct inhibitory action.

Excised tissues exhibit variable characteristics, depending on source and age. When plant tissues are excised and cultured for a few hours in solution, there is usually a decrease in the RNA. In the case of many excised tissues, there may be no overall increase in protein during elongation.

J. L. Key and associates have done extensive work on the effects of auxin on nucleic acid and protein metabolism in elongating and maturing sections of hypocotyls of soybean (*Glycine max*) seedlings. Much of the following discussion is based on their studies. In their work, auxin brings about a general enhancement of the synthesis of all species of RNA, with the major increase in rRNA. However, since rRNA and tRNA have a longer mean life than mRNA, it is to be expected that they would be the species that would accumulate in response to auxin over long labeling periods. In excised elongating soybean hypocotyl, the RNA decreases during incubation. Growth-promoting concentrations of auxin cause the maintenance of RNA at or near the initial level. The enhanced incorporation of ^{14}C-labeled precursor into RNA in response to auxin indicates that auxin maintains the higher RNA level by enhancing synthesis. Growth-inhibiting concentrations of auxin inhibit ^{14}C-labeled precursor incorporation without appreciably affecting the RNA content relative to the control sections. In excised maturing soybean hypocotyl, auxin causes a large net synthesis of RNA, while the RNA content of the control tissues does not change. The net accumulation of RNA in response to auxin was associated with about a twofold enhancement of incorporation of [14]C- or ^{32}P-labeled precursor, and was linear after an initial lag of 2 to 3 hours. A definite but small enhancement of incorporation of ^{14}C-labeled precursor occurred during this lag period.

Incidentally, it should be noted that in many, if not most, investigations of auxin action, 2,4-D or some other synthetic auxin is used rather than the natural

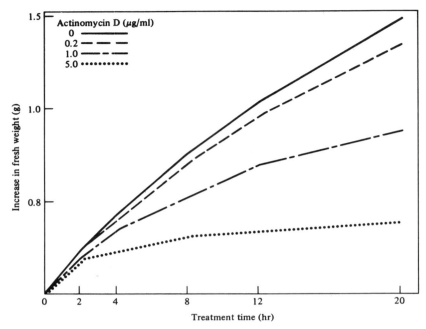

FIGURE 2.26. Inhibition of auxin-induced growth in elongating hypocotyl sections by actinomycin D. (Redrawn, with permission, from Key et al., 1967.)

auxin IAA. This is because synthetic auxins such as 2,4-D are chemically more stable than IAA, more resistant also than IAA to IAA oxidase, and less susceptible than IAA to conjugation with particular amino acids. Yet the available evidence indicates that the several synthetic auxins that are commonly used behave in a manner identical to the hormonal action of IAA.

The evidence for a requirement for continued RNA and protein synthesis for continued cell elongation is derived from studies of inhibitors of RNA and protein synthesis, or of gene transcription and translation. Actinomycin D, an inhibitor of DNA-dependent RNA synthesis, inhibits both growth of elongating hypocotyl sections (Fig. 2.26) and incorporation of radioactive precursor into total RNA (Fig. 2.27). There is a lag period of 1 or 2 hours after presentation of auxin and inhibitor before inhibition is evident. Furthermore, much of the observed RNA synthesis apparently is not directly essential to growth, since low concentrations of actinomycin D inhibited RNA synthesis by as much as 30% without inhibiting growth (Fig. 2.28).

To determine the nature of the RNA required for cell elongation, Key et al. used the more specific inhibitor 5-fluorouracil, which selectively inhibits the synthesis of all types of RNA except messenger RNA (mRNA). 5-Fluorouracil inhibited total RNA synthesis up to 70% without affecting elongation of the hypocotyl sections (Fig. 2.29). These data indicate that the RNA required most

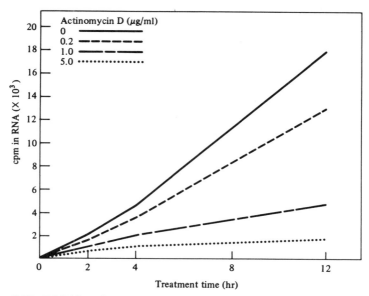

FIGURE 2.27. Inhibition of RNA synthesis in soybean hypocotyl sections by actinomycin D. (Redrawn, with permission, from Key et al., 1967.)

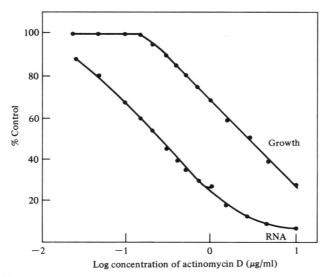

FIGURE 2.28. Inhibition of auxin-induced growth and RNA synthesis in soybean hypocotyl sections by actinomycin D. Low concentrations of actinomycin D inhibited RNA synthesis by as much as 30% without altering growth. (Redrawn, with permission, from Key et al., 1967.)

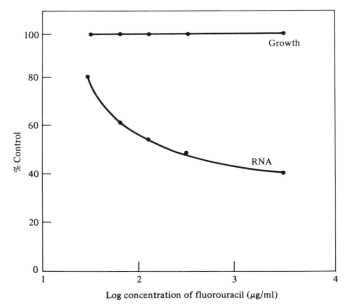

FIGURE 2.29. Inhibition of RNA synthesis but not auxin-induced growth by 5-fluorouracil. RNA synthesis was inhibited up to 60–70% without growth being affected. (Redrawn, with permission, from Key et al.,1967.)

immediately or directly to support protein synthesis and cell elongation over a period of several hours is mRNA.

When the relationship between the auxin-induced growth response and protein synthesis was examined, it was readily seen that protein synthesis, measured as incorporation of [^{14}C]leucine, was inhibited as predicted when growth was inhibited by actinomycin D (Fig. 2.30). When elongating hypocotyl sections were treated with cycloheximide, an inhibitor of cytoplasmic protein synthesis, a parallel inhibition of growth and protein synthesis also was observed (Fig. 2.31). These data show a close association between RNA and protein synthesis and the regulation of cell elongation by auxin. Moreover, in the experiments by J. L. Key and associates, a more specific correlation was observed between inhibition of auxin-induced growth by actinomycin D and putative mRNA synthesis (Fig. 2.32). In those experiments, samples of elongating soybean hypocotyl sections were first preincubated for 4 hours in solutions containing 5-fluorouracil and various concentrations of actinomycin D. Then auxin and [8-^{14}C]ADP were added to the incubation solutions and auxin-induced growth and RNA synthesis were measured during the next 4 hours.

Since cell wall loosening is essential to auxin-induced growth, it was important to ascertain whether the auxin effect on wall loosening was inhibited to the same extent as inhibition of auxin-induced growth by inhibitors of RNA and protein synthesis. The results, displayed in Fig. 2.33, were positive.

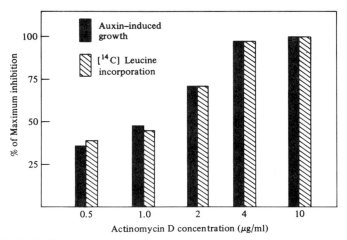

FIGURE 2.30. Inhibition of auxin-induced growth and protein synthesis in soybean hypocotyl sections by actinomycin D. (Redrawn, with permission, from Key et al., 1967.)

The experimental data just discussed show that there is a general requirement for continued synthesis of RNA and protein, very probably mRNA and particular proteins, for continued cell enlargement in excised tissues over a period of several hours. The collective observations indicate that the ability of auxin to enhance the rate of cell enlargement is dependent on new RNA and protein synthesis, and that auxin cannot cause a growth response by utilizing preexisting RNA or protein. The simplest interpretation of these observations, although not necessarily the correct one, is that the hormones are involved in the regulation of

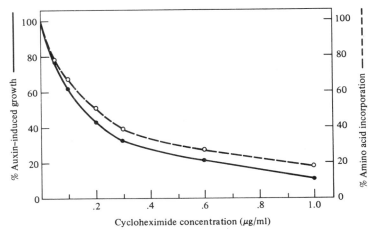

FIGURE 2.31. Parallel inhibition of auxin-induced growth and protein synthesis by cycloheximide. (Redrawn, with permission, from Key et al., 1967.)

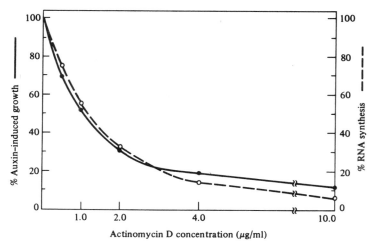

FIGURE 2.32. Parallel inhibition of auxin-induced growth and "D-RNA" (putative mRNA) synthesis by actinomycin D. 5-Fluorouracil also was present in the first incubation medium. (Redrawn, with permission, from Key et al., 1967.)

the synthesis of specific RNAs, probably mRNAs, that is, that the action of auxin is specific gene activation at the transcriptional level. These RNAs would then serve as templates for the synthesis of the proteins required for the physiological response in question. Let us examine additional evidence before drawing any such conclusions.

Refining the experimental system, investigators turned to investigations of the

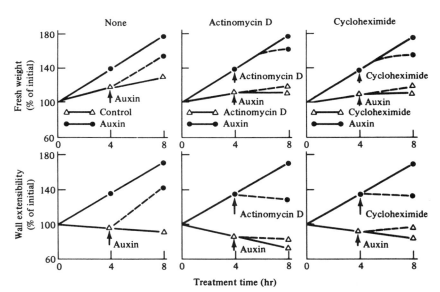

FIGURE 2.33. Effects of auxin, cycloheximide, and actinomycin D on growth and cell wall extensibility in soybean hypocotyl sections. (Redrawn, with permission, from Key et al., 1967.)

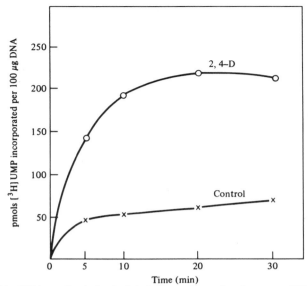

FIGURE 2.34. RNA synthesis by isolated chromatin and endogenous RNA polymerase from hypocotyls of control and 2,4-D-treated soybean seedlings. Seedlings were sprayed with a 2,4-D solution (1 mg/ml) 12 hours prior to chromatin isolation. (Redrawn, with permission, from O'Brien et al., 1968.)

effects of auxin on chromatin-directed RNA synthesis using isolated nuclei and chromatin. It was observed very quickly in such investigations that chromatin isolated from 2,4-D-treated soybean hypocotyls had a greater than twofold higher RNA polymerase activity than control chromatin (Fig. 2.34). RNA polymerase of chromatin preparations showed no response to 2,4-D added to the chromatin directly. RNA polymerase from *Escherichia coli* greatly augmented RNA synthesis of chromatin preparations from both control and 2,4-D-treated plants, and both chromatin preparations exhibited similar saturation curves (Fig. 2.35). The saturation studies with *E. coli* RNA polymerase showed that only a small proportion of the available templates of chromatin from control (7%) and 2,4-D-treated (16%) seedlings was transcribed *in vitro* by endogenous RNA polymerase. This result caused O'Brien et al. (1968) to suggest that the major influence of auxin was to increase the endogenous RNA polymerase of the isolated chromatin. However, the actual amount of RNA synthesis (picomoles) occurring in response to auxin was as great in the presence of added *E. coli* polymerase as with endogenous polymerase. This would seem to indicate that some difference exists between chromatin from control and auxin-treated tissue in addition to, or instead of, an effect on RNA polymerase. Additional evidence, obtained in 1970, showed that the RNA that was synthesized by chromatin from control and auxin-treated hypocotyl tissues was different, indicating that there is some difference in the portion of the genome being read. This could result from a change in the DNA template available for transcription or from association of RNA polymerase with template previously available but not being transcribed because of deficiency of RNA polymerase.

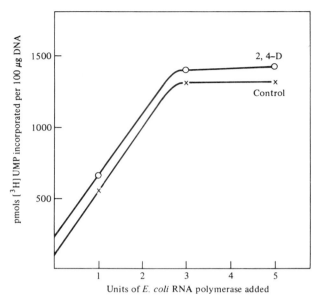

FIGURE 2.35. Experiment as in Fig. 2.34 except the chromatin preparations from control and 2,4-D-treated seedlings were saturated with exogenous RNA polymerase. (Redrawn, with permission, from O'Brien et al., 1968.)

Related investigations were conducted by Matthysse and Phillips in 1969 on effects of auxin on RNA synthesis by isolated nuclei and isolated chromatin. They introduced for the first time, as regards plant hormones, evidence for a hormone receptor. Specifically, they found evidence for a nuclear protein factor that is required as an intermediate in the effect of auxin on RNA synthesis. If nuclei from tobacco and soybean were isolated in the presence of auxin, an auxin enhancement of RNA synthesis was obtained, upon treating nuclei with 3×10^{-7} M IAA. In the absence of auxin, a protein substance required for the hormone response apparently was lost from the nucleus. In a pea chromatin system fortified with $E.$ $coli$ polymerase, neither the factor nor auxin alone affected RNA synthesis. But when both auxin and the protein factor were added together, the chromatin showed a twofold increase in capacity for RNA synthesis. In the presence of this mediator or hormone receptor, auxin increased the rate of RNA synthesis by both isolated plant nuclei and by isolated chromatin; this increased rate of RNA synthesis occurred even in the presence of saturating amounts of ($E.$ $coli$) polymerase. The hormone and protein did not affect the rate of RNA synthesis if pure DNA was used as the template. Matthysse and Phillips concluded that auxin interacts with a binding protein, the IAA–protein complex then interacting with chromatin to cause an increase in DNA template available for transcription. More will be said later about auxin receptors.

More recently Zurfluh and Guilfoyle (1982) and Theologis and Ray (1982) reported that certain mRNA sequences appeared following application of IAA and 2,4-D to soybean and pea seedling stem segments. A few mRNA sequences increased in amount or translation activity within 15 to 20 minutes of exposure of pea and soybean tissues to IAA or 2,4-D, that is, within about the same time as the auxin stimulation of cell enlargement. In Zurfluh's and Guilfoyle's work, the levels of translatable mRNA for at least 10 *in vitro* translation products were increased by 2,4-D in sections of soybean hypocotyl. The induction by auxin occurred rapidly (within 15 minutes), and the amounts of the induced *in vitro* translation products increased with time of auxin treatment. Theologis and Ray stated that, "Although for several reasons, it seems unlikely that those mRNAs are actually causative in the auxin induction of cell enlargement, their increase seems to be relatively close to primary auxin action and might well serve in maintaining a steady rate of cell enlargement over the longer term ('second phase' of auxin action in cell enlargement). Other mRNAs increase substantially subsequent to 0.5, 1 and 2 hours of auxin treatment, and beyond about 2 hours certain other mRNA sequences become repressed by auxins." Obviously, it will be very important ultimately to elucidate the physiological roles of auxin-regulated mRNAs and the mechanism of their regulation.

Auxin and Gene Expression

As we have seen, evidence that the mechanism of auxin action may involve alterations in nucleic acid metabolism was first presented over 30 years ago. Additional, more recent results from several laboratories using sensitive, molecular techniques also strongly support the more specific hypothesis that auxin can rapidly and specifically affect gene expression.

Gretchen Hagen et al. at the University of Missouri have investigated the rapid regulation of specific soybean RNAs by auxin. They have isolated several cDNAs corresponding to auxin-induced RNAs, and used these as probes to show that auxin specifically and rapidly (in minutes) induces synthesis of these sequences in soybean. Several small (500 bases), highly homologous, auxin-induced RNAs have been described. The genes coding for these RNAs are clustered within about 7 kilobase pairs of soybean genomic DNA. DNA sequence analysis of this locus revealed that the genes do not contain introns, are transcribed in opposite orientations, and have conserved sequence elements in the flanking region. This cluster of genes would appear to present a unique, relatively simple system to study the regulation of gene expression by auxin.

Walker et al. (1985) at the University of Georgia investigated the effects of cytokinin, fusicoccin, and ethylene on auxin-induced changes in gene expression during auxin-promoted cell elongation in soybean hypocotyl using cloned cDNAs to two auxin-responsive mRNAs. They found, using RNA blot analyses, that under conditions of cytokinin inhibition of auxin-promoted cell elongation the levels of those two auxin-responsive mRNAs were unaltered. Fusicoccin-promoted elongation was not associated with an enhanced expression of those two mRNAs, which suggested that the increased levels of those mRNAs observed during cell elongation were not due simply to enhanced rates of cell elongation. Finally, they determined that ethylene played no apparent role in the regulation of expression of those mRNAs. However, IAA, 2,4-D, and NAA all enhanced an accumulation of the auxin-responsive mRNAs. They concluded that the regulation of those mRNAs is directly dependent upon auxin. It has not yet been determined whether auxin-induced cell elongation is dependent on the accumulation of those mRNAs.

Auxin Receptor

The auxin receptor of maize (*Zea mays*) coleoptile membranes is one of the most widely investigated and thoroughly characterized plant hormone receptors at present, mainly because of the intensive effort by Michael A. Venis and his associates at the Institute of Horticultural Research, East Malling, Maidstone, Kent, United Kingdom.

Auxin binding sites in maize coleoptile membranes appear to fulfill many of the criteria expected of genuine hormone receptors. The binding protein is readily solubilized from the membranes and purified. It has been found to be a glycosylated homodimer having a molecular mass of 22 kDa.

Venis et al. developed a protocol for purifying (minimum 50%) sufficient auxin receptor for immunization and subsequent screening of hybridoma cultures. The protocol involved DEAE anion exchange, gel filtration, and FPLC mono Q. Five monoclonal antibodies were produced, two of which recognized only the intact 22-kDa receptor polypeptide. These monoclonal antibodies have been used to probe the intracellular distribution of the receptor and how it functions.

Interestingly, using the monoclonal antibodies to assay receptor abundance, Venis et al. showed that maize roots contain 20-fold less auxin receptor per milligram membrane protein than coleoptile. Moreover, they found that the auxin receptor of maize coleoptile membranes is homologous to the receptor in several other species, both monocots and dicots.

To further expedite identification and characterization of the auxin receptor, Venis et al. used the photoaffinity labeling agent [7-^{3}H], 5-aziindole-3-acetic acid to identify the auxin-binding protein in a solubilized membrane preparation that was highly enriched for NAA binding. Only one protein having a molecular mass of 22 kDa was labeled in this preparation. Labeling occurred at a high specific activity and was shown to be saturable. Furthermore, IAA at a concentration 10 times lower than the calculated K_d prevented photolabeling, indicating that labeling occurs at the active site. Specificity was determined by comparing the effectiveness of several compounds to prevent labeling.

In 1988 a group of investigators at the State University of Leiden, The Netherlands, reported on a cytoplasmic/nuclear protein that binds auxin with high affinity and that they considered to be an auxin receptor protein. Addition of auxin to an auxin-starved, stationary-phase cell suspension reportedly induced a rapid (5 minutes) increase of receptor in the nuclei. In parallel at least 7 cDNA clones corresponding to auxin-induced mRNAs appear within 15 minutes. Further research will be required to reveal the significance of this auxin-binding protein.

Rapid Responses to Auxin

Now let us turn to consideration of some responses of tissues to auxins that occur too rapidly and under circumstances that preclude a direct dependence on any primary gene activation effect of auxin. Examples of very rapid responses to auxin include (1) elongation of coleoptile and stem sections, which can be observed to start in less than 15 minutes after presentation of a growth-promoting concentration of auxin; (2) protoplasmic streaming, which, in response to IAA, appears in 2 minutes or less; and (3) increase in respiration rate, which often occurs about 30 minutes after addition of auxin.

The so-called "rapid responses" to auxin generally are believed to be evident in less time than is required for transcription, translation, and action of a gene product (enzyme). Transcription alone requires on the order of 2 to 10 minutes. And translation, that is, polypeptide assembly on the ribosomes, takes an additional 1.5 to 5 minutes.

M. L. Evans and P. M. Ray published a paper in 1969 which provided impetus for a surge of interest in the rapid responses to auxin. They devised an experimental apparatus (Fig. 2.36) with which elongation of a file of several sections of stem or coleoptile could be measured accurately in short periods of time, and which permitted rapid changes of solutions bathing the sections. The results are recorded automatically on photographic paper attached to a rotating cylinder by a shadowgraph technique.

A typical result is illustrated in Fig. 2.37. Immediately after being mounted in the apparatus, the sections showed a burst of rapid elongation, due to tactile stimulation, which lasted for about half an hour. After the addition of auxin in growth-promoting concentration to the bathing solution, the sections then continued a slow rate of elongation for about 10 minutes (at 23°C), the so-called "lag period," after which time a rather sudden increase in elongation occurred. Within about 3 additional minutes, a steady-state rate of elongation in response to auxin developed. Upon withdrawal of auxin (at $5 \times 10^{-5} M$ IAA), a decline in the rate of elongation began in approximately 10 minutes, and after 50 minutes the control rate was resumed. Later investigations have showed that the lag time in response to auxin could be reduced to less that 1 minute if the methyl ester of IAA, which penetrates cell membranes more rapidly than IAA, was used instead of IAA.

Very interestingly, neither actinomycin D nor cycloheximide at concentrations that partially inhibit the elongation response caused an extension of the lag period in the response to auxin, as would be expected if the rapid elongation was directly dependent upon *de novo* synthesis of RNA or protein.

Evans and Ray concluded in their 1969 paper that auxin probably does not act

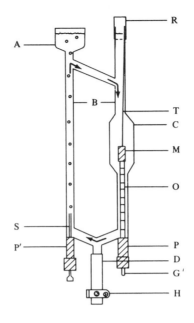

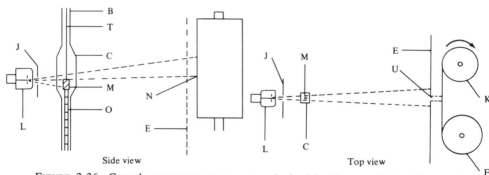

Side view   Top view

FIGURE 2.36. Growth measurement apparatus devised by Evans and Ray for use in experiments on rapid cell elongation responses. A, overflow reservoir, gas exit vent, and funnel for filling chamber; B, 7-mm-o.d. glass tubing; C, spectrophotometer cuvette (10-mm square); D, drain outlet, Tygon tubing; E, baffle, with slit (U); F, roll of photographic paper; G, glass plug; H, screw clamp; J, diaphragm; K, kymograph drum; L, zirconium arc lamp; M, weight; N, upper edge of shadow of M; O, row of coleoptile segments; P, P', Teflon plugs; S, No. 25 syringe needle; T, thread. (Redrawn, with permission, from Evans and Ray, 1969.)

on the elongation of the tissues studied, oat and corn coleoptiles, by promoting the synthesis of informational RNA or enzymatic protein. They did not exclude the possibility that auxin acts at the translational level to induce synthesis of a structural protein, such as cell wall protein or membrane protein. However, they stressed that a number of alternatives to the gene-activation hypothesis of auxin

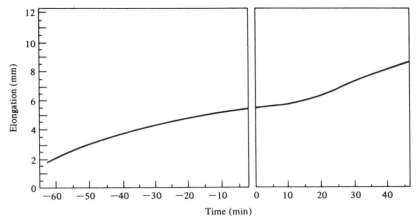

FIGURE 2.37. Record produced by the apparatus shown in Fig. 2.36. Medium was changed from water to a solution of IAA (3 μg/ml) at zero time. the elongation that was occurring at the beginning of the record was the result of tactile stimulation of the coleoptile sections as they were mounted in the apparatus. (Redrawn, with permission, from Evans and Ray, 1969.)

action exist, any of which might depend indirectly on continual protein synthesis in order to remain functional. Serious consideration should be given to possible effects of auxin on intracellular transport processes or membrane permeability properties, for example, they suggested. And they emphasized another important point which should be borne in mind when using inhibitors of RNA and protein synthesis, namely, that antibiotics will inhibit the developmental process if it depends in any way upon RNA or protein synthesis, whether or not the hormone works by gene activation. It is only to be expected that enhanced growth will in general be accompanied by increased synthesis of RNA and protein, even if the hormone does not act directly upon these processes.

The "Acid-Growth" Theory of Auxin Action

Briefly stated, the "acid-growth" theory says that auxin initiates an acidification mechanism, possibly a membrane-bound H^+ pump, with the result that the pH of the solution in the matrix of the cell wall decreases. Some still hypothetical enzyme or enzymes reponsible for wall loosening are activated by the lowered pH, the wall is loosened, and cell enlargement takes place under the direct force of the cell's own turgor pressure.

This concept that hydrogen ions act as a second messenger for auxin-promoted cell elongation, formulated first by David L. Rayle and Robert Cleland in 1970, is supported by such evidence as (1) segments of *Avena* coleoptile and pea and soybean stems elongate more rapidly at low pH; (2) wall extensibility in *Avena* and sunflower (*Helianthus*) is increased at low pH; (3) auxin-promoted elongation is preceded by auxin-induced H^+ extrusion in

Avena, pea, and corn; and (4) certain wall-bound enzymes, glycosidases, have relatively low pH optima.

Auxin-induced H^+ excretion and the capacity of H^+ to promote growth of *Avena* coleoptile segments actually had been known for some time prior to formulation of the "acid-growth" theory. In 1934 James Bonner reported that the growth of coleoptile sections was eight times greater at pH 4.1 than at pH 7.2, and he also observed that a low pH induced a rather large increase in extensibility of the cell walls. And Kenneth V. Thimann noted in 1956 that elongation of coleoptile segments in response to auxin was accompanied by an acidification of the incubation medium. In the same year, 1956, Jean and Collette Nitsch reported that hydrogen ions had a stimulatory effect on cell enlargement both in the presence and absence of auxin.

If the wall-loosening factor is indeed H^+, certain criteria obviously must be met. One is that there ought to be similar growth responses to exogenous auxin and exogenous hydrogen ions. In the case of *Avena* coleoptile sections, this indeed is the case (Fig. 2.38). Sections begin to elongate rapidly about 10 minutes after the addition of auxin and reach a maximum rate in another 10–15 minutes. Likewise, the same maximum rate of elongation is obtained by an optimal H^+ concentration (pH 3.0), but the lag period is only about 1 minute. Both auxin and H^+ induce similar increases in wall extensibility, both agents are effective only when the cells' turgor pressure is nearly maximal, and both show the same unusual temperature dependence for elongation ($Q_{10} \simeq 5$ at 15–25°C and $Q_{10} \simeq 1$ between 25 and 35°C). Moreover, there are differences between the growth responses to auxin and H^+. Metabolic inhibitors (e.g., cyanide and dinitrophenol) prevent auxin-induced elongation but are without effect on the acid-induced response, as the theory would predict. Furthermore, acidic

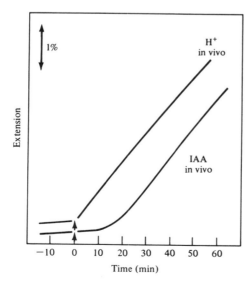

FIGURE 2.38. Kinetics of elongation of *Avena* coleoptile segments in response to 10^{-5} *M* IAA (pH 6.5 solution) and 10 m*M* citrate buffer at pH 3.0.(Redrawn, with permission, from Rayle, D. L. and R. Cleland. 1977. Control of plant cell enlargement by hydrogen ions. In: Moscona, A. A. and A. Monroy, eds. *Current Topics in Developmental Biology. Vol. 11. Pattern Development.* Pp. 187–214. Copyright by Academic Press, Inc., New York.)

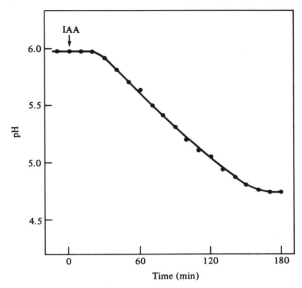

FIGURE 2.39. Auxin-induced acidification of a weakly buffered incubation medium ($10^{-5} M$ IAA in 1 mM K-P$_i$, pH 6.0). (Redrawn, with permission, from Rayle, D. L. and R. Cleland. 1977. Control of plant cell enlargement by hydrogen ions. In: Moscona, A. A. and A. Monroy, eds. *Current Topics in Developmental Biology. Vol. 11. Pattern Development.* Pp. 187–214. Copyright by Academic Press, Inc., New York.)

solutions but not auxin cause wall loosening even when added to isolated cell walls.

Determination of the optimum concentration of H^+ for elongation presented some problems. The answer was found to depend really on whether or not the cuticle of the coleoptile segments was intact. With cuticle intact, sections extended only when the pH of the incubation solution was below 4.5, and the maximum response required a pH of 3. However, if the waxy cuticle (a barrier to H^+) was removed or abraded, cell wall loosening was induced whenever the pH was below about 5.8, and a maximum response occurred at pH 4.8 or below, a much more realistic physiological pH.

A second major criterion that must be satisfied to establish the validity of the acid-growth theory is that growth-promoting concentrations of auxin should cause excretion of H^+ from coleoptile cells. If coleoptile segments are "peeled" to remove the cuticle and floated on a small volume of a weakly buffered solution (e.g., 1 mM potassium phosphate, pH 6.0), and the pH is monitored with time, auxin-induced acidification can readily be detected within approximately 20 minutes (Fig. 2.39). If a flat surface electrode is placed directly on the *Avena* sections, acidification can be detected more rapidly. The acidification and growth responses induced by exogenous auxin ($10^{-5} M$ IAA) obviously are directly correlated (Fig. 2.40).

More evidence for the role of H^+ as the wall-loosening factor came from

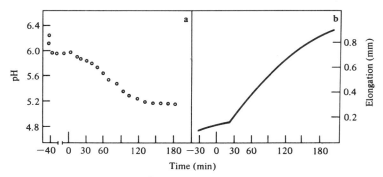

FIGURE 2.40. Effects of auxin (10^{-5} M IAA, added at zero time) on acidification of the incubation medium, **a**, and on elongation, **b**, of *Avena* coleoptile segments. (Redrawn with modifications, by permission, from Rayle, 1973.)

studies with a fungal toxin called fusicoccin. Fusicoccin acts as a "super auxin," causing pea stem sections and *Avena* coleoptile sections to grow and excrete H^+ at rates greater even than those obtained with maximal auxin (Fig. 2.41). Thus two very different agents, IAA and fusicoccin, with different modes of action, both induce acidification with kinetics that closely approximate their relative speeds of action with respect to extension growth. The important conclusion, of course, is that both cause H^+ extrusion, and it is the H^+ that evidently lead to wall loosening and thus to cell elongation.

The mechanism by which auxin causes wall acidification is unknown, but there are some hypothetical models worthy of consideration. According to one model, it is speculated that the primary action of auxin is at the plasmalemma. In

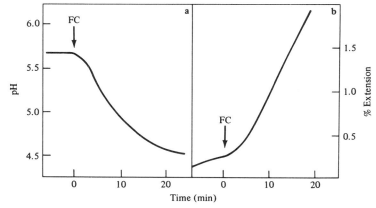

FIGURE 2.41. Kinetics of fusicoccin (FC)-initiated proton excretion, **a**, and growth, **b**, in *Avena* coleoptile sections. (Redrawn, with permission, from Rayle, D. L. and R. Cleland. Control of plant cell enlargement by hydrogen ions. In: Moscona, A. A. and A. Monroy, eds. *Current Topics in Developmental Biology. Vol. 11. Pattern Development.* Pp. 187–214. Copyright by Academic Press, Inc., New York).

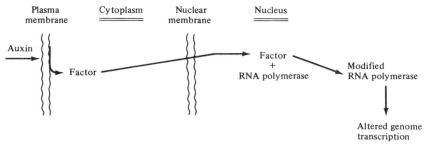

FIGURE 2.42. Hypothesis for auxin-regulated nucleic acid biosynthesis. It is proposed that auxin interacts with a factor within the plasma membrane. The factor is then transported through the nuclear membrane into the nucleus where it regulates the activity of RNA polymerase. (Redrawn, with permission, from Hardin et al., 1972.)

1972, J. W. Hardin, J. H. Cherry, D. J. Morré, and C. A. Lembi at Purdue University proposed that plasma membranes of soybean cells contain a regulatory factor that is specifically released by auxin, and that enhances the activity of purified soybean hypocotyl RNA polymerase. While by this model, the primary interaction of auxin would be with the plasmalemma, the Purdue group devised their hypothesis so as to account also for auxin-regulated nucleic acid biosynthesis (Fig. 2.42). Although still highly speculative, this model provides a possible mechanism whereby the interaction of auxin with the plasmalemma could be transmitted to the nucleus via the release of a receptor that could interact with a specific RNA polymerase. A result of this action would be altered DNA transcription and quantitative and qualitative changes in RNA synthesis. With a bit more speculation one could hypothesize that interaction of auxin with the receptor is accompanied by a release of H^+ from the plasmalemma into the cell wall, which effect seems to be directly related to cell wall loosening.

Ray (1977) reported on binding sites with high affinity for auxin on rough endoplasmic reticulum (ER) of maize coleoptiles, and he suggested an alternative model of auxin action in which the primary auxin action occurs at ER membranes. He reasoned that combination of auxin with its receptor sites on the ER could induce hydrogen ion transport from the cytoplasm into the cisternal space of the ER. The H^+ contained therein could be transported, along with secretory proteins contained in the ER space, to the cell wall, probably via the Golgi system. Assuredly this model of auxin action via receptors on the ER suggests a possible connection between rapid and longer term effects of the hormone. As Ray (1977) stated, "Longer term auxin effects that involve stimulation of protein synthesis might stem from improved delivery of nascent secretory protein into the ER lumen, which transport might be coupled to the secretion of H^+ across the same membranes."

No discussion of the acid-growth theory would be complete, at the present state of our knowledge, without mention that there are some yet unresolved reports which are not in agreement. Two reports in particular, by L. N. Vanderhoef and associates in 1977, will be mentioned. Using excised elongating soybean (*Glycine max*) hypocotyl segments, Vanderhoef et al. (1977, 1977) reported that elongating cells actively adjusted the external pH, from any pH in the range of 4 to 8, to pH 5.4. Auxin, at growth-promoting concentration, had no effect on the cellular pH adjustment of the external medium. From a comparison of acid-induced growth and auxin-induced growth, it was concluded that H^+ do not mediate in long-term auxin-induced elongation in soybean hypocotyl segments (Vanderhoef et al., 1977, 1977). Only time and more research will enable possible reconciliation of these observations with the acid-growth theory.

A model presented by Vanderhoef and Dute (1981) summarizes some earlier evidence reconciling the effects of wall acidification and RNA and protein synthesis as a two-phase action of auxin following addition of the hormone to auxin-depleted, responsive tissues. Vanderhoef and Dute pointed out that the two hypotheses—gene activation and wall acidification—are not incompatible if, as they postulated, auxin regulates both wall loosening and the supply of new wall materials that accompanies sustained cell elongation. According to their model, in the *intact seedling* a continuous supply of auxin keeps the cell wall loose by maintaining a low wall pH and the cells growing by maintaining a supply of material for wall development. There can be a steady-state growth rate of the seedling over prolonged periods of several hours and longer.

However, when exogenous auxin is added to *auxin-depleted, excised stem sections* (e.g., soybean hypocotyl) two auxin-regulated components of elongation can be observed. In this case, the reintroduction of hormone to the excised sections causes first wall loosening and a subsequent burst of turgor-driven elongation. But this initial burst of elongation of excised sections is transient, lasting only some 30 to 90 minutes, depending on the species and conditions. The rate of elongation first increases, after a lag period of 15 minutes or less, then begins to decrease, with kinetics resembling (perhaps identical to) acid-induced growth.

Then with exogenous auxin still present, the postulated auxin-caused insertion of new cell wall material begins in the excised sections about 50 minutes after auxin was first added, and the rate of elongation increases once again and eventually reaches a steady-state. The difference in lag times for the two postulated effects of auxin, 15 minutes or less and approximately 50 minutes, respectively, allows the experimental separation of the two postulated components of auxin action.

This is an attractive model, since regulation of the two major cell elongation events, wall loosening and supply of wall materials, does not necessarily require two auxin activities. Both conceivably could be initiated by a single action of auxin. According to this model, the "rapid response" experiments first conducted by Evans and Ray (1969) do not refute the gene activation hypothesis, since those experiments are applicable only to the first elongation response. Moreover, there is much evidence, discussed previously, for a gene activation action of auxin. It now seems likely that future work will yield more evidence for a unified concept of a single fundamental mechanism of auxin action.

References

Anthon, G. E. and R. M. Spanswick. 1986. Purification and properties of the H^+-translocating ATPase from the plasma membrane of tomato roots. *Plant Physiol.* **81**: 1080–1085.

Ashton, F. M. and A. S. Crafts. 1973. *Mode of Action of Herbicides.* John Wiley & Sons, New York.

Audus, L. J. 1972. *Plant Growth Substances. Vol. 1. Chemistry and Physiology.* Leonard Hill Books, London.

Bailey, H. M., E. J. D. Barker, K. R. Libbenga, P. C. G. Van der Linde, A. M. Mennes, and M. C. Elliott. 1985. An auxin receptor in plant cells. *Biol. Plant.* **27**: 105–109.

Bandurski, R. S. 1980. Homeostatic control of concentrations of indole-3-acetic acid. In: Skoog, F., ed. *Plant Growth Substances 1979.* Springer-Verlag, Berlin, Heidelberg, New York. Pp. 37–49.

Bandurski, R. S. and A. Schulze. 1974. Concentrations of indole-3-acetic acid and its esters in *Avena* and *Zea. Plant Physiol.* **54**: 257–262.

Bandurski, R. S., A. Schulze, P. Dayanandan, and P. B. Kaufman. 1984. Response to gravity by *Zea mays* seedlings. I. Time course of the response. *Plant Physiol.* **74**: 284–288.

Bandurski, R. S., A. Schulze, and D. M. Reincke. 1986. Biosynthetic and metabolic aspects of auxins. In: Bopp, M., ed. *Plant Growth Substances 1985.* Springer-Verlag, Berlin, Heidelberg, New York, Tokyo. Pp. 83–91.

Beyer, E. M., Jr. 1972. Auxin transport: A new synthetic inhibitor. *Plant Physiol.* **50**: 322–327.

Bonner, J. 1934. The relation of hydrogen ions to the growth rate of the *Avena* coleoptile. *Protoplasma* **21**: 406–423.

Bonner, J. 1961. On the mechanics of auxin-induced growth. In: *Plant Growth Regulation.* Iowa State University Press, Ames. Pp. 307–328.

Bonner, J. and A. W. Galston. 1952. *Principles of Plant Physiology.* W. H. Freeman and Company, San Francisco.

Bower, P. J., H. M. Brown, and W. K. Purves. 1978. Cucumber seedling indoleacetaldehyde oxidase. *Plant Physiol.* **61**: 107–110.

Chang, Y. P. and W. P. Jacobs. 1972. The contrast between active transport and diffusion of indole-3-acetic acid in *Coleus* petioles. *Plant Physiol.* **50**: 635–639.

Chisnell, J. R. 1984. *Myo*-inositol esters of indole-3-acetic acid are endogenous components of *Zea mays* L. shoot tissue. *Plant Physiol.* **74**: 278–283.

Chisnell, J. R. and R. S. Bandurski. 1988. Translocation of radio-labeled indole-3-acetic acid and indole-3-acetyl-*myo*-inositol from kernel to shoot of *Zea mays* L. *Plant Physiol.* **86**: 79–84.

Cleland, R. 1971. Cell wall extension. *Annu. Rev. Plant Physiol.* **22**: 197–222.

Cleland, R. E. 1987. Auxin and cell elongation. In: Davies, P. J., ed. *Plant Hormones and Their Role in Plant Growth and Development.* Martinus Nijhoff Publishers, Dordrecht, The Netherlands. Pp. 132–148.

Cleland, R. E. 1976. Kinetics of hormone-induced H^+ excretion. *Plant Physiol.* **58**: 210–213.

Cohen, J. D. 1982. Identification and quantitative analysis of indole-3-acetyl-L-aspartate from seeds of *Glycine max* L. *Plant Physiol.* **70**: 749–753.

Corcuera, L. J. and R. S. Bandurski. 1982. Biosynthesis of indol-3-yl-acetyl-*myo*-inositol arabinoside in kernels of *Zea mays* L. *Plant Physiol.* **70**: 1664–1666.

Cosgrove, D. J. 1981. Analysis of the dynamic and steady-state responses of growth rate and turgor pressure to changes in cell parameters. *Plant Physiol.* **68**: 1439–1446.

Darwin, C. 1880. *The Power of Movements in Plants.* D. Appleton and Company, New York.

Davidonis, G. H., R. H. Hamilton, and R. O. Mumma. 1980. Metabolism of 2,4-dichlorophenoxyacetic acid (2,4-D) in soybean root callus. Evidence for the conversion of 2,4-D amino acid conjugates to free 2,4-D. *Plant Physiol.* **66**: 537–540.

Davidonis, G. H., R. H. Hamilton, R. P. Vallejo, R. Buly, and R. O. Mumma. 1982. Biological properties of D-amino conjugates of 2,4-D. *Plant Physiol.* **70**: 357–360.

Davies, P. J. 1973. Current theories on the mode of action of auxin. *Bot. Rev.* **39**: 139–171.

dela Fuente, R. K. and A. C. Leopold. 1966. Kinetics of polar auxin transport. *Plant Physiol.* **41**: 1481–1484.

dela Fuente, R. K. and A. C. Leopold. 1972. Two components of auxin transport. *Plant Physiol.* **50**: 491–495.

Epstein, E., B. G. Baldi, and J. D. Cohen. 1986. Identification of indole-3-acetylglutamate from seeds of *Glycine max* L. *Plant Physiol.* **80**: 256–258.

Epstein, E., J. D. Cohen, and R. S. Bandurski. 1980. Concentration and metabolic turnover of indoles in germinating kernels of *Zea mays* L. *Plant Physiol.* **65**: 415–421.

Epstein, E., O. Sagee, J. D. Cohen, and J. Garty. 1986. Endogenous auxin and ethylene in the lichen *Ramalina duriaei*. *Plant Physiol.* **82**: 1122–1125.

Erdmann, N. and U. Schiewer. 1971. Tryptophan dependent indoleacetic-acid biosynthesis from indole, demonstrated by double-labelling experiments. *Planta* **97**: 135–141.

Evans, M. L. 1974. Rapid responses to plant hormones. *Annu. Rev. Plant Physiol.* **25**: 195–223.

Evans, M. L. 1985. The action of auxin on plant cell elongation. *Crit. Rev. Plant Sci.* **2**: 317–365.

Evans, M. L. and P. M. Ray. 1969. Timing of the auxin response in coleoptiles and its implications regarding auxin action. *J. Gen. Physiol.* **53**: 1–20.

Feldman, L. J. and W. R. Briggs. 1987. Light-regulated gravitropism in seedling roots of maize. *Plant Physiol.* **83**: 241–243.

Gabathuler, R. and R. E. Cleland. 1985. Auxin regulation of a proton translocating ATPase in pea root plasma membrane vesicles. *Plant Physiol.* **79**: 1080–1085.

Galston, A. W. 1967. Regulatory systems in higher plants. *Am. Scientist* **55**: 144–160.

Galston, A. W. and P. J. Davies. 1970. *Control Mechanisms in Plant Development.* Prentice-Hall, Englewood Cliffs, New Jersey.

Goldsmith, M. H. M. 1968. The transport of auxin. *Annu. Rev. Plant Physiol.* **19**: 347–360.

Goldsmith, M. H. M. 1977. The polar transport of auxin. *Annu. Rev. Plant Physiol.* **28**: 439–478.

Gordon, S. A. 1961. The biogenesis of auxin. In: Ruhland, W., ed. *Handbuch der Pflanzenphysiologie. Vol. XIV.* Springer-Verlag, Berlin. Pp. 620–646.

Gove, J. P. and M. C. Hoyle. 1975. The isozymic similarity of indoleacetic acid oxidase to peroxidase in birch and horseradish. *Plant Physiol.* **56**: 684–687.

Guilfoyle, T. J., C. Y. Lin, Y. M. Chien, R. T. Nagao, and J. L. Key. 1975. Enhancement of soybean RNA polymerase I by auxin. *Proc. Natl. Acad. Sci. U.S.A.* **72**: 69–72.

Haagen-Smith, A. J., W. B. Dandliker, S. H. Wittwer, and A. E. Murneek. 1946. Isolation of 3-indoleacetic acid from immature corn kernels. *Am. J. Bot.* **33**: 118–120.

Hagen, G. 1987. The control of gene expression by auxin. In: Davies, P. J., ed. *Plant Hormones and Their Role in Plant Growth and Development.* Martinus Nijhoff Publishers, Dordrecht, The Netherlands. Pp. 149–163.

Hagen, G. and T. J. Guilfoyle. 1985. Rapid induction of selective transcription by auxins. *Mol. Cell Biol.* **5**: 1197–1203.

Hager, A., H. Menzel, and A. Krauss. 1971. Versuche and Hypothese zur Primärwirkung des Auxins beim Streckungswachstum. *Planta* **100:** 47–75.

Hangarter, R. P. and N. E. Good. 1981. Evidence that IAA conjugates are slow-release sources of free IAA in plant tissues. *Plant Physiol.* **68**: 1424–1427.

Hardin, J. W., J. H. Cherry, D. J. Morré, and C. A. Lembi. 1972. Enhancement of RNA polymerase activity by a factor released by auxin from plasma membrane. *Proc. Natl. Acad. Sci. U.S.A.* **69**: 3146–3150.

Harrison, M. A. and P. B. Kaufman. 1980. Hormonal regulation of lateral bud (tiller) release in oats (*Avena sativa* L.). *Plant Physiol.* **66**: 1123–1127.

Hasenstein, K.-H. and M. L. Evans. 1986. Calcium dependence of rapid auxin action in maize roots. *Plant Physiol.* **81**: 439–443.

Hasenstein, K.-H. and D. Rayle. 1984. Cell wall pH and auxin transport velocity. *Plant Physiol.* **76**: 65–67.

Hatfield, R. D. and C. E. LaMotte. 1985. Gravitropic responses of partially decapitated corn coleoptiles with and without applied [^{14}C]indoleacetic acid. *Plant Physiol.* **77**: 475–480.

Hertel, R. 1986. Two comments on auxin transport: The uptake/efflux-mechanism and the problem of adaptation. In Bopp, M., ed. *Plant Growth Substances 1985.* Springer-Verlag, Berlin, Heidelberg, New York, Tokyo. Pp. 214–217.

Hinman, R. L. and J. Lang. 1965. Peroxidase-catalyzed oxidation of indole-3-acetic acid. *Biochemistry* **4**: 144–158.

Iino, M. and D. J. Carr. 1982. Sources of free IAA in the mesocotyl of etiolated maize seedlings. *Plant Physiol.* **69**: 1109–1112.

Jacobs, M. and S. F. Gilbert. 1983. Basal localization of the presumptive auxin transport carrier in pea stem cells. *Science* **220**: 1297–1300.

Jacobs, W. P., K. Falkenstein, and R. H. Hamilton. 1985. Nature and amount of auxin in algae. IAA from extracts of *Caulerpa paspaloides* (Siphonales). *Plant Physiol.* **78**: 844–848.

Kasamo, K. 1986. Purification and properties of the plasma membrane H$^+$-translocating adenosine triphosphatase of *Phaseolus mungo* L. *Plant Physiol.* **80**: 818–824.

Kateckar, G. F. and A. E. Geissler. 1980. Auxin transport inhibitors. IV. Evidence of a common mode of action for a proposed class of auxin transport inhibitors: The phytotropins. *Plant Physiol.* **66**: 1190–1195.

Keegstra, K., K. W. Talmadge, W. D. Bauer, and P. Albersheim. 1973. The structure of plant cell walls. III. A model of the walls of suspension-cultured sycamore cells based on the interconnections of the macromolecular components. *Plant Physiol.* **51**: 188–196.

Kende, H. and G. Gardner. 1976. Hormone binding in plants. *Annu. Rev. Plant Physiol.* **27**: 267–290.

Key, J. L. 1969. Hormones and nucleic acid metabolism. *Annu. Rev. Plant Physiol.* **20**: 449–474.

Key, J. L., N. M. Barnett, and C. Y. Lin. 1967. RNA and protein biosynthesis and the regulation of cell elongation by auxin. *Ann. New York Acad. Sci.* **144**: 49–62.

Kögl, F. and A. J. Haagen-Smit. 1931. Über die Chemie des Wuchsstoffs. K. Akad. Wetenschap. Amsterdam. *Proc. Sect. Sci.* **34**: 1411–1416.

Krul, W. R. 1972. Polar indole-3-acetic acid diffusion in nonliving and model systems. *Plant Physiol.* **50**: 784–787.

Kutschera, U. and P. Schopfer. 1985. Evidence against the acid growth theory of auxin action. *Planta* **163**: 483–493.

Labarca, C., P. B. Nicholls, and R. S. Bandurski. 1966. A partial characterization of indoleacetylinositols from *Zea mays*. *Biochem. Biophys. Res. Commun.* **20**: 641–646.

Lamport, D. T. A. 1970. Cell wall metabolism. *Annu. Rev. Plant Physiol.* **21**: 235–270.

Larsen, P. 1951. Formation, occurrence, and inactivation of growth substances. *Annu. Rev. Plant Physiol.* **2**: 169–198.

Lee, J. S. and M. L. Evans. 1985. Polar transport of auxin across gravistimulated roots of maize and its enhancement by calcium. *Plant Physiol.* **77**: 824–827.

Leopold, A. C. 1955. *Auxins and Plant Growth*. University of California Press, Berkeley and Los Angeles.

Leopold, A. C. and O. F. Hall. 1966. Mathematical model of polar auxin transport. *Plant Physiol.* **41**: 1476–1480.

Leopold, A. C. and P. E. Kriedemann. 1975. *Plant Growth and Development*. 2nd ed. McGraw-Hill Book Company, New York.

Libbenga, K. R. and A. M. Mennes. 1987. Hormone binding and its role in hormone action. In: Davies, P. J., ed. *Plant Hormones and Their Role in Plant Growth and Development*. Martinus Nijoff Publishers, Dordrecht, The Netherlands. Pp. 194–221.

Libbert, E., S. Wichner, U. Schiewer, H. Risch, and W. Kaiser. 1966. The influence of epiphytic bacteria on auxin metabolism. *Planta* **68**: 327–334.

Löbler, M. and D. Klämbt. 1985. Auxin-binding protein from coleoptile membranes of corn (*Zea mays* L.). II. Localization of a putative receptor. *J. Biol. Chem.* **260**: 9854–9859.

Lomax, T. L. 1986. Active auxin uptake by specific plasma membrane carriers. In: Bopp, M., ed. *Plant Growth Substances 1985*. Springer-Verlag, Berlin, Heidelberg, New York, Tokyo. Pp. 209–213.

MacDonald, I. R. and J. W. Hart. 1987. New light on the Cholodny-Went Theory. *Plant Physiol.* **84**: 568–570.

Marré, E., P. Lado, F. Rasi-Caldogno, R. Colombo, M. Cocucci, and M. I. DeMichelis. 1975. Regulation of proton extrusion by plant hormones and cell elongation. *Physiol. Vég.* **13**: 797–811.

Masuda, Y. 1969. Auxin-induced cell expansion in relation to cell wall extensibility. *Plant Cell Physiol.* **10**: 1–9.

Matheron, M. E. and T. C. Moore. 1973. Properties of an aminotransferase of pea (*Pisum sativum* L.). *Plant Physiol.* **52**: 63–67.

Matthysse, A. G. and C. Phillips. 1969. A protein intermediary in the interaction of a hormone with the genome. *Proc. Natl. Acad. Sci. U.S.A.* **63**: 897–903.

Migliaccio, F. and D. L. Rayle. 1984. Sequence of key events in shoot gravitropism. *Plant Physiol.* **75**: 78–81.

Miura, G. A. and S. E. Mills. 1971. The conversion of D-tryptophan to L-tryptophan in cell cultures of tobacco. *Plant Physiol.* **47**: 483–487.

Moore, T. C. 1969. Comparative net biosynthesis of indoleacetic acid from tryptophan in cell-free extracts of different parts of *Pisum sativum* plants. *Phytochemistry* **8**: 1109–1120.

Moore, T. C. and C. A. Shaner. 1967. Biosynthesis of indoleacetic acid from tryptophan-¹⁴C in cell-free extracts of pea shoot tips. *Plant Physiol.* **42**: 1787–1796.

Moore, T. C. and C. A. Shaner. 1968. Synthesis of indoleacetic acid via indolepyruvic acid in cell-free extracts of pea seedlings. *Arch. Biochem. Biophys.* **127**: 613–621.

Morré, D. J. and J. H. Cherry. 1977. Auxin hormone-plasma membrane interactions. In: Pilet, P. E., ed. *Plant Growth Regulation.* Springer-Verlag, New York. Pp. 35–43.

Murray, A. K. and R. S. Bandurski. 1975. Correlative studies of cell wall enzymes and growth. *Plant Physiol.* **56**: 143–147.

Nelson, N. 1988. Structure, function, and evolution of proton-ATPases. *Plant Physiol.* **86**: 1–3.

Newman, I. A. 1963. Electric potentials and auxin translocation in *Avena. Aust. J. Biol. Sci.* **16**: 629–646.

Newman, I. A. 1970. Auxin transport in *Avena.* I. Indoleacetic acid-¹⁴C distributions and speeds. *Plant Physiol.* **46**: 263–272.

Nishitani, K. and Y. Masuda. 1983. Auxin-induced changes in cell wall xyloglucans. *Plant Cell Physiol.* **24**: 345–355.

Nitsch, J. P. and C. Nitsch. 1956. Studies on the growth of coleoptile and first internode sections. A new, sensitive, straight-growth test for auxins. *Plant Physiol.* **31**: 94–111.

Nowacki, J. and R. S. Bandurski. 1980. *Myo*-inositol esters of indole-3-acetic acid as seed auxin precursors of *Zea mays* L. *Plant Physiol.* **65**: 422–427.

O'Brien, T. J., B. C. Jarvis, J. H. Cherry, and J. B. Hanson. 1968. Enhancement by 2,4-dichlorophenoxyacetic acid of chromatin RNA polymerase in soybean hypocotyl tissue. *Biochim. Biophys. Acta* **169**: 35–43.

Paleg, L. G. 1965. Physiological effects of gibberellins. *Annu. Rev. Plant Physiol.* **16**: 291–322.

Park, R. D. and C. K. Park. 1987. Oxidation of indole-3-acetic acid-amino acid conjugates by horseradish peroxidase. *Plant Physiol.* **84**: 826–829.

Percival, F. W. and R. S. Bandurski. 1976. Esters of indole-3-acetic acid from *Avena* seeds. *Plant Physiol.* **58**: 60–67.

Phillips, I. D. J. 1971. *Introduction to the Biochemistry and Physiology of Plant Growth Hormones.* McGraw-Hill Book Company, New York.

Phillips. I. D. J. 1975. Apical dominance. *Annu. Rev. Plant Physiol.* **26**: 341–367.

Pilet, P.-E. and M. Saugy. 1987. Effect on root growth of endogenous and applied IAA and ABA. A critical reexamination. *Plant Physiol.* **83**: 33–38.

Pillay, I. and I. D. Railton. 1983. Complete release of axillary buds from apical dominance in intact, light-grown seedlings of *Pisum sativum* L. following a single application of cytokinin. *Plant Physiol.* **71**: 972–974.

Pless, T., M. Böttger, P. Hedden, and J. Graebe. 1984. Occurrence of 4-Cl-indoleacetic acid in broad beans and correlation of its levels with seed development. *Plant Physiol.* **74**: 320–323.

Price, C. A. 1970. *Molecular Approaches to Plant Physiology.* McGraw-Hill Book Company, New York.

Rasi-Caldogno, F., M. I. DeMichelis, M. C. Pugliarello, and E. Marré. 1986. H⁺-pumping driven by the plasma membrane ATPase in membrane vesicles from radish:

Stimulation by fusicoccin. *Plant Physiol.* **82**: 212–125.

Raven, J. A. 1975. Transport of indoleacetic acid in plant cells in relation to pH and electrical potential gradients, and its significance for polar IAA transport. *New Phytol.* **74**: 163–172.

Ray, P. M. 1958. Destruction of auxin. *Annu. Rev. Plant Physiol.* **9**: 81–118.

Ray, P. M. 1960. The destruction of indoleacetic acid. III. Relationships between peroxidase action and indoleacetic acid oxidation. *Arch. Biochem. Biophys.* **87**: 19–30.

Ray, P. M. 1974. The biochemistry of the action of indoleacetic acid on plant growth. In: Runeckles, V. C., E. Sondheimer, and D. C. Walton, eds. *The Chemistry and Biochemistry of Plant Hormones. Vol. 7. Recent Advances in Phytochemistry.* Academic Press, New York, Pp. 93–122.

Ray, P. M. 1977. Auxin-binding sites of maize coleoptiles are localized on membranes of the endoplasmic reticulum. *Plant Physiol.* **59**: 594–599.

Ray, P. M. 1985. Auxin and fusicoccin enhancement of β-glucan synthase in peas. An intracellular enzyme activity apparently modulated by proton extrusion. *Plant Physiol.* **78**: 466–472.

Ray, P. M. 1987. Involvement of macromolecule biosynthesis in auxin and fusicoccin enhancement of β-glucan synthase activity in pea. *Plant Physiol.* **85**: 523–528.

Ray, P. M., P. B. Green, and R. Cleland. 1972. Role of turgor in plant cell growth. *Nature (London)* **239**: 163–164.

Rayle, D. L. 1973. Auxin-induced hydrogen-ion secretion in *Avena* coleoptiles and its implications. *Planta* **114**: 63–73.

Rayle, D. L. and R. Cleland. 1970. Enhancement of wall loosening and elongation by acid solutions. *Plant Physiol.* **46**: 250–253.

Rayle, D. L. and R. Cleland. 1972. The *in vitro* acid-growth response: relation to *in vivo* growth responses and auxin action. *Planta* **104**: 282–296.

Rayle, D. L. and R. Cleland. 1977. Control of plant cell enlargement by hydrogen ions. In: Moscona, A. A. and A. Monroy, eds. *Current Topics in Developmental Biology. Vol. 11. Pattern Development.* Academic Press, New York. Pp. 187–214.

Rayle, D. L. and R. E. Cleland. 1980. Evidence that auxin-induced growth of soybean hypocotyls involves proton excretion. *Plant Physiol.* **66**: 433–437.

Rayle, D. L., M. L. Evans, and R. Hertel. 1970. Action of auxin on cell elongation. *Proc. Natl. Acad. Sci. U.S.A.* **65**: 184–191.

Rayle, D. L. and W. K. Purves. 1967a. Isolation and identification of indole-3-ethanol (tryptophol) from cucumber seedlings. *Plant Physiol.* **42**: 520–524.

Rayle, D. L. and W. K. Purves. 1967b. Conversion of indole-3-ethanol to indole-3 acetic acid in cucumber seedling shoots. *Plant Physiol.* **42**: 1091–1093.

Reinecke, D. M. and R. S. Bandurski. 1981. Metabolic conversion of ^{14}C-indole-3-acetic acid to ^{14}C-oxindole-3-acetic acid. *Biochem. Biophys. Res. Commun.* **103**: 429–433.

Reinecke, D. M. and R. S. Bandurski. 1983. Oxindole-3-acetic acid, an indole-3-acetic acid catabolite in *Zea mays. Plant Physiol.* **17**: 211–213.

Reinecke, D. M. and R. S. Bandurski. 1987. Auxin biosynthesis and metabolism. In: Davies, P. J., ed. *Plant Hormones and Their Role in Plant Growth and Development.* Martinus Nijhoff Publishers, Dordrecht, The Netherlands. Pp. 24–42.

Reinecke, D. M. and R. S. Bandurski. 1988. Oxidation of indole-3-acetic acid to oxindole-3-acetic acid by an enzyme preparation from *Zea mays. Plant Physiol.* **86**: 868–872.

Rubery, P.H. 1981. Auxin receptors. *Annu. Rev. Plant Physiol.* **32**: 569–596.

Rubery P. H. 1987. Auxin transport. In: Davies, P. J., ed. *Plant Hormones and Their Role in Plant Growth and Development*. Martinus Nijhoff Publishers, Dordrecht, The Netherlands. Pp. 341–362.

Rubery, P. H. and A. R. Sheldrake. 1974. Carrier-mediated auxin transport. *Planta* **118**: 101–121.

Salisbury, F. B. and C. W. Ross. 1985. *Plant Physiology*. 3rd ed. Wadsworth Publishing Company, Belmont, California.

Schneider, A. and F. Wightman. 1974. Metabolism of auxin in higher plants. *Annu. Rev. Plant Physiol.* **25**: 487–513.

Schrank, A. R. 1951. Electrical polarity and auxins. In: Skoog, F., ed. *Plant Growth Substances*. University of Wisconsin Press, Madison. Pp. 123–140.

Scott, T. K. 1972. Auxins and roots. *Annu. Rev. Plant Physiol.* **23**: 235–258.

Scott, T. K. and W. R. Briggs. 1960. Auxin relationships in the Alaska pea (*Pisum sativum*). *Am. J. Bot.* **47**: 492–499.

Scott, T. K. and W. R. Briggs. 1963. Recovery of native and applied auxin from the dark-grown 'Alaska' pea seedling. *Am. J. Bot.* **50**: 652–657.

Scott, T. K. and M. B. Wilkins. 1968. Auxin transport in roots. II. Polar flux of IAA in *Zea* roots. *Planta* **83**: 323–334.

Senn, A. P. and M. H. M. Goldsmith. 1988. Regulation of electrogenic proton pumping by auxin and fusicoccin as related to the growth of *Avena* coleoptiles. *Plant Physiol.* **88**: 131–138.

Sherwin, J. E. 1970. A tryptophan decarboxylase from cucumber seedlings. *Plant Cell Physiol.* **11**: 865–872.

Shinkle, J. R. and W. R. Briggs. 1984. Auxin concentration/growth relationship for *Avena* coleoptile sections from seedlings grown in complete darkness. *Plant Physiol.* **74**: 335–339.

Sonner, J. M. and W. K. Purves. 1985. Natural occurrence of indole-3-acetylaspartate and indole-3-acetylglutamate in cucumber shoot tissue. *Plant Physiol.* **77**: 784–785.

Spanswick, R. M. 1981. Electrogenic ion pumps. *Annu. Rev. Plant. Physiol.* **32**: 267–289.

Taiz, L. 1984. Plant cell expansion: regulation of cell wall mechanical properties. *Annu. Rev. Plant Physiol.* **35**: 585–657.

Talbott, L. D., P. M. Ray, and J. K. M. Roberts. 1988. Effect of indoleacetic acid- and fusicoccin-stimulated proton extrusion on internal pH of pea internode cells. *Plant Physiol.* **87**: 211–216.

Talmadge, K. W., K. Keegstra. W. D. Bauer, and P. Albersheim. 1973. The structure of plant cell walls. I. The macromolecular components of the walls of suspension-cultured sycamore cells with a detailed analysis of the pectic polysaccharides. *Plant Physiol.* **51**: 158–173.

Theologis, A., T. V. Huynh, and R. W. Davis. 1985. Rapid induction of specific mRNAs by auxin in pea epicotyl tissue. *J. Mol. Biol.* **183**: 53–68.

Theologis, A. and P. M. Ray. 1982. Changes in messenger RNAs under the influence of auxins. In: Wareing, P. F., ed. *Plant Growth Substances 1982*. Academic Press, London. Pp. 43–57.

Thimann, K. V. 1934. Studies on the growth hormone of plants. VI. The distribution of the growth substance in plant tissues. *J. Gen. Physiol.* **18**: 23–34.

Thimann, K. V. 1937. On the nature of inhibitions caused by auxins. *Am. J. Bot.* **24**: 407–412.

Thimann, K. V. 1956. Studies on the growth and inhibition of isolated plant parts. V. The effects of cobalt and other metals. *Am. J. Bot.* **43**: 241–250.

Thimann, K. V. 1963. Plant growth substances; past, present and future. *Annu. Rev. Plant Physiol.* **14**: 1–18.

Thimann, K. V. 1969. The auxins. In: Wilkins, M. B., ed. *The Physiology of Plant Growth and Development*. McGraw-Hill Publishing Company Limited, London. Pp. 1–45.

Thimann, K. V. 1977. *Hormone Action in the Whole Life of Plants*. University of Massachusetts Press, Amherst.

Thimann, K. V. and F. Skoog. 1934. On the inhibition of bud development and other functions of growth substance in *Vicia fava*. *Proc. Royal Soc., Ser. B., Biol. Sci. London* **114**: 317–339.

Torrey, J. G. 1976. Root hormones and plant growth. *Annu. Rev. Plant Physiol.* **27**: 435–459.

Travis, R. L. and J. L. Key. 1976. Auxin-induced changes in the incorporation of ^{3}H-amino acids into soybean ribosomal proteins. *Plant Physiol.* **57**: 936–938.

Tuli, V. and H. S. Moyed. 1967. Inhibitory oxidation products of indole-3-acetic acid: 3-hydroxymethyloxindole and 3-methyleneoxindole as plant metabolities. *Plant Physiol.* **42**: 425–430.

Ueda, M., A. Ehmann, and R. S. Bandurski. 1970. Gas-liquid chromatographic analysis of indole-3-acetic acid myoinositol esters in maize kernels. *Plant Physiol.* **46**: 715–719.

Vanderhoef, L. N. and R. R. Dute. 1981. Auxin-regulated wall loosening and sustained growth in elongation. *Plant Physiol.* **67**: 146–149.

Vanderhoef, L. N., J. S. Findley, J. J. Burke, and W. E. Blizzard. 1977. Auxin has no effect on modification of external pH by soybean hypocotyl cells. *Plant Physiol.* **59**: 1000–1003.

Vanderhoef, L. N., T. S. Lu, and C. A. Williams. 1977. Comparison of auxin-induced and acid-induced elongation in soybean hypocotyl. *Plant Physiol.* **59**: 1004–1007.

Varner, J. E. and D. T. Ho. 1976. Hormones. In: Bonner. J. and J. E. Varner, eds. *Plant Biochemistry*. 3rd ed. Academic Press, New York. Pp. 713–770.

Venis, M. A. 1972. Auxin-induced conjugation systems in peas. *Plant Physiol.* **49**: 24–27.

Venis, M. A. 1985. *Hormone Binding in Plants*. Longman Inc., New York, London.

Verma, D. P. S., G. A. Maclachlan, H. Byrne, and D. Ewings. 1975. Regulation and *in vitro* translation of messenger ribonucleic acid for cellulase from auxin-treated pea epicotyls. *J. Biol. Chem.* **250**: 1019–1026.

Vickery, L. E. and W. K. Purves. 1972. Isolation of indole-3-ethanol oxidase from cucumber seedlings. *Plant Physiol.* **49**: 716–721.

Vijayaraghavan, S. J. and W. L. Pengelly. 1986. Bound auxin metabolism in cultured crown-gall tissues of tobacco. *Plant Physiol.* **80**: 315–321.

Walker, J. C., J. Legocka, L. Edelman, and J. L. Key. 1985. An analysis of growth regulator interactions and gene expression during auxin-induced cell elongation using cloned complementary DNAs to auxin-responsive messenger RNAs. *Plant Physiol.* **77**: 847–850.

Weaver, R. J. 1972. *Plant Growth Substances in Agriculture*. W. H. Freeman and Company, San Francisco.

Went, F. W. 1928. Wuchstoff und Wachstum. *Rec. Trav. Bot. Néer.* **25**: 1–116.

Went, F. W. and K. V. Thimann. 1937. *Phytohormones*. Macmillan Company, New York.

Wightman, F. and B. S. Rauthan. 1974. Evidence for the biosynthesis and natural occurrence of the auxin, phenylacetic acid, in shoots of higher plants. In: *Plant Growth Substances 1973*. Hirokawa Publishing Company, Tokyo. Pp. 15–27.

Winter, A. 1966. A hypothetical route for the biogenesis of IAA. *Planta* **71**: 229–239.

Wright, L. Z. and D. L. Rayle. 1983. Evidence for a relationship between H^+ excretion and auxin in shoot gravitropism. *Plant Physiol.* **72**: 99–104.

Zenk, M. H. and H. Scherf, 1963. D-Tryptophan in höheren Pflanzen. *Biochim. Biophys. Acta* **71**: 737–738.

Zurfluh, L. L. and T. J. Guilfoyle. 1982a. Auxin-induced changes in the population of translatable messenger RNA in elongating sections of soybean hypocotyl. *Plant Physiol.* **69**: 332–337.

Zurfluh, L. L. and T. J. Guilfoyle. 1982b. Auxin-and ethylene-induced changes in the population of translatable messenger RNA in basal sections and intact soybean hypocotyl. *Plant Physiol.* **69**: 338–340.

CHAPTER 3

Gibberellins

Brief History of Discovery

All research on the gibberellins (GAs) actually stems from the work of E. Kurosawa, a Japanese plant pathologist working in Formosa, who generally is credited with having discovered GA in 1926. However, the gibberellin story actually had its beginning in the last decade of the nineteenth century.

In 1898, Konishi, a semiliterate Japanese farmer, dictated an agricultural book in which was contained the earliest known description of what is now called the "bakanae" ("foolish seedling") disease of rice. The most characteristic symptom of the disease is the appearance of tall, spindly plants. Also, leaf sheaths are longer; leaves are longer, narrower, and thinner; root growth and tillering are reduced; plants are chlorotic; and severe cases cause death.

From 1898 to 1910, Hori provided the first description of the disease with its causal fungus. In 1926, Kurosawa, after several failures, succeeded in producing the bakanae effect in rice and maize seedlings by treating them with cell-free culture medium in which *Gibberella fujikuroi* had been grown (Fig. 3.1). In 1931, Wollenweber corrected the taxonomy of the pathogen and established that bakanae is caused by *Fusarium moniliforme* Sheld., the asexual or imperfect stage of the ascomycete *Gibberella fujikuroi* (Saw.) Wr. Actually, *F. moniliforme* is known to attack a wide variety of hosts, among the most important being maize, cotton, rice, and sugar cane.

T. Yabuta, at the University of Tokyo, assigned the name "gibberellin" to the active factor in *G. fujikuroi* culture filtrates in 1935, and in 1938 Yabuta and Y. Sumiki announced the isolation of two crystalline, biologically active substances, which they named "gibberellins A and B."

It was nearly 10 years before Western abstracts of these papers became available. So actually, the existing GA story remained unknown in the Western world until about 1950, when several groups of workers, both in the United States and in England, discovered the existing Japanese papers.

In 1955, F. H. Stodola et al. of the U. S. Department of Agriculture undertook the purification of GAs from fungal filtrates. Apparently, almost simultaneously with the work of Stodola et al., scientists at Imperial Chemical Industries in England (including P. W. Brian, A. Borrow, G. W. Elson, B. E. Cross, and others) began large scale preparation of GAs from fungus cultures. Both the American and British work resulted in the isolation of an entirely new compound from *G. fujikuroi* culture filtrates, named "gibberellin X" by the USDA team (1955) and "gibberellic acid" by the I.C.I. group (1954). The latter

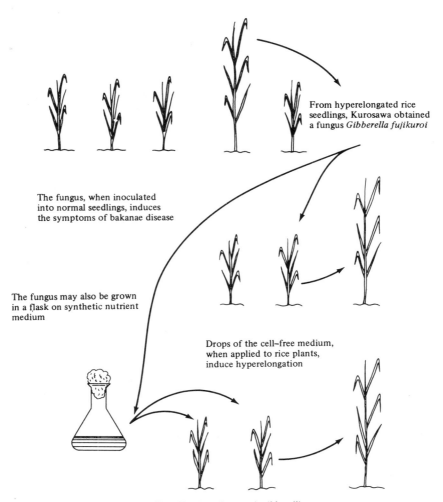

From hyperelongated rice
seedlings, Kurosawa obtained
a fungus *Gibberella fujikuroi*

The fungus, when inoculated
into normal seedlings, induces
the symptoms of bakanae disease

The fungus may also be grown
in a flask on synthetic nutrient
medium

Drops of the cell–free medium,
when applied to rice plants,
induce hyperelongation

The effective substance is gibberellin

FIGURE 3.1. Discovery of GA through investigations of bakanae, a disease of rice caused by *Gibberella fujikuroi* (*Fusarium moniliforme*). (Redrawn from *The Life of the Green Plant* by A. W. Galston. p. 74. Copyright © 1961 by Prentice-Hall, Inc. By permission of Prentice-Hall, Inc.)

name has now been universally accepted, and gibberellic acid is now also known as GA_3.

Intensive research on the effects of gibberellic acid, of fungal origin, on higher plants was conducted in the 1950s, and in the same decade GA-like substances were discovered to occur naturally in higher plants.

The discovery that GAs are natural products of higher plants was made independently in 1956 by C. A. West and B. O. Phinney at the University of

California at Los Angeles and by Margaret Radley of the Akers Research Laboratories of Imperial Chemical Industries in England (Radley, 1956; West and Phinney, 1956). GA-like substances were isolated from seeds and fruits of plants of nine genera representing seven families of angiosperms by Phinney and associates (West and Phinney, 1956; Phinney et al., 1957) and from pea (*Pisum sativum* L.) seedling shoots (Radley, 1956), and later from other parts of pea seedlings and mature seeds of several species, by Radley (1958). Actually, in 1951, J. W. Mitchell, D. P. Skaggs, and W. P. Anderson of the U. S. Department of Agriculture prepared extracts from immature bean seeds that had activity that can now be interpreted as GA activity, but at the time they did not know about GAs.

    MacMillan and Suter apparently were the first to isolate and chemically identify a natural GA from higher plants when they extracted and identified GA_1 from immature seeds of *Phaseolus coccineus* L. in 1958. In the same year, West and Murashige isolated and identified Ga_1 from seed of *Phaseolus vulgaris* L. Now it is firmly established that GAs occur naturally in a large number of species, and it is reasonable to conclude that these substances are ubiquitous in all angiosperms and gymnosperms, and plants of at least some other phyla in addition. The number of GAs now known from all sources is 76.

Chemical Characterization of GAs

GAs were defined by L. G. Paleg in 1965 as compounds having an *ent*-gibberellane skeleton (Fig. 3.2) and biological activity in stimulating cell division or cell elongation, or both, or such other biological activity as may be specifically associated with this type of naturally occurring substance.

    GAs occur naturally in three chemical forms or states, two of which are chemically defined and the third of which is hypothetical: (1) "free GAs," (2) "conjugated GAs," and (3) other "water-soluble" or "bound GAs."

    Free GAs occur as the 19-carbon or 20-carbon, mono-, di-, or tricarboxylic acids unassociated by any detectable form of bonding to other substances (Fig.

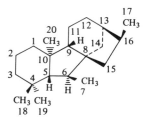

ent-gibberellane

FIGURE 3.2. Structural formula of *ent*-gibberellane.

3.3). The terms C_{20}- and C_{19}-GAs denote compounds that have retained and lost, respectively, carbon atom 20. Other differences among the GAs involve the presence or absence of a lactone configuration in ring A and the number and position of hydroxyl groups.

Many workers have compared the plant growth-promoting properties of numerous GAs and some related compounds in a variety of bioassays. It is now established that, besides differing in potency in any one test system, the GAs also exhibit a certain amount of plant species specificity. P. W. Brian and associates at Imperial Chemical Industries in England examined 134 compounds related to GAs A_1–A_9 for growth-promoting activity in four bioassays in 1967, and they discussed extensively relationships between structure and activity. The general conclusions at that time were (1) no one particular feature of the active compounds could be regarded as essential for activity, but there appeared to be requirements for the higher grades of activity; (2) the highly active compounds all contained an intact gibberellane ring system (rings A, B, C, D), and it seems safe to conclude that the intact gibberellane skeleton is essential for growth-promoting activity; (3) the carboxyl group at C-7 is essential for activity; and (4) the most active GAs tested all possess a lactone ring in ring A.

There are numerous reports of the occurrence in fruits, seeds, potato tubers, tulip bulbs, tomato seedlings, and tobacco shoot tips of GA-like substances that are more polar than the GAs known prior to 1970. These substances are more readily extracted with water or aqueous buffers than with organic solvents such as methanol, which is commonly used to extract free GAs. Unlike free GAs generally, these substances cannot be efficiently extracted from acidified aqueous solution with ethyl acetate but can be with n-butanol. In the early literature, such substances were referred to as "water-soluble GAs" or, more commonly, "bound GAs."

Several substances fitting the foregoing description have been isolated and their structures determined. It is now recognized that these comparatively polar forms of GA are chemically diverse, representing at least two, perhaps more, major types of substances. At least one substance is, in fact, a highly polar "free GA." This is GA_{32}, which was isolated from immature seeds of *Prunus persica* L. and *Prunus armeniaca* L. and its structure determined. GA_{32} is reportedly active in eight biological systems. It was the most potent GA tested in the barley endosperm bioassay, being three times as active as GA_3.

Other relatively polar forms of GA that have been chemically characterized as natural constituents, specifically of certain seeds, are conjugates of β-D-glucose and particular GAs (Fig. 3.4). From immature seeds of Japanese morning glory (*Pharbitis nil* Chois.) six GA glucosides have been isolated and their structures determined. The GA_8 glucoside has also been isolated from immature seeds and seedlings of *Phascolus coccineus* L. There have also been isolated at least seven GA glucosyl esters (Fig. 3.4). Sembdner et al. termed the GA-glucosides "conjugated GAs" in 1968, and this name is likely to be widely adopted for the glucosides, glucosyl esters, and other conjugated forms of GAs that may be

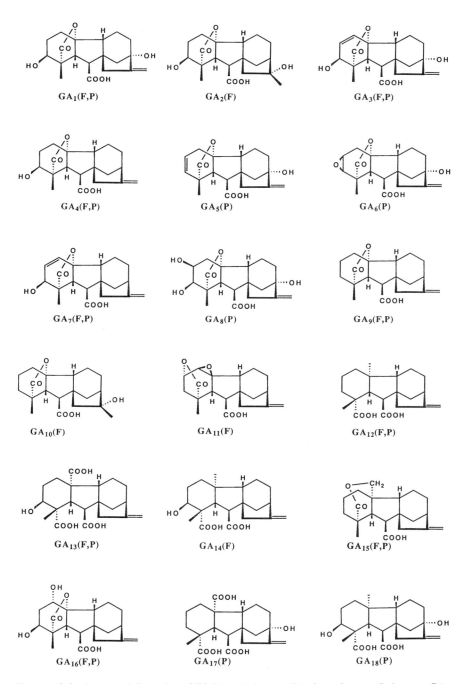

FIGURE 3.3. Structural formulas of 76 GAs. F denotes GA from fungus; P denotes GA from plant. (Courtesy of N. Takashashi, 1988.)

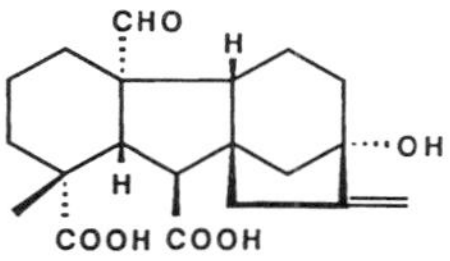

GA$_{19}$(P) 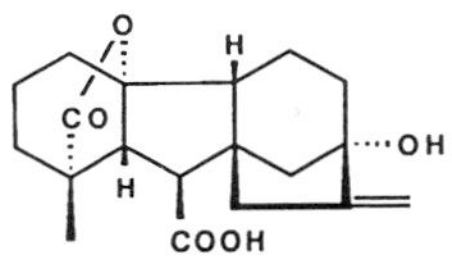 GA$_{20}$(P) 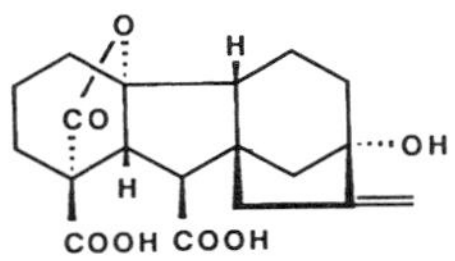 GA$_{21}$(P)

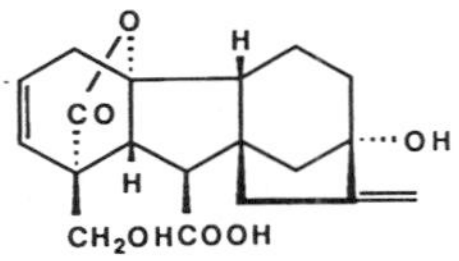

GA$_{22}$(P) 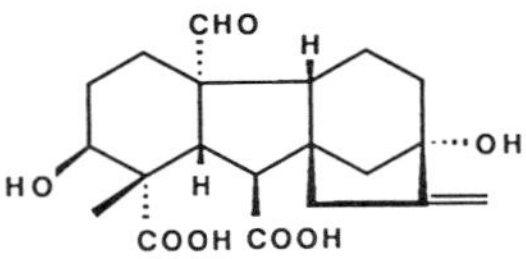 GA$_{23}$(P) 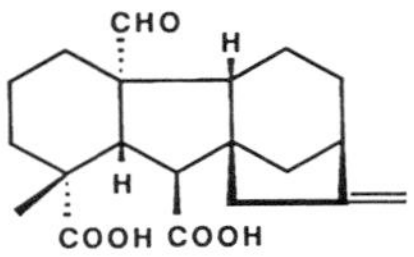 GA$_{24}$(F,P)

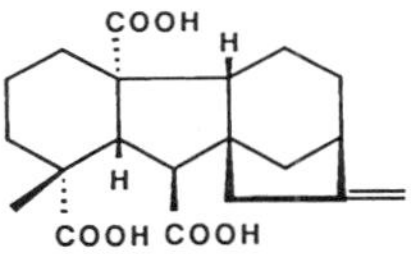

GA$_{25}$(F,P) 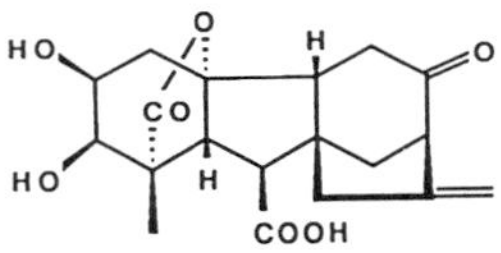 GA$_{26}$(P) 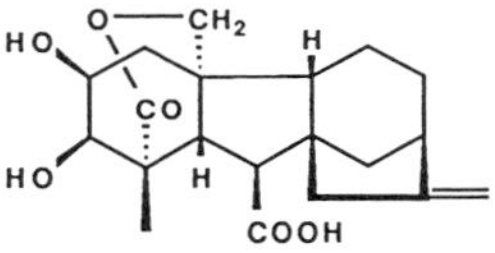 GA$_{27}$(P)

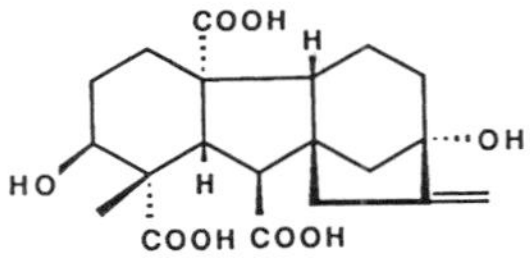

GA$_{28}$(P) 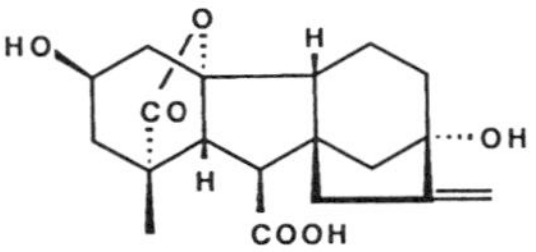 GA$_{29}$(P) 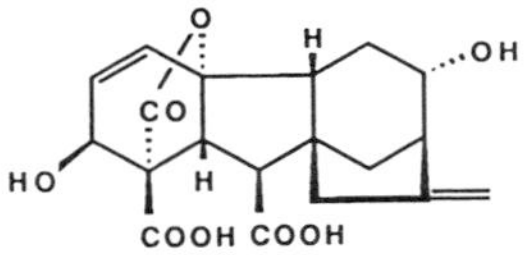 GA$_{30}$(P)

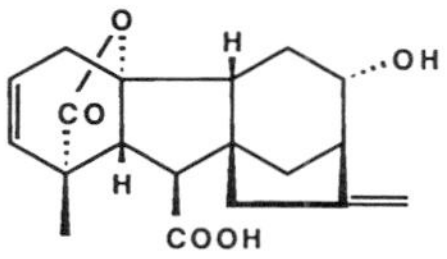

GA$_{31}$(P) 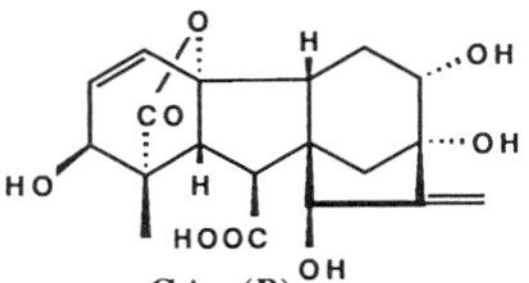 GA$_{32}$(P) 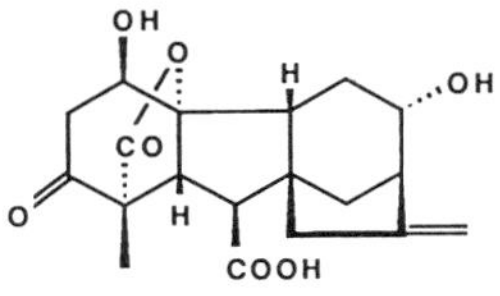 GA$_{33}$(P)

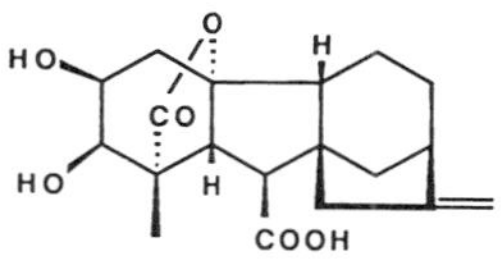

GA$_{34}$(P) 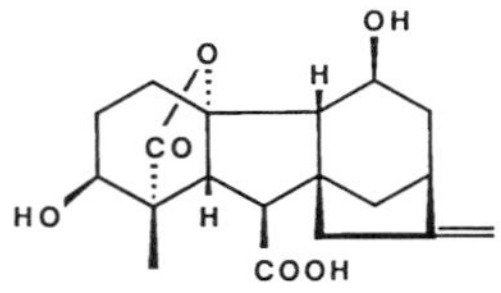 GA$_{35}$(P) 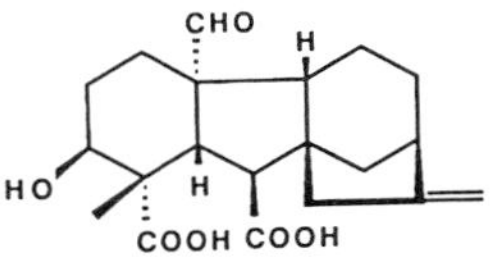 GA$_{36}$(F,P)

FIGURE 3.3. *Continued.*

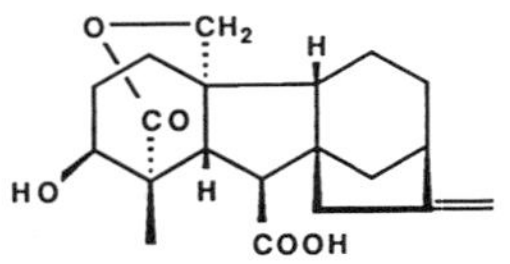

GA$_{37}$(F,P)

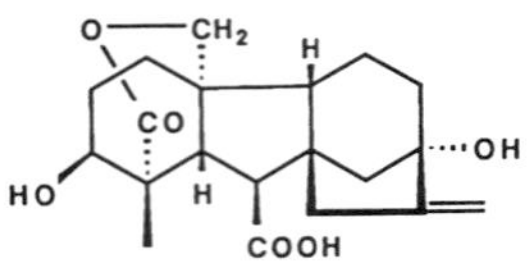

GA$_{38}$(P)

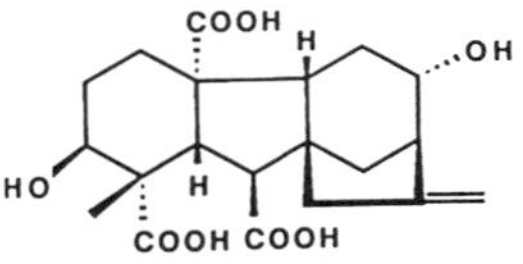

GA$_{39}$(P)

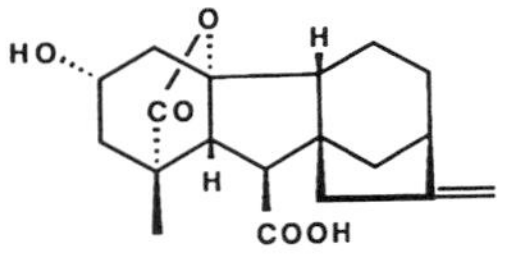

GA$_{40}$(F)

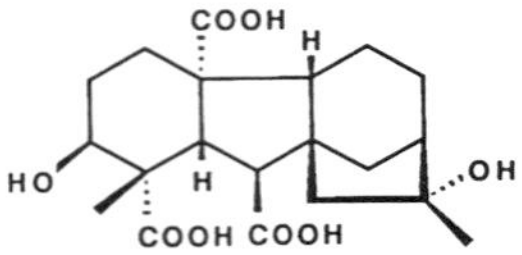

GA$_{41}$(F)

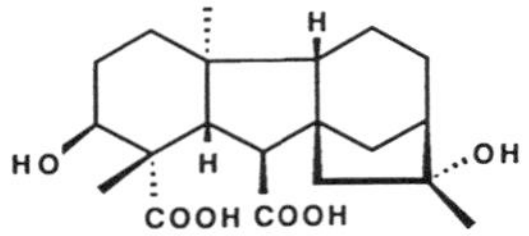

GA$_{42}$(F)

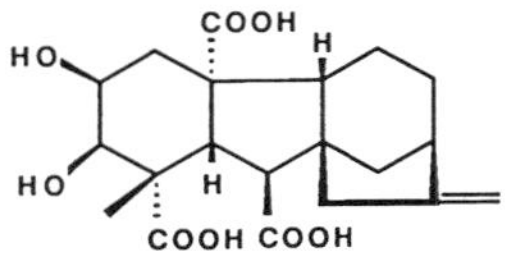

GA$_{43}$(P)

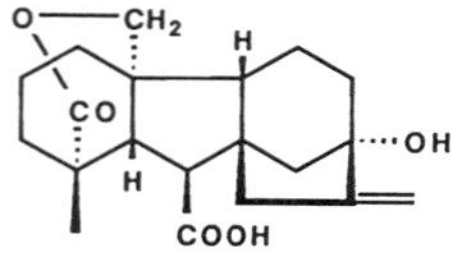

GA$_{44}$(P)

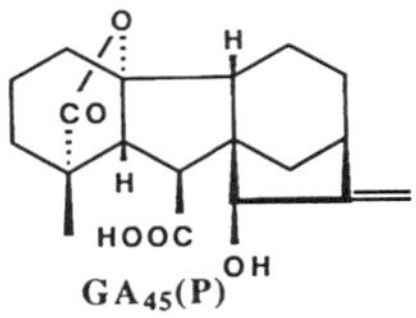

GA$_{45}$(P)

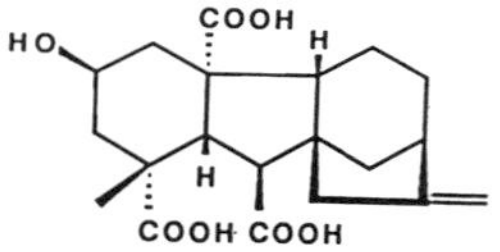

GA$_{46}$(P)

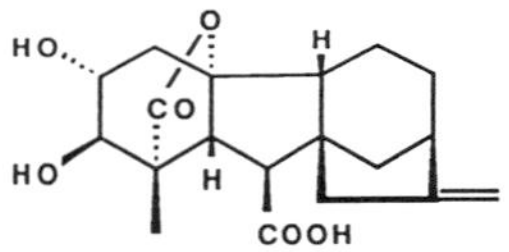

GA$_{47}$(F)

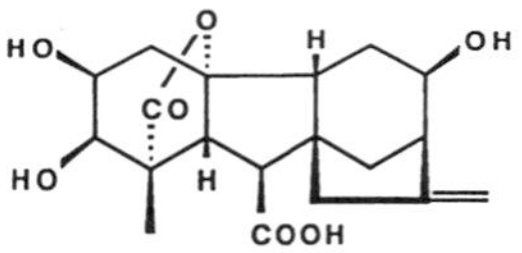

GA$_{48}$(P)

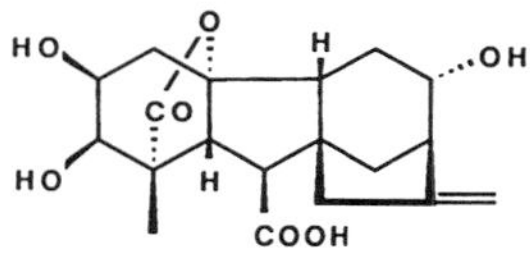

GA$_{49}$(P)

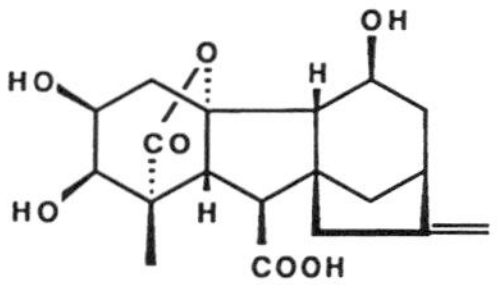

GA$_{50}$(P)

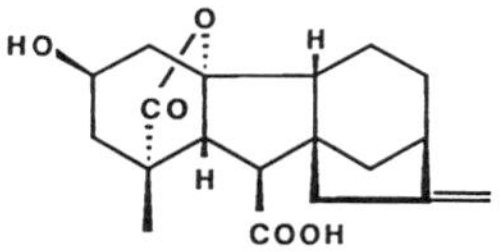

GA$_{51}$(P)

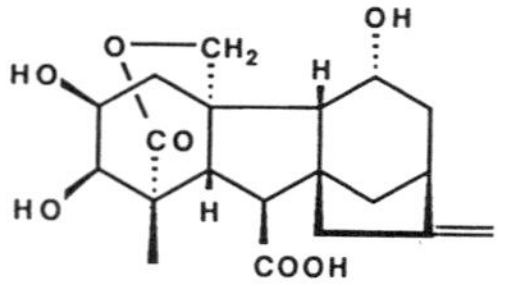

GA$_{52}$(P)

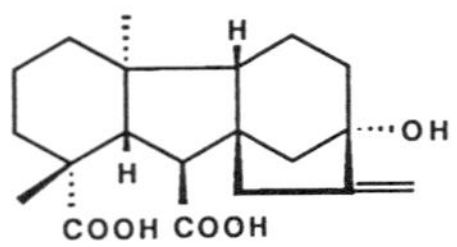

GA$_{53}$(P)

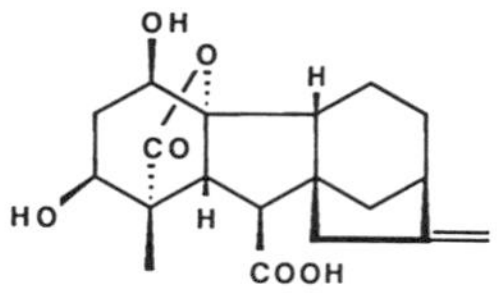

GA$_{54}$(F,P)

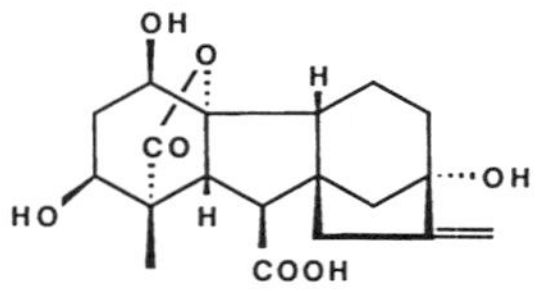

GA$_{55}$(F,P)

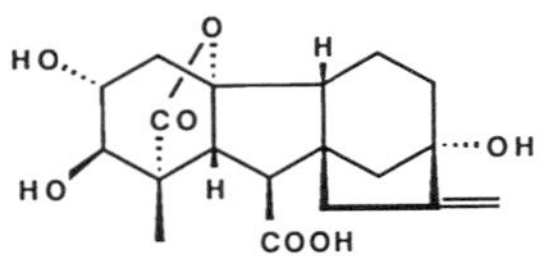

GA$_{56}$(F)

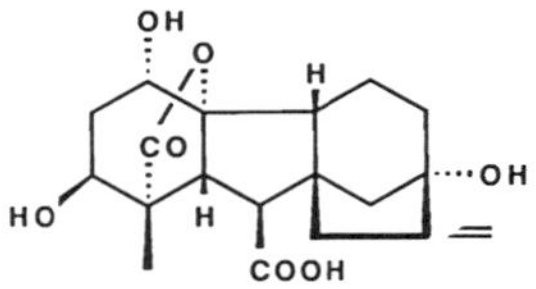

GA$_{57}$(F)

FIGURE 3.3. *Continued.*

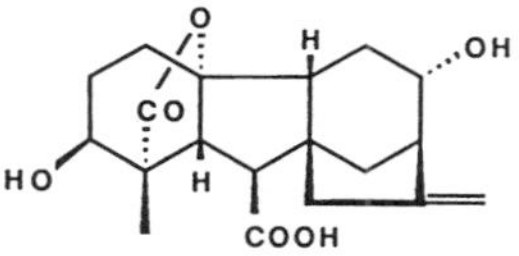

GA$_{58}$(P)

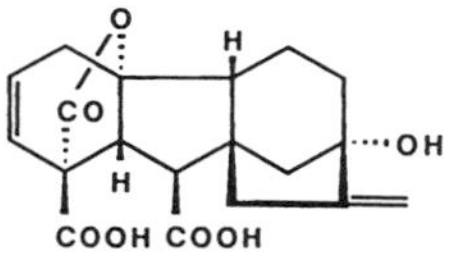

GA$_{59}$(P)

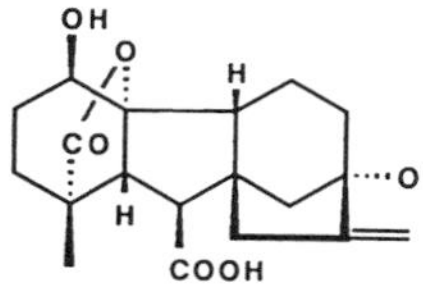

GA$_{60}$(P)

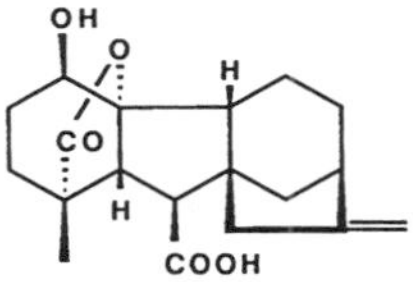

GA$_{61}$(P)

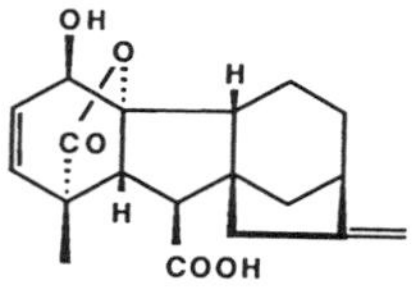

GA$_{62}$(P)

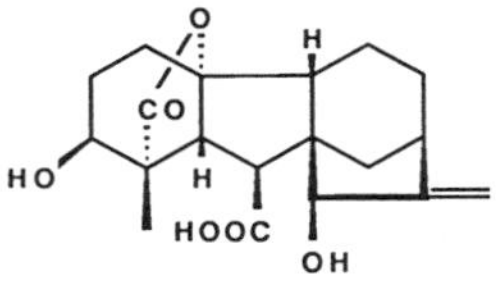

GA$_{63}$(P)

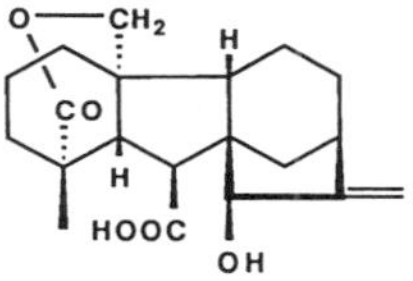

GA$_{64}$(P)

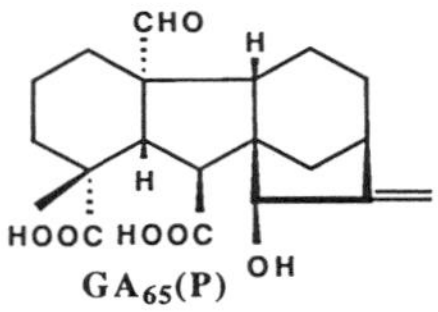

GA$_{65}$(P)

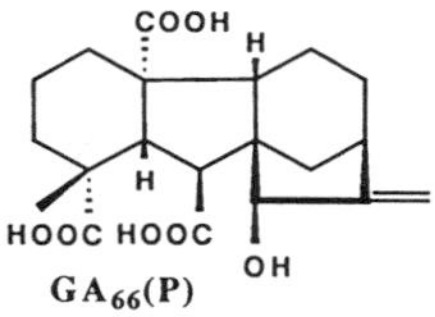

GA$_{66}$(P)

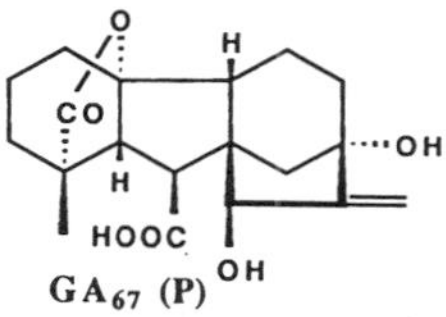

GA$_{67}$ (P)

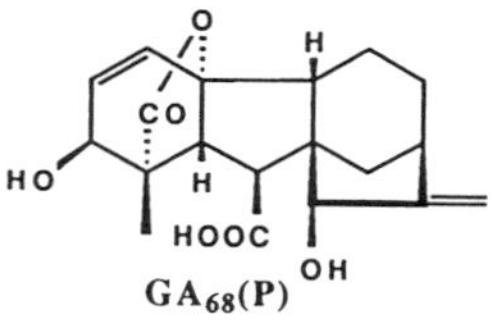

GA$_{68}$(P)

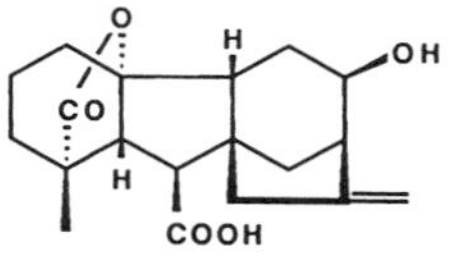

GA$_{69}$(P)

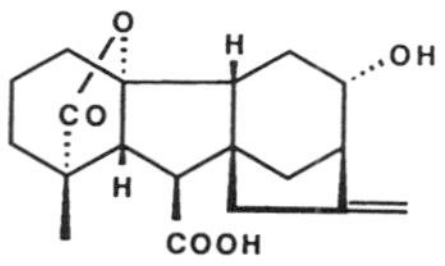

GA$_{70}$(P)

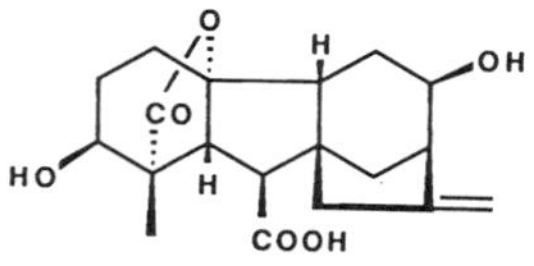

GA$_{71}$(P)

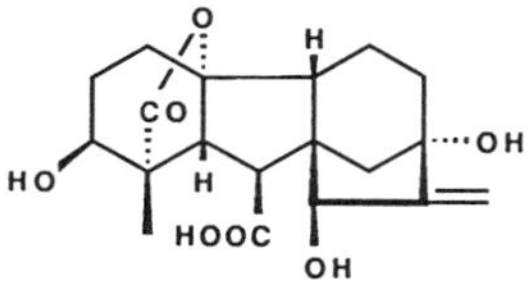

GA$_{72}$(P)

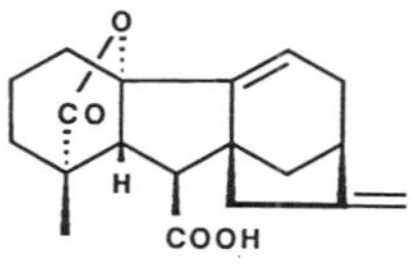

GA$_{73}$(P)

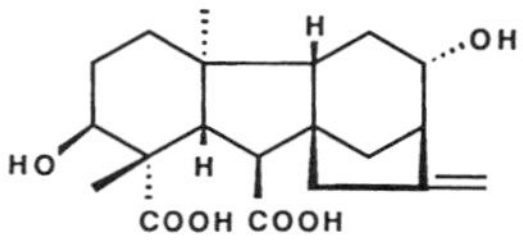

GA$_{74}$(P)

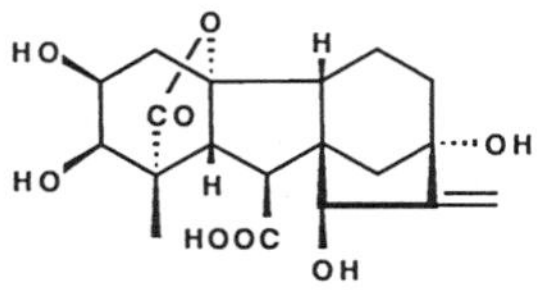

GA$_{75}$(P)

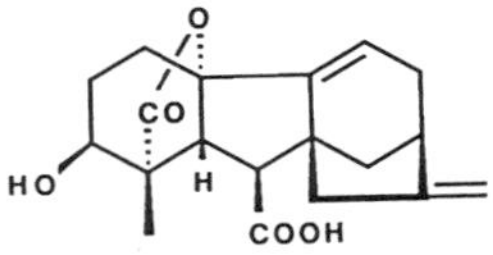

GA$_{76}$(P)

FIGURE 3.3. *Continued.*

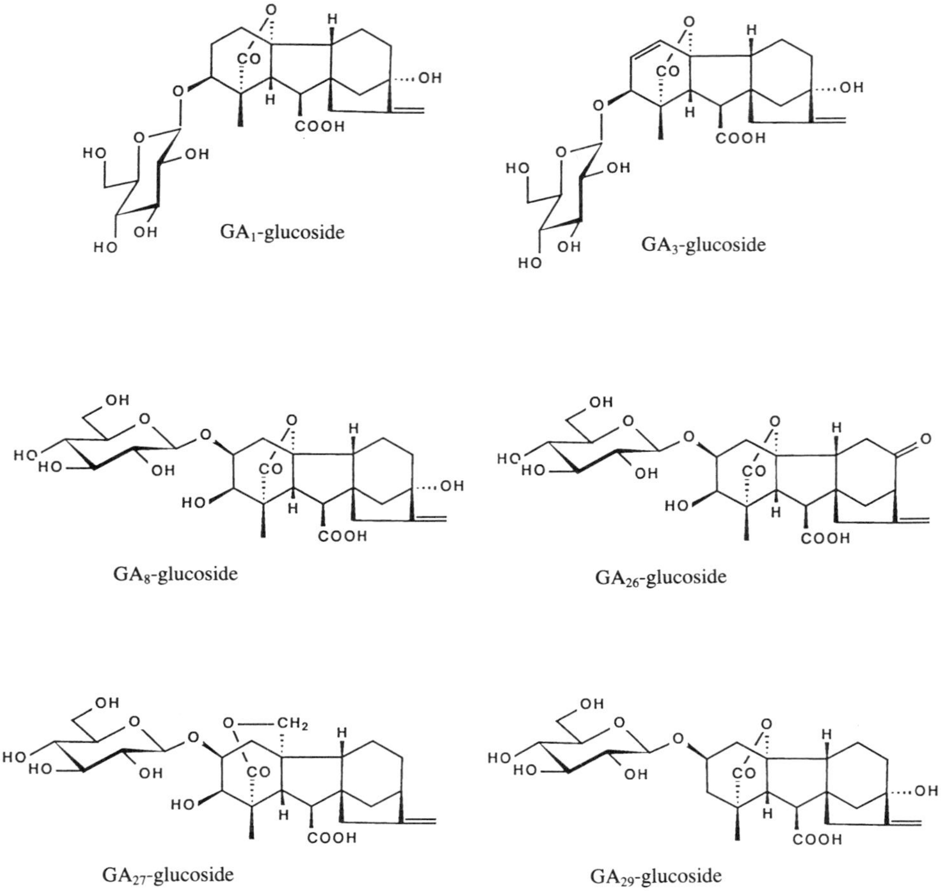

FIGURE 3.4. Structural formulas of GA glucosides and glucosyl esters. (Courtesy of N. Takahashi, 1988.)

found. GA glucosides per se probably are inactive in growth regulation. Yokota and associates compared the growth-promoting effects of GA_3, GA_8, GA_{26}, GA_{29}, and their glucosides in six bioassays in 1971. They found that the activities of the GA glucosides were much less than that of their aglycones, and concluded that the growth-promoting effects of the glucosides observed in some bioassay systems are probably due to the aglycones liberated by hydrolysis in plant tissue.

The term "bound GAs" continues to be used for as yet unidentified GA-like substances more polar than known GAs. Since free GAs reportedly are released in extracts containing unidentified "bound GAs" by treatment of extracts with preparations of proteinases, ficin and papain, for example, and by acid hydrolysis, some authors have suggested the existence of "protein-bound

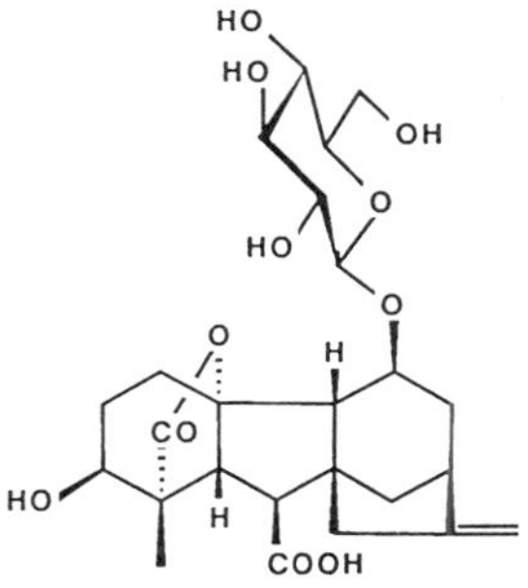

GA$_{35}$-glucoside 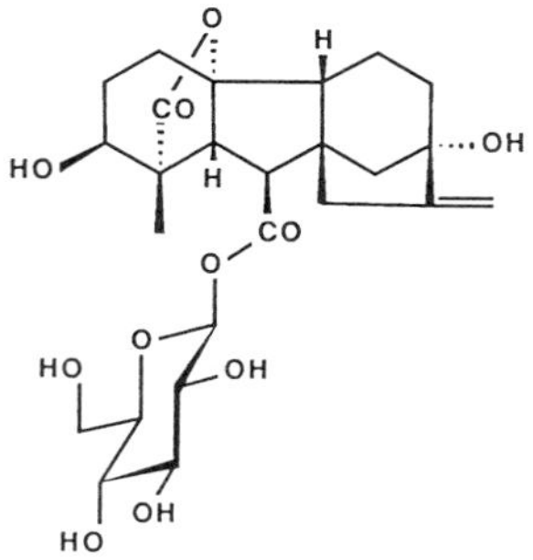   GA$_1$-glucosyl ester 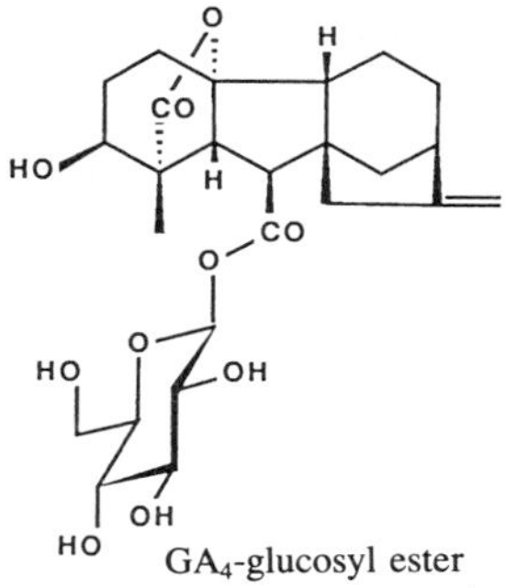   GA$_4$-glucosyl ester

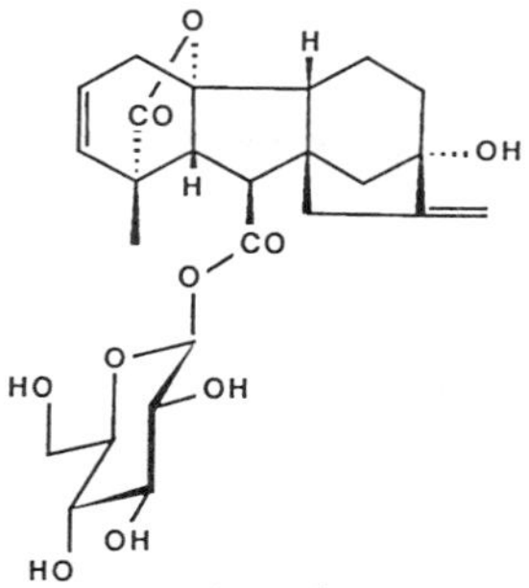

GA$_5$-glucosyl ester 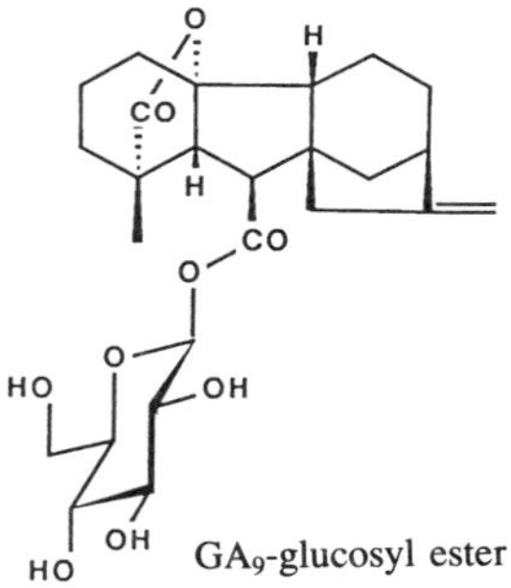   GA$_9$-glucosyl ester 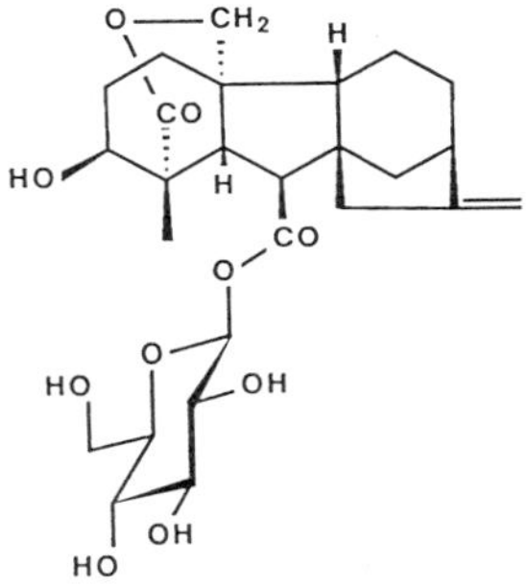   GA$_{37}$-glucosyl ester

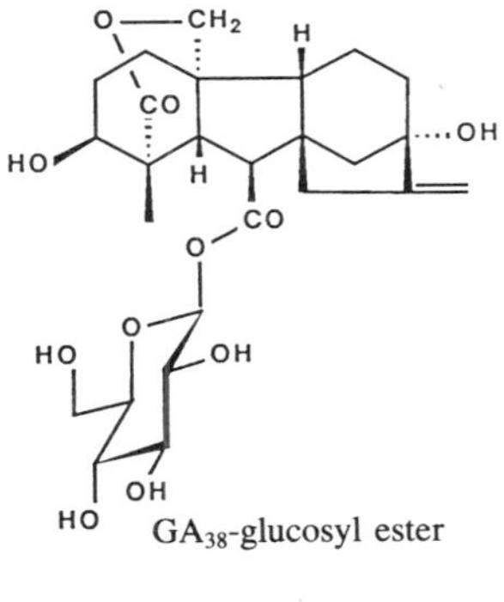

GA$_{38}$-glucosyl ester 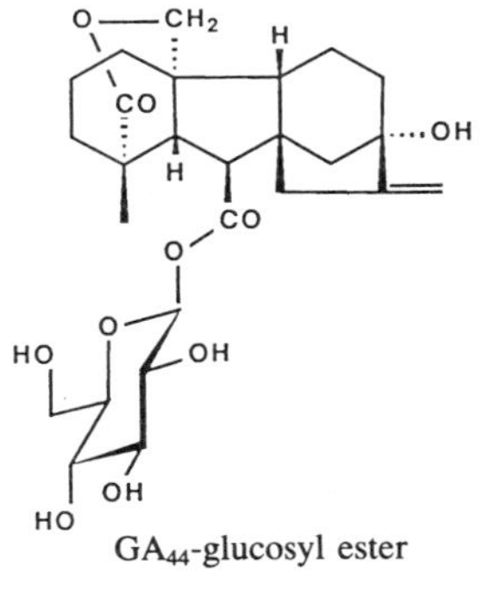   GA$_{44}$-glucosyl ester

FIGURE 3.4. *Continued.*

GAs.'' The reality of natural occurrence of GAs in protein-bound states remains uncertain, however. Whereas glucosyl-GAs have been isolated from tissues following treatment with free GAs, no such evidence for the formation of GA-protein complexes has been reported.

Currently, the major source of GAs for experimental and practical use is biosynthesis by large-scale cultures of various genetic strains of *Fusarium moniliforme.* However, progress has been and continues to be made to perfect organic synthesis of GAs.

Natural Occurrence of GAs

According to Anton Lang, by 1970, GAs and GA-like substances had been found in representatives of angiosperms, gymnosperms, ferns, brown algae, green algae, fungi, and bacteria. It would not be surprising, therefore, to ultimately discover that GAs are ubiquitous in the plant kingdom.

Selected examples of specific fruits and seeds from which particular GAs have been isolated are presented in Table 3.1. In general, it appears that a given organ or tissue contains two to several GAs. Furthermore, the kinds, amounts, and states (free or bound) of the GAs are not constant during development, as later discussion will show.

Attempts to determine the amounts of GAs in higher plant tissues prior to the use of gas chromatography-mass spectrometry and GC-selected ion monitoring generally yielded at best only semiquantitative results. The bioassays employed often yielded inexact data, and relatively few analyses measured the amounts of identified GAs. Rather, the general practice was to estimate the amounts of unidentified GA-like substances present in an extract by comparison with response of test material to standard dosages of GA_3, or, less commonly, some other specific GA. Representative such data are presented in Table 3.2. A general conclusion from the available data is that the concentrations of GAs in fruits and seeds, particularly immature seeds, are much higher, approximately two orders of magnitude in the latter, than in vegetative organs of established herbaceous angiosperm sporophytes.

GA Biosynthesis in Seeds

Present knowledge concerning GA biosynthesis in higher plants, is based largely on investigations of immature seeds. The biosynthetic pathways from mevalonate (Fig. 3.5, 3.6 and 3.7), as far as currently known for higher plants, are

TABLE 3.1. Examples of some fruits and seeds in which specific GAs have been identified.[a]

Species	GAs
Marah macrocarpus (Greene) Greene	A_1, A_3
Mature seeds	A_4, A_7
Phaseolus coccineus L.	A_1, A_3, A_5, A_6
Immature seeds	A_8, A_{17}, A_{19}
Pharbitis nil Chois.	A_3, A_5, A_{20}
Immature seeds	A_{26}, A_{27}
Cucumis sativus L.	A_1, A_3, A_4
Immature seeds	A_7
Lupinus luteus L.	A_{18}, A_{19}
Immature fruits and seeds	A_{22}, A_{28}
Malus sylvestris L.	A_3, A_4, A_7
Fruits and seeds	

[a] With permission, from Moore and Ecklund (1975).

TABLE 3.2. Representative data on quantitative analyses of GAs and GA-like substances in plant materials.[a]

Plant material	Concentration of GAs	Assay methods
Immature florets of *Lolium perenne* L.	256 µg GA_3 equivalents per kg dry weight	Chromatography and lettuce hypocotyl bioassay
Immature seeds of *Pisum sativum* L.	1400 µg GA_3 equivalents per kg fresh weight; 0.430 µg per seed	Chromatography and dwarf pea and dwarf maize bioassays
Immature partheno-carpic fruit of *Malus sylvestris* Mill.	143 µg GA_3 equivalents per kg fresh weight	Chromatography, GLC, and dwarf pea and cucumber bioassay
Young fruits of *Citrus sinenis* Osbeck (9 days after petal fall)	Approximatelty 105 µg GA_3 equivalents per kg fresh weight; approx. 4 µg GA_3 equivalents per fruit	Chromatography and fluorometric analysis
7-day-old etiolated dwarf seedlings of *Pisum sativum* L.	88 µg GA_3 equivalents per kg dry weight; 0.033 µg per plant	Chromatography and dwarf maize bioassay

[a] With permission, from Moore and Ecklund (1975).

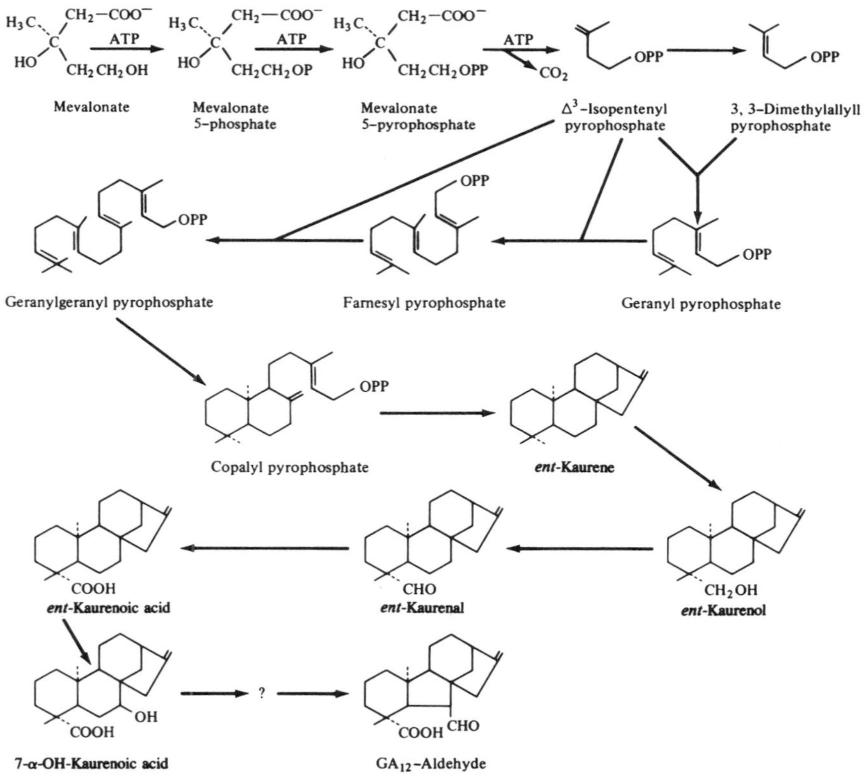

FIGURE 3.5. The early part of the pathway of GA biosynthesis from mevalonate in seed plants.

based predominantly on the work by C. A. West and associates of the University of California at Los Angeles with preparations of liquid (free nuclear) endosperm of immature seeds of wild cucumber (*Marah macrocarpus*), and that of J. E. Graebe and associates at the Pflanzenphysiologisches Institut at Göttingen with endosperm of pumpkin (*Cucurbita maxima*) seeds and immature pea (*Pisum sativum*) seeds. Other systems have been developed subsequently.

Substantial progress toward the development of a cell-free enzyme system from higher plant tissues capable of GA biosynthesis first was made when J. E. Graebe et al., working in C. A. West's laboratories, reported in 1965 on the incorporation of [2-^{14}C]mevalonic acid into geranylgeraniol, kaurene,[1] and kaurenol in extracts of endosperm-nucellus of immature seeds of *Marah macrocarpus*. The intermediates in the formation of geranylgeranyl pyrophosphate from mevalonate appear to be identical to those involved in farnesyl pyrophosphate (hence sesquiterpene and triterpene) synthesis. Thus the isoprenoid pathway contributes to the synthesis of abscisic acid (a sesquiterpenoid compound) and the isopentenyl side chains of the cytokinins, as well as the GAs. In some cases the prenyl transferase specificity determines the 20-C rather than the 15-C prenyl pyrophosphate as the major product of chain elogation. The cyclization of geranylgeranyl pyrophosphate to kaurene occurs in two steps and involves the formation of copalyl pyrophosphate as an intermediate. Both reactions are catalyzed by kaurene synthetase.

Prior to 1975 the enzyme systems for GA biosynthesis were thought to be cytoplasmic-microsomal although, as subsequent discussion will reveal, it is now clear that some GA biosynthesis from mevalonate may be compartmentalized in plastids (see discussion of this topic also in Chapter 8.) The reactions from mevalonate to kaurene are catalyzed by enzymes that are readily solubilized and that are dependent on ATP and a divalent cation in all systems that have been studied. In most systems the most effective cation is Mg^{2+}, but Mn^{2+} is more effective than Mg^{2+} in cell-free extracts of immature pea seeds and pea seedling shoot tips.

As we have seen, mevalonic acid is converted by soluble enzymes in nine steps to the hydrocarbon *ent*-kaurene. The subsequent steps by which GA_{12}-aldehyde is formed are associated with the endoplasmic reticulum. Gibberellin A_{12}-aldehyde is converted to C_{20}- and C_{19}-GAs by soluble enzymes requiring α-ketoglutarate and Fe^{2+} for activity. Figure 3.6 illustrates the biosynthesis of GAs in immature pea seeds. At the bottom is illustrated a pathway that has become known as the "early 13-hydroxylation pathway," which plays an important role in many of the plants investigated to date. The 13-hydroxylation is associated with the endoplasmic reticulum, and it takes place early in the

---

[1] *Kaurene* refers specifically to *ent*-kaur-16-ene [also referred to as (−)-kaur-16-ene], and kaurenol, kaurenal, and kaurenoic acid refer to *ent*-kaur-16-en-19-ol, *ent*-kaur-16-en-19-al, and *ent*-kaur-16-en-19-oic acid, respectively. The trivial designation of the GAs and the *ent*-gibberellane system as described by MacMillan (1971) is employed.

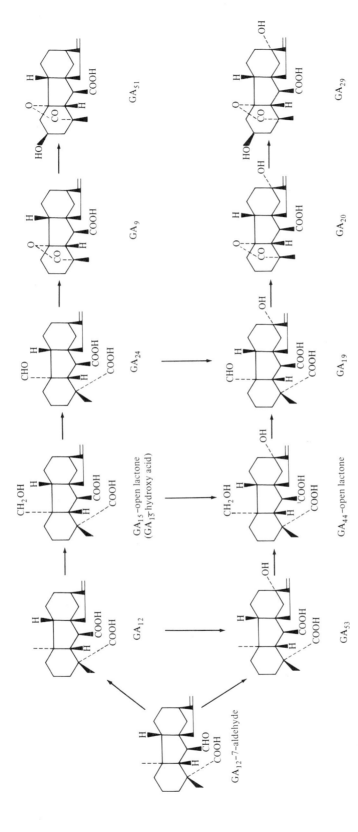

FIGURE 3.6. Pathway of gibberellin biosynthesis from GA_{12}–7-aldehyde in immature pea seeds. The lower sequence is the "early 13-hydroxylation pathway." In shoots there is 3β-hydroxylation of GA_{20} to form GA_1, which in turn is 2β-hydroxylated to form GA_8. (Redrawn, with permission, from Kamiya and Graebe, 1983.)

pathway in conjugation with the microsomal steps by which GA_{12}-aldehyde is formed (Graebe, 1986).

Other patterns of hydroxylation are known. For example, in a system from *Cucurbita maxima* endosperm (Fig. 3.7), there is what has been termed a "late 3β-hydroxylation pathway." In contrast to 13-hydroxylation, 3β-hydroxylation is catalyzed by soluble enzymes, at least in cell-free systems from *Cucurbita maxima* and *Phaseolus vulgaris*. This may explain why 3β-hydroxylation tends to take place at intermediate or late steps rather than at the beginning of the pathway (Graebe, 1986).

GA Biosynthesis in Systems Other Than Seeds

Prior to 1973, most investigations of GA biosynthesis in seed plants utilized cell-free enzyme extracts prepared from developing seeds and fruits. While these systems have yielded abundant valuable information about the biosynthetic pathway, the development of cell-free systems also from vegetative organs of young seedlings and older sporophytes is essential to an ultimate understanding of the factors regulating the kinds and amounts of GAs produced in plant tissues.

Synthesis of kaurene was reported using cell-free extracts of pea (*Pisum sativum* L.) shoots by J. E. Graebe in 1968 and castor bean (*Ricinus communis* L.) seedlings by D. R. Robinson and C. A. West in 1970. In 1973 the development of a cell-free system from pea shoot tips, the site of most abundant GA synthesis in the vegetative sporophyte, was reported that synthesized kaurene from MVA.

Kaurene is a key intermediate in the GA biosynthetic pathway, and it was considered plausible from previous results with developing pea seeds that the capacity of cell-free extracts of pea shoot tips to synthesize kaurene might be indicative of the potential of shoot tips to produce GAs. Accordingly, investigations were undertaken in an effort to ascertain possible correlations between rate of kaurene biosynthesis and rates of growth and other physiological parameters in etiolated and light-grown dwarf and tall cultivars of pea.

It was discovered in 1974 that cell-free extracts from shoot tips of light-grown Alaska pea (*Pisum sativum*) seedlings had a fivefold greater capacity for synthesizing *ent*-kaurene from mevalonate than extracts from shoot tips of etiolated seedlings of the same age (Table 3.3). On continuous irradiation of 10-day-old etiolated seedlings with high-intensity white light, activity increased approximately exponentially between the third and twelfth hours, attaining a level equal to that of light-grown plants of the same age, and remained nearly constant during the succeeding 24 hours of development (Fig. 3.8). Although the increase in capacity for *ent*-kaurene synthesis and chloroplast development occurred concurrently, there was no evidence then that the two processes were related.

These results reinforced the growing suspicion that at least some, if not all, of the observed synthesis of kaurene might be occurring in the chloroplasts. The

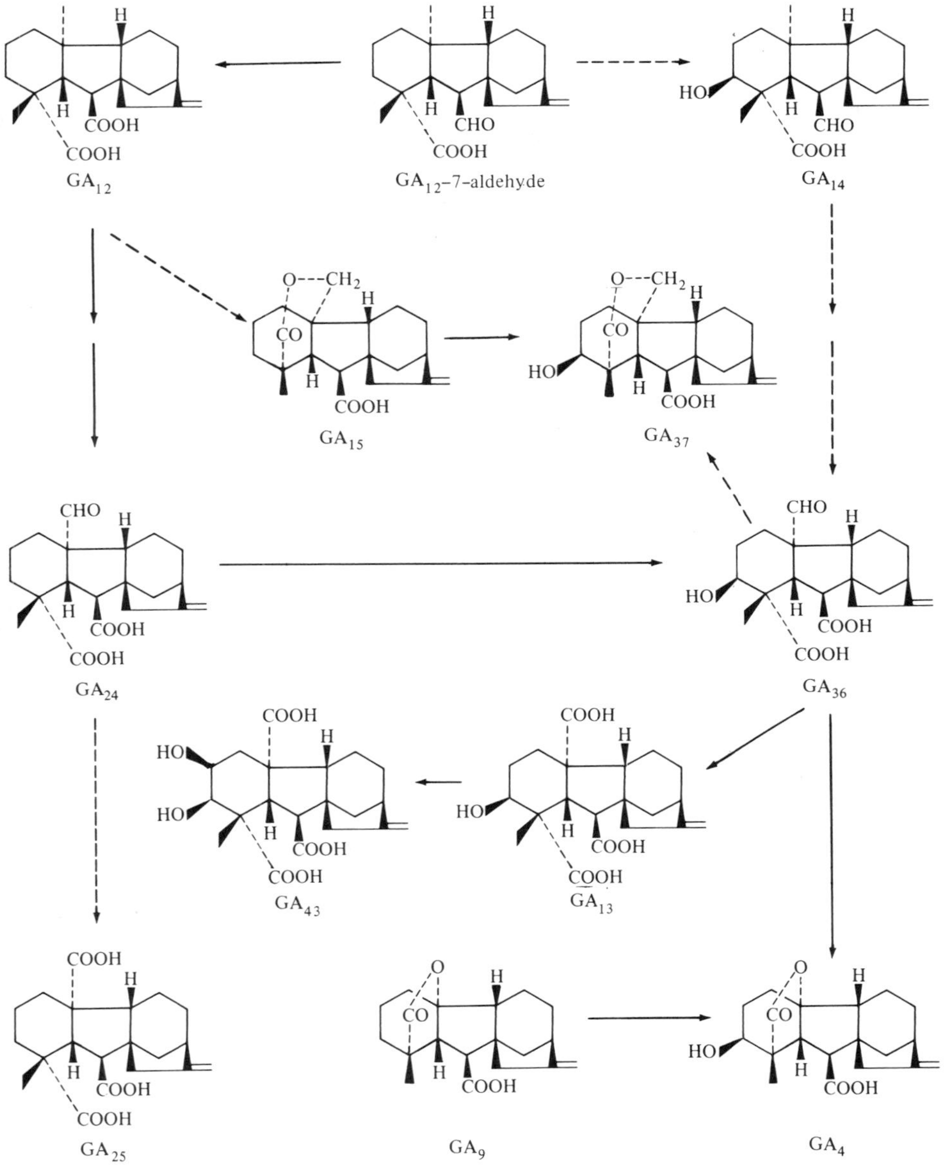

FIGURE 3.7. Pathway from GA_{12}-7-aldehyde to GAs in a cell-free system from *Cucurbita maxima* endosperm. Broken-line arrow pathways were observed when an excess of GA_{12}-7-aldehyde was used as substrate. GA_9 was not found in the system, but was readily converted to GA_4. (Revised, with permission, from Hedden and Graebe, 1982.)

TABLE 3.3. Comparisons of kaurene-synthesizing capacities of cell-free extracts from shoot tips of light-and dark-grown Alaska and Progress No. 9 peas.[a]

Alaska		Progress No. 9	
Light-grown[b]	Etiolated[c]	Light-grown[d]	Etiolated[e]
	dpm/mg protein		
78	17	40	35
	dpm/g fresh wt		
944	170	558	350

[a] With permission, from Moore and Ecklund (1974).
[b-e] Means of 5, 3, 4, and 1 independent experiments, respectively.

GA Biosynthesis in Systems Other Than Seeds

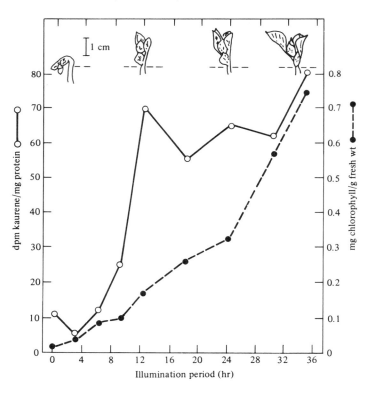

FIGURE 3.8. Change in chlorophyll content and in kaurene-synthesizing capacity of Alaska pea shoot tips during 36 hour irradiation of 10-day-old seedlings with high-intensity white light, as measured in cell-free enzyme extracts. The drawings illustrate the shoot tips from which enzyme and chlorophyll extracts were prepared. (Redrawn, with permission, from Moore and Ecklund, 1974.)

author and associates had noted as early as 1974 that the procedures utilized to prepare enzyme extracts from pea shoot tips did not preclude contamination with enzymes or other substances of chloroplast origin, since the procedures resulted in extensive breakage of chloroplasts.

Hence, investigations were initiated to attempt to ascertain whether the enzyme kaurene synthetase occurs in pea chloroplasts. All the enzymes necessary for the synthesis of geranylgeranyl pyrophosphate from mevalonate were known to occur in chloroplasts *in vivo,* since the former compound is also a precursor of carotenoids. Thus the enzymic competence to cyclize geranylgeranyl pyrophosphate to kaurene was all that had to be shown to prove that chloroplasts possess complete metabolic competence for kaurene synthesis. It was reported in 1976 that enzyme extracts from sonicated pea chloroplasts could indeed enzymically convert geranylgeranyl pyrophosphate to kaurene in very low yield. While that was in progress, P. D. Simcox, D. T. Dennis, and C. A. West reported in 1975 on the conversion of copalyl pyrophosphate to *ent*-kaurene in extracts of proplastids from etiolated pea shoot tips, developing castor bean endosperm extracts, and wild cucumber endosperm. Only the wild cucumber endosperm extracts were capable of converting GGPP to *ent*-kaurene. Further evidence for *ent*-kaurene synthesis in chloroplasts was reported by Railton et al. (1984).

The cumulative evidence to date seems clearly to warrant the conclusion that some GA biosynthesis in higher plants, and perhaps some metabolic interconversions of GAs as well, may be compartmentalized in chloroplasts and precursor organelles.

Interestingly, B. V. Milborrow reported in 1974 that abscisic acid, a sesquiterpenoid hormone (see Chapter 5), is synthesized from mevalonate by a preparation of lysed chloroplasts of ripening avocado fruit. Indeed in the avocado fruit, it appeared that the majority of the abscisic acid is synthesized within the chloroplasts, although the occurrence of synthesis outside chloroplasts was not discounted. T. W. Goodwin and colleagues have shown that there is a chloroplastic and an extrachloroplastic system for the conversion of mevalonate into terpenoid compounds. The chloroplast membrane is highly impermeable to mevalonate, and the mevalonate kinase of chloroplasts is distinct from the cytoplasmic mevalonate kinase. These observations reinforce the importance of plastids as compartments for hormone biosynthesis.

One naturally wonders why, during the course of evolution, synthesis of GAs and other terpenoid compounds might come to be compartmentalized in chloroplasts. All we can do at the present state of our knowledge is speculate. One possible explanation is that photosynthesis produces abundant adenosine triphosphate (ATP) and reduced pyridine nucleotide (NADPH), relatively large quantities of which are required for the biosynthesis of terpenoid compounds. Hence it would appear to be a positive adaptive feature for GA biosynthesis to be localized in an organelle where the levels of ATP and NADPH are comparatively high. There may also be a relationship between stimulation of GA biosynthesis by light and compartmentation in the chloroplasts.

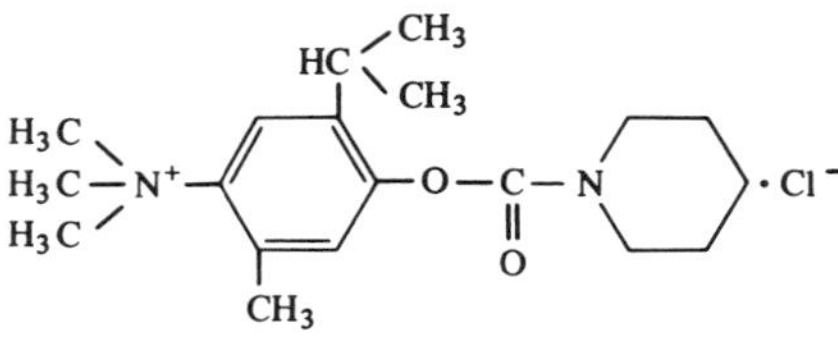

AMO–1618

[2′–Isopropyl–4′–(trimethylammonium chloride)–5′–methylphenyl piperidine–1–carboxylate]

$$CH_2Cl-CH_2-N^+\underset{CH_3}{\overset{CH_3}{\underset{\textstyle}{\diagdown}}}CH_3 \cdot Cl^-$$

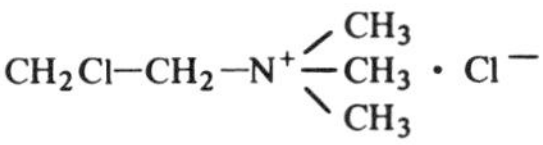

CCC (Cycocel)

(2–Chloroethyl) trimethylammonium chloride

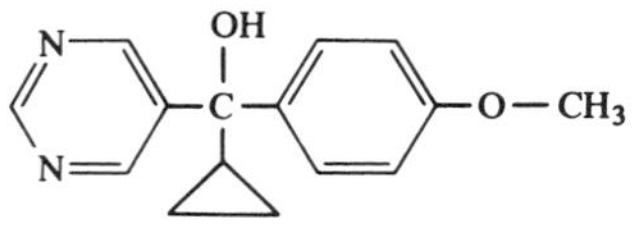

Ancymidol

[α–Cyclopropyl–α–(p–methoxyphenyl)–5–pyrimidine methyl alcohol]

FIGURE 3.9. Structural formulas of three synthetic inhibitors of GA biosynthesis.

Before leaving the subject of GA biosynthesis, mention should be made of some purely synthetic plant growth regulators (growth retardants). Several compounds have been produced that inhibit the growth of at least some seed plants, and that have been demonstrated to inhibit GA biosynthesis. Three compounds in particular have received attention both for their practical utility in inhibiting growth and for inhibiting GA biosynthesis. They are Amo-1618 [2′-isopropyl-4′-(trimethylammonium chloride)-5′-methylphenyl piperidine-1-carboxylate], CCC or Cycocel [(2-chloroethyl) trimethylammonium chloride], and ancymidol [α-cyclopropyl-α-(p-methoxyphenyl)-5-pyrimidine methyl alcohol] (Fig. 3.9).

Amo-1618 is the most specific inhibitor of GA biosynthesis in that it inhibits specifically the cyclization of geranylgeranyl pyrophosphate to copalyl pyrophosphate (see Fig. 3.5), one of the two reactions catalyzed by kaurene synthetase. However, Amo-1618 also inhibits the synthesis of sterols and other triterpenes. Specifically, it inhibits the cyclization step by which squalene-2,3-epoxide is converted to cyclized triterpene sterols.

CCC inhibits GA biosynthesis somewhere beyond the formation of kaurene. However, like Amo-1618, CCC also inhibits other reactions and in general appears to be the least specific of the two growth retardants. Ancymidol blocks the conversion of kaurene to kaurenol and the succeeding two mixed function oxidase-catalyzed reactions. Other synthetic plant growth regulators reported to inhibit GA biosynthesis are paclobutrazol and tetracyclacis. The latter also inhibits sterol biosynthesis. Recently, it was reported that Amo-1618, ancymidol, and CCC, among other growth retardants, inhibited ABA biosynthesis in the fungus *Cercospora rosicola*.

Effects of Light on GA Biosynthesis

There have been numerous reports of stimulation of GA biosynthesis by light, and in some cases phytochrome (see Chapter 8) has been implicated or directly shown to be involved as the photoreceptor. In the case of etiolated wheat leaf sections, an increase in GA-like substances occurred shortly after 5 minutes of red light treatment, which appeared to be the result of a release of GA from bound forms. However, in other work GA biosynthesis, as well as RNA and protein synthesis, appeared to be implicated in an observed increase in GA-like substances that occurred subsequent to a 30-minute red light treatment of wheat leaves. A transient increase in GA-like substances in etiolated barley leaf sections occurred within 15 minutes after a 30-minute exposure to red light and also appeared to require GA biosynthesis and RNA and protein synthesis. Additionally, D. M. Reid et al. reported in 1972 that homogenates of etiolated barley leaves, when exposed to brief periods (20 minutes or less) of red irradiation, displayed an immediate increase in the levels of GA-like substances as compared to dark controls, and that the red light-induced increase in GA-like activity could be explained partially by the greater conversion of $[^3H]GA_9$ into other GA-like substances with high biological activity.

Previously an increase in kaurene-synthesizing activity in shoot tips of Alaska pea seedings during de-etiolation was described (Fig. 3.8.) Since kaurene is a key intermediate in GA biosynthesis, and rate of kaurene biosynthesis may be correlated closely with rate of GA biosynthesis, the light-induced increase in kaurene-synthesizing capacity reflects a potential light-induced increase in rate of GA biosynthesis. In those investigations, continuous illumination for at least 12 hours was required to attain a maximum rate of kaurene synthesis in greening shoot tips. This is a relatively sluggish light-induced change, compared to the results described in the preceding paragraph. However, there also are reports of less rapid effects with which these observations agree more closely. D. Köhler reported in 1966 that exposure to red light [either continuous or intermittent (2.5 minutes/hour)] resulted in an increase in extractable GA from shoots of a normal (tall) pea cultivar, and that the increase, susceptible to inhibition by CCC, appeared about 1 day after light exposure. Y. Ogawa showed in 1964 that several hours of irradiation of etiolated *Pharbitis nil* seedlings with white, red, or far-red light resulted in an increase in extractable GA-like substances, which was correlated with the opening of the hypocotyl hook and expansion of cotyledons. On the basis of all available reports it seems certain that light may affect diverse reactions involved in GA metabolism. A more detailed discussion of these effects, and the involvement of phytochrome in them, is presented in Chapter 8.

Whether there are significant differences in the levels of GAs in etiolated and light-grown seedlings of a given species and variety remains uncertain. R. L. Jones and A. Lang reported in 1968 that they detected no significant difference between tall and dwarf peas or between etiolated and light-grown peas with respect to their contents of extractable or diffusible GA. On the contrary, D. Köhler found that the extractable GA content of light-grown tall or dwarf peas was considerably greater than that of etiolated seedlings of the tall cultivar. Work by the author and associates showed (Table 3.3) that enzyme extracts of the shoot tips of light-grown seedlings of both a tall pea cultivar and a dwarf pea cultivar possessed greater capacities for kaurene synthesis than enzyme extracts from their etiolated counterparts. These results are in agreement with those of Köhler and, in certain respects also, with those of G. W. M. Barendse and A. Lang who found in 1972 that a dwarf strain of *Pharbitis nil* contained less GA than a tall strain, and that dark-grown plants of both strains contained less GA than light-grown ones. Notably, in this case the differences in GA content were attributable mainly to differences in levels of bound GA.

Role of GAs in Dwarfism

For numerous species, particularly some of horticultural importance, there are genetic dwarf strains or cultivars and normal or tall types. In the case of the garden pea (*Pisum sativum* L.), for example, there are dozens of varieties or cultivars. Dwarf cultivars typically mature at heights of 30 cm or less (Fig. 3.10), whereas some tall or normal cultivars exceed a meter in shoot height at maturity.

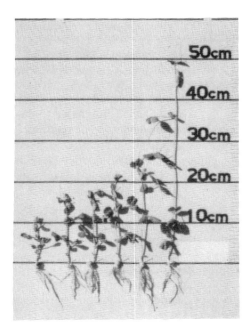

FIGURE 3.10. Growth of dwarf peas (*Pisum sativum*) sprayed with GA 10 days after planting. The plant on the left was left untreated; the others were sprayed with increasing concentrations (10-fold increases), from 0.0015 to 15.0 mg/liter, progressing toward the right. Photographed 18 days after treatment. (Unpublished data of Moore, 1958.)

Interestingly, dwarfism generally is readily apparent only in light-grown plants. When etiolated the difference in stem elongation between dwarf and tall cultivars is greatly reduced or nonexistent (see, e.g., Fig. 2.22). Furthermore, etiolated pea plants show little if any response to exogenous GA, whereas light-grown dwarf cultivars of peas and many other species often respond with dramatic stem elongation. Indeed it was in large part on the basis of this observation that it was first postulated that GAs are natural hormones of seed plants, and that GA deficiency might be causally related to genetic dwarfism.

In some cases, but not all, dwarfism does in fact seem to be correlated with GA deficiency. The case of dwarfism that is most completely understood at the molecular level is that of a dwarf mutant of corn (*Zea mays*) known as the dwarf-5 (d_5) mutant (Fig. 3.11). The d_5 phenotype is due to a single gene mutation, which results in such a marked reduction in internode elongation that mature dwarf plants are only about 20% the height of normal plants. Dwarfism is expressed from the seedling stage to maturity and, in contrast to dwarf types of some other species, is expressed in etiolated as well as light-grown plants. The dwarf-5 mutant is deficient in endogenous GA, compared to normal genetic strains, and responds by normal growth to appropriate dosages of exogenous GAs and such GA precursors as *ent*-kaurene, *ent*-kaurenol, and *ent*-kaurenoic acid.

In 1976, Peter Hedden and B. O. Phinney of the University of California at Los Angeles showed that the d_5 genetic lesion controls the cyclization of copalyl

FIGURE 3.11. The effect of GA$_3$ on normal and dwarf corn (*Zea mays*). Left to right: normal control, normal plus GA, dwarf control, dwarf plus GA. (Courtesy of B. O. Phinney, 1978.)

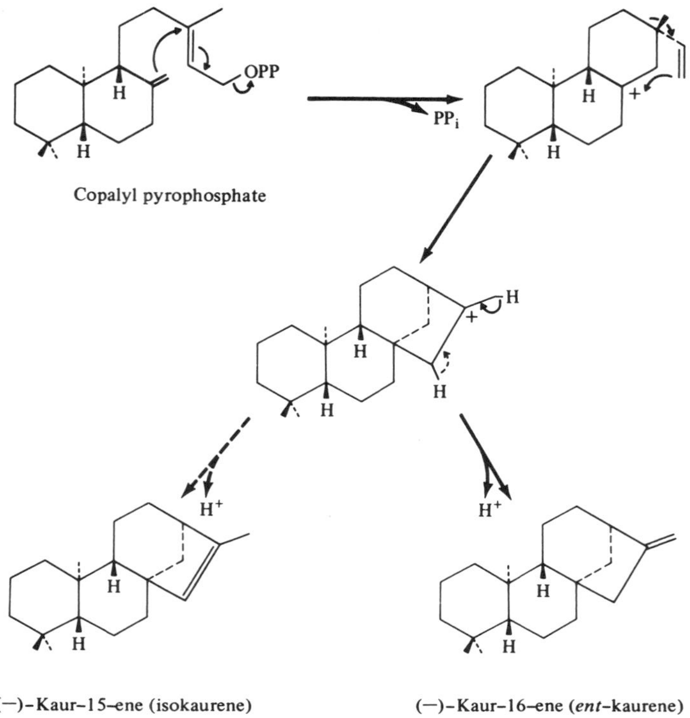

(−)-Kaur-15-ene (isokaurene)   (−)-Kaur-16-ene (ent-kaurene)

FIGURE 3.12. Cyclization of copalyl pyrophosphate to the normal (−)-kaur-16-ene (ent-kaurene) and to (−)-kaur-15-ene (isokaurene), a deadend intermediate as in dwarf-5 mutant of corn (*Zea mays*). (Courtesy of Peter Hedden and B. O. Phinney, 1976.)

pyrophosphate (B activity of kaurene synthetase) to kaurene. The d_5 mutant produces mainly isokaurene (Fig. 3.12), an intermediate that is not converted to GAs, and only a comparatively small amount of the GA precursor *ent*-kaurene. Normal plants, on the other hand, efficiently synthesize *ent*-kaurene and only very small amounts of isokaurene (Fig. 3.13). Hence, dwarfism in the d_5 mutant is correlated, in a true cause and effect manner, with failure to carry out efficiently one reaction in GA biosynthesis.

Actually more than 20 dwarf mutants of maize have been catalogued by maize geneticists. Additional dwarf mutants are also available from recent investigations by geneticists using chemical mutagens and transposable elements. The phenotypes of these mutants vary from examples 10 cm in shoot height at maturity to those that are 1 m in height (50% the height of normals). The reduced height is due primarily to the length of internodes, not the number of internodes. There is also wide variation in the morphology (shape) of the dwarf plants. Some are miniatures of the normal; some have shortened and pointed leaves; others (GA mutants) have short, crinkly, relatively wide leaves and anthers in the ears. All dwarf mutants are simple recessives, with one exception,

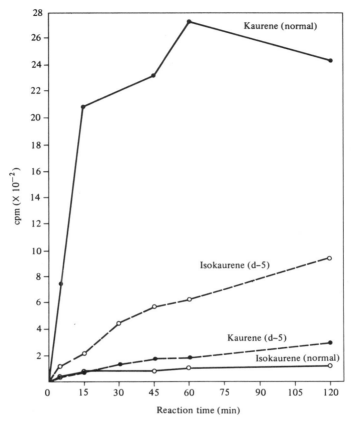

FIGURE 3.13. Comparative biosynthesis of *ent*-kaurene and isokaurene from [³H]copalyl pyrophosphate by cell-free enzyme extracts prepared from normal and dwarf-5 mutants of *Zea mays*. (Courtesy of Peter Hedden and B. O. Phinney, 1976.)

Dominant Dwarf. The GA dwarf mutants lack one or more endogenous GAs. They respond by normal growth to exogenous GA and to no other growth regulators.

Bernard O. Phinney of the University of California at Los Angeles and his associates have identified the precise location in the GA metabolic pathways where lesions (mutations) occur for several different dwarf mutants (Fig. 3.14). Four of the GA-deficient and GA-sensitive mutants of *Zea mays* have received particular attention by Phinney and associates, namely d_1, d_2, d_3, and d_5, and they have determined with varying degrees of certainty precisely where each dwarfing gene works (blocks) in the GA biosynthetic pathway (Fig. 3.14). Thus, as we have seen, gene d_5 blocks the cyclization of copalyl pyrophosphate to *ent*-kaurene by reducing the B activity of *ent*-kaurene synthetase. Gene d_1 blocks the 3β-hydroxylation of GA_{20} to GA_1. Gene d_2 probably controls the carbon-7-

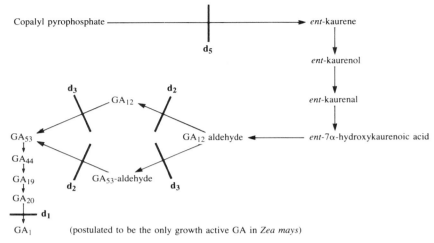

FIGURE 3.14. Proposed sites of action of certain dwarfing genes in mutants of *Zea mays*. Based on the extensive research of Bernard O. Phinney, University of California, Los Angeles [see, e.g., Phinney et al. (1986)].

oxidation step between GA_{12}-aldehyde and GA_{12} and between GA_{53}-aldehyde to GA_{53}. Gene d_3 probably controls the C-13 hydroxylation step (Phinney and Spray, 1982; Phinney et al., 1986). Throughout their investigations Phinney and associates have utilized extraction and bioassay procedures, feeding of putative radioactive substrates for specific reactions of interest, and analysis by GC-MS. Thus, these four mutations probably are the best understood genes controlling plant hormone biosynthesis. Similar investigations are underway with dwarf cultivars of peas (*Pisum sativum* L.) and some other higher plants, and important new data are emerging rapidly (Reid, 1987).

Genetic dwarfs of some other species also appear to be deficient in GAs, compared to their normal or tall counterparts. Five dwarf cultivars of rice and one dwarf mutant of bean are reported to be GA-deficient. Yet in 10 other dwarf rice cultivars no differences in GA content, compared to normal plants, could be found.

In 1972 G. W. M. Barendse and A. Lang reported on a comparison of the levels of endogenous free and bound GAs in the normal strain Violet and the dwarf strain Kidachi (Fig. 3.15) of Japanese morning glory. They found that the dwarf strain contained less GA than the tall counterpart, and that etiolated plants of both strains contained less GA than their light-grown counterparts (Fig. 3.16). In this case the differences in GA content reportedly were attributable mainly to differences in levels of bound GA.

The situation in peas is complex, according to the collective investigations reported on to date. H. Kende and A. Lang in 1964 found no difference in the extractable GA content in light-grown and dark-grown dwarf peas. Furthermore, R. L. Jones and A. Lang detected no significant difference between tall

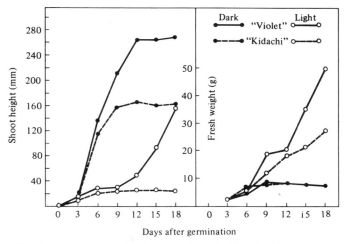

FIGURE 3.15. Growth curves of the dwarf strain Kidachi and of the normal or tall strain Violet of Japanese morning glory (*Pharbitis nil*). Left, total height of shoot; right, fresh weight (of the whole plants). (Redrawn, with permission, from Barendse and Lang, 1972.)

and dwarf peas or between light-grown and dark-grown peas with respect to their extractable or diffusible GA contents in a study reported on in 1968. These authors described dwarfism in peas in terms of a light-induced difference between tall and dwarf cultivars in their sensitivity to a certain endogenous GA. However, this explanation is somewhat doubtful in view of other reports such as those of Köhler, who found that light-grown tall peas contained much more of a

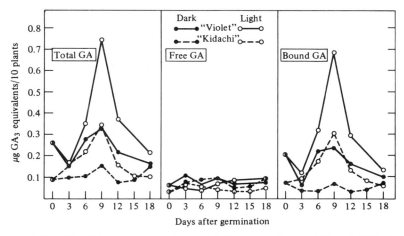

FIGURE 3.16. The GA content of light- and dark-grown plants of (dwarf) Kidachi and (normal or tall) Violet varieties of *Pharbitis nil* determined at 3-day intervals after germination. (Redrawn, with permission, from Barendse and Lang, 1972.)

certain extractable GA fraction than did light-grown dwarfs, and that the extractable GA content of light-grown tall or dwarf peas was considerably greater than that of the tall cultivar grown in the dark. P. R. Ecklund and T. C. Moore reported in 1974 that cell-free extracts from shoot tips of light-grown Alaska peas repeatedly possessed greater kaurene-synthesizing activity (1.4–5 times) than that found in extracts from light-grown Progress No. 9 shoot tips (Table 3.3). Furthermore, cell-free extracts from light-grown shoot tips of either the tall or the dwarf cultivar exhibited greater kaurene-synthesizing activity than was found in such extracts from etiolated shoot tips.

Chung and Coolbaugh (1986) assayed *ent*-kaurene biosynthesis in cell-free enzyme extracts of excised parts of tall (Alaska) and dwarf (Progress No. 9) pea seedlings. They found that *ent*-kaurene-synthesizing activity in extracts of dwarf shoot tips and internodes was generally lower than in equivalent tall plants, but the activity in dwarf leaflets and stipules was somewhat higher than in tall plants. With the exception of root tips, they found a strong correlation between growth potential of a tissue and the rate of *ent*-kaurene biosynthesis in extracts from that tissue.

Moore et al. (1988) analyzed *ent*-kaurene in extracts of the shoots of one normal and two dwarf cultivars of rice (*Oryza sativa* L.) of 14 and 28 days of age by gas chromatography–mass spectrometry (GC-MS) and GC-selected ion monitoring (GC-SIM). They found that the dwarf cultivar Waito-C contained much less *ent*-kaurene than the other two cultivars at both developmental stages (Table 3.4). The level of *ent*-kaurene in the dwarf cultivar Tan-ginbozu was similar to that in the normal cultivar Nihonbare at 14 days but was lower than in Nihonbare at 28 days. These data may be compared to those of Murakami (1972), who postulated that (1) Waito-C plants have a genetic block in the conversion of GA_{20} to GA_1, the latter being considered to be the GA active in the growth of rice; and (2) dwarfism in Tan-ginbozu is attributable to a deficiency in the biosynthesis of *ent*-kaurene. Obviously, further research will be required to elucidate fully the hormonal basis of dwarfism in the Waito-C and Tan-ginbozu cultivars of rice.

In view of the findings in recent years concerning free and bound GAs, and considering the extraction and bioassay techniques used in many studies in the past, many early investigations on levels of endogenous GAs in dwarf and normal plants need to be repeated, taking into account modern techniques and

TABLE 3.4. Concentrations of *ent*-kaurene in extracts of rice shoots.[a]

		ent-kaurene (ng/g fresh wt)	
Cultivar	Phenotype	14 days	28 days
Nihonbare	Normal	10	147
Tan-ginbozu	Dwarf	13	102
Waito-C	Dwarf	< 1	26

[a] With permission, from Moore et al. (1988).

current knowledge. Until this is done, many conclusions about the role of GAs in dwarfism can be regarded only as tentative.

Other Aspects of GA Metabolism

Very little is known about the origin and the physiological significance of conjugated or bound GAs. With respect to physiological role, A. Lang (1970) stated that bound GAs can be considered either as reserve or storage forms, from which GAs are released, or as inert GAs which have no function, or as transport forms. The evidence for a role for conjugated GAs in GA transport rests on their occurrence in the bleeding sap of some trees. In bleeding sap from *Acer platanoides* and *Ulmus glabra,* treatment with a β-glucosidase caused a release of free GAs. Actually the two possible functions, serving as storage and as transport forms, are not mutually exclusive. In fact, the bleeding saps of maple and elm in which the conjugated GAs were found were collected in early spring, and the conjugates may represent storage GAs being transported to growth centers.

Only meager evidence is available concerning other metabolic conversions, including metabolic fate, of GAs in plants. Obviously, a comprehensive understanding of metabolic conversions of GAs is essential to an ultimate full understanding of the GA relationships of plants.

H. Kende prepared tritiated GA_1 and investigated its metabolism in dwarf peas in 1967. He applied $[^3H]GA_1$ to etiolated and light-grown dwarf peas and extracted the seedling shoots 24, 48, and 72 hours after treatment. There was some conversion of $[^3H]GA_1$ to an acidic, biologically active compound (approximately 40% of administered label after 72 hours), and a slight amount was converted to a neutral (water-soluble) fraction (about 10% after 72 hours).

There is abundant evidence for metabolic interconversions among free GAs and reversible formation of GA conjugates and other bound GAs during development of fruits, seeds, and other storage organs. Thus a free GA that is present at one time or at a particular developmental stage may not be present at some other time or stage of development.

It has been reported that developing seeds of *Phaseolus coccineus* L. ($= Ph.$ *multiflorus* Lam.) contain GA_1, GA_3, GA_5, GA_6, GA_8, GA_{17}, GA_{19}, and GA_{20}, but so far none of these GAs has been demonstrated in mature seeds. Furthermore, the two major GAs that have been isolated from *P. coccineus* L. seedlings apparently are not identical with any of them. The 20-carbon GAs may in many cases be precursors of 19-carbon GAs, and interconversions among some 19-carbon GAs have been demonstrated.

Other evidence for metabolic interconversions or degradation of free GAs is scant. L. G. Paleg and B. G. Coombe found no evidence in 1967 for utilization of exogenous GA_3 in embryoless barley half-grains during the induction of hydrolases in a barley endosperm bioassay procedure. A. J. McComb applied radioactive GA_3 to pea seedlings in 1964 and found no changes in the amount of recoverable GA_3 during a succeeding 94 hours. Some evidence has been reported for the metabolism of some exogenous GA_1 in seeds and seedlings, however, by H. Kende in 1967 and G. W. M. Barendse et al. in 1968.

The postulated storage function of bound GAs is indicated by evidence for interconversions between free and "bound" GAs during seed maturation and subsequent germination. G. W. M. Barendse et al., in a study published in 1968, injected radioactive GA_1 ($[^3H]GA_1$) into excised immature fruits of pea (*Pisum sativum* L.) and Japanese morning glory (*Pharbitis nil* Chois.) and investigated the fate of the applied GA_1 during further seed development and subsequent germination. During maturation of both kinds of seeds, part of the radioactivity became associated with the aqueous fraction, while another part of the $[^3H]GA_1$ was apparently converted to two other unidentified biologically active acidic compounds (Fig. 3.17). During germination of pea and morning glory seeds, part of the bound radioactivity was released in the form of $[^3H]GA_1$ and the two unidentified growth-active, acidic compounds. During rapid seedling growth, a further conversion of $[^3H]$, mainly to one of the unidentified acidic substances occurred.

It appears from these data that during seed development a part of the exogenous $[^3H]GA$ was converted to a bound form, while during germination

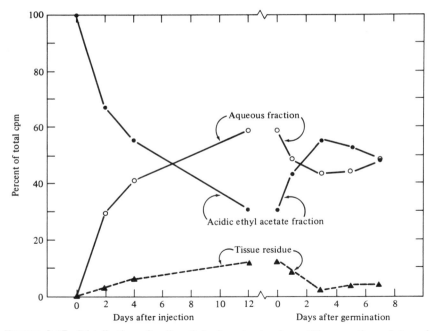

FIGURE 3.17. Distribution of radioactivity in extracts of pea (*Pisum sativum* L.) seeds developing in pods which were excised 13 days after anthesis, injected with $[^3H]GA_1$, and grown *in vitro*, and in seedlings that subsequently developed from such seeds. The acidic ethyl acetate fraction contained free GAs and the aqueous fraction "bound GAs." (Redrawn, with permission, from Barendse et al., 1968.)

part of this bound GA (approximately 10–15%) was reconverted to free GA. The latter process was accompanied by transfer of radioactive material from the cotyledons to the shoot–root axis. Whatever contribution of free GA may be accounted for by bound GA in early growth of pea seedlings, it appears that GA biosynthesis is initiated during early seedling growth and also that endogenous GA is present in growth rate-limiting quantity during early seedling development.

G. Sembdner et al. in 1968 applied radioactive GA_8 to developing *Phaseolus coccineus* L. pods and observed a decrease of radioactivity in water-soluble substances and an increase in more ethyl acetate-soluble substances as the seeds germinated. These authors also observed transformations of labeled GA_3 and GA_6 following their application to maturing *Phaseolus coccineus* L. pods, the former to a hexapyranosyl-GA_3 and glucosyl-GA_8, the latter to $GA_{1/3}$, GA_8, and glucosyl-GA_8. Barendse (1971) applied $[^3H]GA_1$ to young *Pharbitis nil* Chois. plants and found that part of the exogenous GA was converted to two water-soluble compounds. The rate of conversion was enhanced when both $[^3H]GA_1$ and $[^{14}C]$ glucose were applied to the same plants.

T. Hashimoto and L. Rappaport analyzed endogenous GAs in developing bean (*Phaseolus vulgaris* L.) seeds in 1966. They found that as the seeds matured the acidic ethylacetate-soluble GA-like substances initially decreased and then almost disappeared, while neutral substances and one of the acidic butanol-soluble substances increased. The neutral GA-like substances (bound GAs) were present in fairly large amounts in mature dry seeds. Related work by the same authors, involving the application of GA_1, further indicated formation of bound GAs during seed maturation, possibly products of conversion from acidic substances. In 1966 G. F. Pegg extracted GA-like substances possessing acidic, basic, and neutral properties from mature, dry tomato (*Lycopersicon esculentum* Mill.) seeds and young seedlings. He suggested that protein hydrolysis during germination is an important means of liberating free GAs from bound GAs.

Quantitative Changes in GA Content during Development

Development of Shoots

The author and P. R. Ecklund conducted studies in 1967 and 1968 designed in part to ascertain whether there are quantitative changes in sensitivity of shoots of pea plants of the normal or tall Alaska cultivar to exogenous GA and Amo-1618. It was reasoned that the relative rate of GA biosynthesis and the level of active, endogenous GA could be estimated indirectly by relating the growth rates to the growth responses of the plants to standard dosages of the two compounds.

In the initial phase of the investigations seedlings were treated with single growth-saturating doses of 10 μg of GA_3, which were applied to the shoot tips, at intervals of 3, 6, and 9 days after planting. Shoot height was measured daily from the third through the thirteenth day after planting. Seedlings treated at 3 days

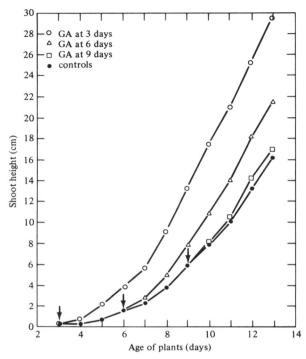

FIGURE 3.18. Time-course of stem elongation in pea seedlings to which single growth-saturating doses of GA, 10 μg per plant, were applied on the shoot tips on the days indicated by arrows. (Redrawn, with permission, from Moore, 1967.)

exhibited a pronounced stem elongation response (Fig. 3.18). Sensitivity of the seedlings to exogenous GA decreased as development progressed, to the extent that seedlings treated at 9 days of age were only slightly taller than the controls on the thirteenth day of growth. The plants completely failed to respond to exogenous GA when treated at 14 days of age. Hence, unlike dwarf varieties of peas, Alaska pea shoots become insensitive to exogenous GA at about 2 weeks of age.

Interestingly, however, Alaska pea plants again become sensitive to exogenous GA_3 after the linear phase of growth is passed and senescence begins. This was seen in experiments wherein plants were treated by applying 1 μg of GA_3 or 50 μg of Amo-1618. On some plants flowers and fruits were left on as they formed; the flowers were removed as they formed on other groups. Sensitivity to exogenous GA was measured as the percentage increase in stem elongation over that of untreated plants 1 week after application. Sensitivity to Amo-1618 was determined as the percentage inhibition in stem elongation compared with untreated plants 1 week after application. In Fig. 3.19 sensitivity to a growth regulator is plotted at the age of the plants when the regulator was applied. It was assumed that the growth response of a plant to exogenous GA would be inversely

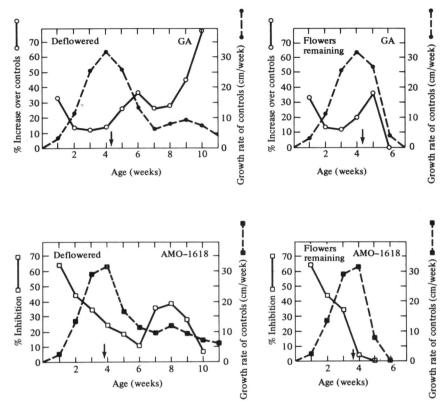

FIGURE 3.19. Ontogenetic changes in sensitivity of fruit-bearing and deflowered Alaska peas to a single shoot-tip application of 1 μg of GA₃ per plant or 50 μg of Amo-1618 per plant. Vertical arrows indicate the time of anthesis. Deflowered control plants were those from which flowers were removed as they formed and that received no growth regulator. Controls for experiments using plants with flowers remaining were flower-and fruit-bearing plants that received no growth regulator. Sensitivity to a growth regulator is plotted at the plant age at which the regulator was applied. Sensitivity to exogenous GA was measured as the percentage increase in stem elongation over that of untreated plants 1 week after the application. Sensitivity to Amo-1618 was measured as the percentage inhibition in stem elongation compared with untreated plants 1 week after the application. (Redrawn, with permission, from Ecklund and Moore, 1968.)

related to the level of active, endogenous GA in the shoot during the time the response is measured and that a specific amount of Amo-1618 would inhibit the biosynthesis of a specific amount of endogenous GA. Therefore, the magnitude of growth inhibition by Amo-1618 would depend on the rate of GA biosynthesis throughout the time interval that inhibition is measured.

The plants displayed high sensitivity to both exogenous GA and Amo-1618 during the exponential phase of growth (Fig. 3.19), suggesting that endogenous GA was at a relatively low level, but a fairly high rate of GA biosynthesis was occurring. During the linear phase of growth, sensitivity to both exogenous GA and Amo-1618 declined, presumably because the rate of GA biosynthesis was rapidly increasing, thereby increasing the endogenous GA to a level that was practically growth-saturating.

In the case of plants with flowers and fruits intact, a second period of high sensitivity to exogenous GA appeared near the time of anthesis (Fig. 3.19) while sensitivity to Amo-1618 continued to decline, or remained quite low. These results suggest that subsequent to the linear growth phase the level of endogenous GA was decreasing because the rate of GA biosynthesis was rapidly declining. A lack of response to exogenous GA or Amo-1618 was associated with the rapid decline in growth rate, presumably because as apex senescence progressed, irreversible catabolic processes in the apex resulted in a loss of growth capability.

The decline in growth rate following anthesis was not as precipitous, the second period of high sensitivity to exogenous GA was delayed, and inhibition by Amo-1618 did not decrease as rapidly with deflowered plants as with flower- and fruit-bearing plants. These differences are assumed to result from a slower decline in the rate of GA biosynthesis following anthesis in deflowered plants. Throughout the period of prolonged growth due to flower removal, the sensitivity to exogenous GA was inversely related to the growth rate of the plants, while inhibition by Amo-1618 was directly related to the growth rate. The troughs in the GA-sensitivity curves correspond with the peaks in the Amo-1618-sensitivity curves for deflowered plants, suggesting that a second period of increasing GA biosynthesis occurred in the shoot tips of plants in which apex senescence was delayed. As deflowered plants approached the stage of determinate growth with the apparent disappearance of the shoot apex, the growth potential of the shoots was lost and the sensitivity to exogenous GA or Amo-1618 disappeared.

It appears from these experiments that light-grown Alaska peas exhibit two periods of sensitivity to exogenous GA in their ontogeny, the first being the growth period prior to the linear phase of growth, and the second being the period of declining growth rate and onset of apex senescence. Thus it seems that endogenous GA is rate-limiting for stem elongation in light-grown Alaska peas during early seedling growth and during the period of decline. It may be speculated that the level of endogenous GA in Alaska peas is regulated primarily by the rates of synthesis and utilization in growth processes. Stem elongation in light-grown Alaska peas may be specifically limited by the rate of endogenous GA biosynthesis prior to and subsequent to the linear phase of growth. Insensitivity to exogenous GA during the linear phase of growth probably is correlated with the presence of growth-saturating levels of endogenous GA. A direct approach toward confirmation of this hypothesis would be to conduct time-course studies of the rate of GA biosynthesis and the concentration of extractable GA in pea shoots throughout ontogeny.

To date precisely those types of investigations have not been reported. However, the author and P. R. Ecklund did conduct studies on ontogenetic

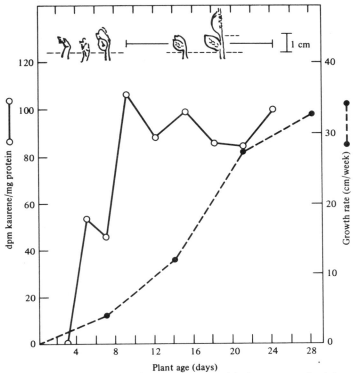

FIGURE 3.20. Ontogenetic changes in growth rate and in kaurene-synthesizing capacity of Alaska pea shoot tips, as measured in cell-free enzyme extracts. The drawings illustrate the shoot tips from which enzyme extracts were prepared. The data represent means based on three independent experiments. (Redrawn, with permission, from Moore and Ecklund, 1974.)

variations in capacity for kaurene synthesis in enzyme extracts of Alaska pea plants. The results of that effort are illustrated in Fig. 3.20. Kaurene synthesis was not detectable in epicotyls of 3-day-old seedlings. However, activity was detectable in extracts of plants older than 3 days and became maximal in shoot tips of 9-day-old plants. Thereafter, kaurene-synthesizing activity remained relatively constant until the twenty-fourth day, during which interval the growth rate was increasing. The same relationship can be seen when activity is expressed on a fresh weight basis. These results reveal a direct correlation between the apparent rate of kaurene synthesis in pea shoot tips and the apparent rate of GA production, as assessed by indirect procedures in previous investigations. Interestingly, the results are also analogous to those that were reported previously by R. C. Coolbaugh and the author for developing pea seeds, in which a large increase in capacity for kaurene production appears to slightly precede the period of maximum GA biosynthesis and maturation (see Fig. 3.22).

Coolbaugh (1985) assayed *ent*-kaurene biosynthesis from [^{14}C]MVA in cell-free enzyme extracts of shoot tips, leaflet blades, internodes, and root tips of Alaska pea seedlings. He found that the highest rates were obtained with extracts from the fifth (youngest) internode, the fourth (youngest) expanded leaflet, and the shoot tip. The tissues with the greatest potential for growth had the highest apparent rates of *ent*-kaurene biosynthesis. The yields of *ent*-kaurene (pmol/mg protein) were 2.02, 8.78 and 3.32 for shoot tips, fifth internode, and fourth leaf, respectively.

Development of Fruits and Seeds

As in the case of shoots, only fragmentary and incomplete evidence exists regarding changes in GA content, rate of GA biosynthesis, and correlations between GA content and growth rate in fruits, seeds, and other specific organs. However, developing seeds and fruits have been most intensively investigated, primarily because they contain much higher concentrations of GA than vegetative organs. Questions of interest include where in the seeds and fruits GA

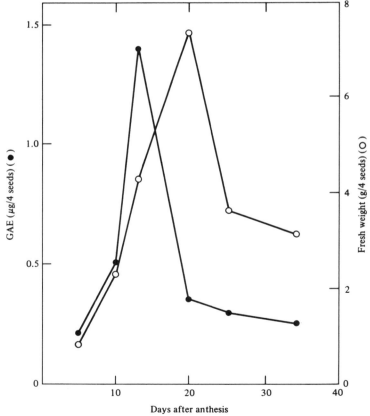

FIGURE 3.21. Changes in GA content and fresh weight in seeds of *Pharbitis nil* Chois. (Japanese morning glory) from anthesis to maturity. GA content is expressed as GA_3 equivalents (GAE). (Redrawn, with permission, from Murakami, 1961.)

synthesis is localized, and how the levels of endogenous GA vary during development.

In immature seeds of many species of angiosperms the maximum amounts of extractable GAs are found when the seeds have attained about half their maximum fresh weights (e.g., Fig. 3.21). The characteristic decrease in extractable GA content as the seed matures quite probably is correlated with the formation of conjugated GAs or other forms of bound GA that are not effectively extracted from acidified aqueous solution with relatively nonpolar solvents. By extraction procedures that used to be commonly employed, such substances, when present, generally were inadvertently discarded in an aqueous fraction.

It is interesting to compare these data with available information from studies with cell-free enzyme extracts. In 1969 R. C. Coolbaugh and T. C. Moore found that the capacity to convert mevalonic acid to kaurene in cell-free enzyme extracts varied markedly with the stage of seed development (Fig. 3.22). Kaurene-synthesizing capacity increased from a barely assayable level in very young ovules to a maximum at about 13 days after anthesis, when the seeds had attained about half-maximal fresh weight.

J. E. Graebe in 1968 observed comparatively low incorporation of MVA into kaurene in cell-free extracts prepared from very young pea fruits and even lower

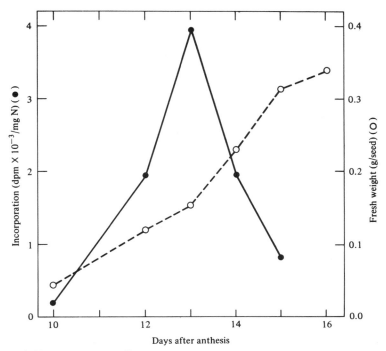

FIGURE 3.22. Net rates of [^{14}C]kaurene biosynthesis from [2-^{14}C]mevalonate in cell-free enzyme extracts (40,000 g supernatant fraction of homogenates) prepared from pea *(Pisum sativum* L.) seeds at different stages of development, in relation to seed growth. (Redrawn, with permission, from Coolbaugh and Moore, 1969.)

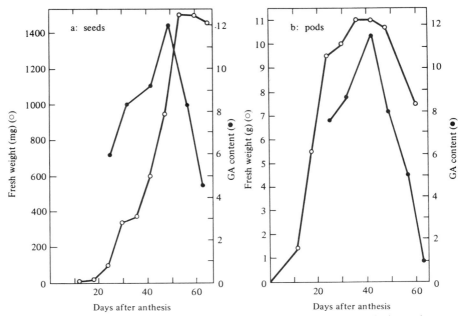

FIGURE 3.23. Changes in amount of GA-like substances and fresh weight in seeds, **a**, and pods, **b**, of bean (*Phaseolus vulgaris* L.) during development. GA content is expressed as mm of growth in the dwarf (d_1) *Zea mays* leaf section bioassay. (Adapted from Wareing and Seth, 1967, by permission.)

incorporation with immature seeds from almost fully grown pods. The kaurene-synthesizing activity that he observed with very young fruits probably reflected synthesis in the carpel. Hence, just as the growth curves for the legume pod and seed do not coincide in time (Fig. 3.23), maturation of the seed lagging behind that of the pod, it seems likely that the time-courses of kaurene and GA synthesis and GA accumulation in the pod and seed are similarly related.

J. L. Stoddart in 1965 investigated changes in GA content during ripening of seeds of two species of grasses. Floret samples of perennial ryegrass *Lolium perenne* L.) and timothy *(Phleum pratense* L.) were analyzed for GA content at 2–5 day intervals from inflorescence emergence to seed harvest, using chromatographic and bioassay techniques. Both species had relatively high GA content at inflorescence emergence, which diminished during anther formation. During final seed maturation the level of GA decreased steadily to reach a stable, low level prior to harvest.

Changes in endogenous GAs during development of fleshy fruits have also been investigated rather extensively. D. I. Jackson and B. G. Coombe in 1966 compared growth rates of the seed, endocarp, and mesocarp of apricot (*Prunus armeniaca* L.) between anthesis and fruit maturity with the concentrations of GA-like substances extracted from those tissues, and also described some properties of the GA-like substances. A strong correlation was shown between the concentration of GA-like substances in all three tissues and their respective growth rates for the first 60 days after anthesis (Fig. 3.24). Results with

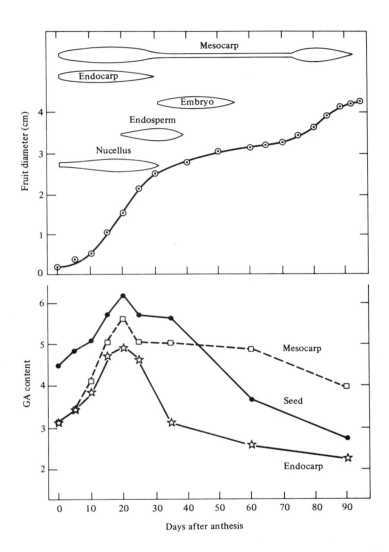

FIGURE 3.24. Concentrations of GA-like substances in methanolic extracts of seed, endocarp, and mesocarp of apricot (*Prunus armeniaca* L.) fruits in relation to their growth rates during development. GA content is expressed as log picogram GA_3 equivalents/10 mg tissue determined by the barley endosperm bioassay. Figures with tissue names inserted in top half of figure represent the comparative volume increases of the tissues. (Redrawn, with permission, from Jackson, D. I. and B. G. Coombe, 1966. Gibberellin-like substances in the developing apricot fruit. *Science* **154**: 277–278, 14 October 1966. Copyright 1966 by the American Association for the Advancement of Science.)

methanol extracts showed that in all three parts of the apricot fruit, GA activity increased after anthesis and reached a maximum 20 days after anthesis. The seed had the greatest GA activity and the endocarp the least. Thereafter concentrations decreased in all three tissues, the decrease being most rapid in the endocarp and slowest in the mesocarp. The time at which GA activity was greatest coincided with the time of greatest growth rate in the mesocarp, endocarp, and nucellus. Activity in the endocarp declined to a low level concurrently with cessation of growth and sclerification of the tissue, 30–45 days after anthesis. Activity in the seed remained high and did not decline until 35–60 days after anthesis during which seed growth stopped. The only tissue that grew after 60 days was the mesocarp, which contained the highest content of GA-like substances at that time. Evidence for the occurrence of bound GAs, that is substances more polar than free GAs, in developing apricot fruits was seen in the fact that methanol extracts of the seed and pericarp were 10 times more active than ethylacetate extracts.

D. I. Jackson reported in 1968 on an analogous, equally impressive study of changes in GA-like substances during development of peach (*Prunus persica* Batsch) fruits. No GA was found in the ovary before bloom. Immediately after full bloom GA activity was found first in the seed and later in the mesocarp and endocarp as well. GA concentration was closely correlated with the rate of cell expansion in each tissue, but not with cell division. Until the final growth phase (in the double-sigmoid growth curve), when activity was found only in the mesocarp, the highest GA level was always found in the seed. W. J. Wiltbank and A. H. Krezdorn reported in 1969 that there was a striking direct correlation between concentration of GA-like substances and mean fresh weight of whole navel orange (*Citrus sinensis* Osbeck) fruits during development (Fig. 3.25).

An interesting question concerns the role of GAs produced in seeds in the development of fleshy fruits. D. I. Jackson stated in 1968 that hormone levels in seeds rarely correlate with the growth of the whole fruit. More commonly, he noted, high levels of hormone correspond to periods of rapid seed growth. This is sometimes interpreted to indicate that growth of other fruit tissue is not directly dependent on supplies of hormone synthesized by the seed. D. I. Jackson and B. G. Coombe found in 1966, in fact that in apricot the GA concentration may at times be higher in the mesocarp than in the seed and that the respective tissues of apricots may be self-sufficient for this hormone. Similar results with peach fruits supported that view and further suggested that GA levels in a given tissue are intimately associated with cell expansion in that tissue.

S. Iwahori et al. in 1968 determined levels of GA-like activity in berries of both seeded Tokay and seedless Tokay grapes (*Vitis vinifera* L.) at different stages of development. In general, they found, as did R. J. Weaver and R. M. Pool in 1965, more GA-like activity in seeded than in seedless berries, indicating that grape seeds may be rich source of GA-like substances. This view is supported by the fact that exogenous GA greatly promotes enlargement of seedless Tokay berries, but not seeded Tokay berries. In seedless Tokay the

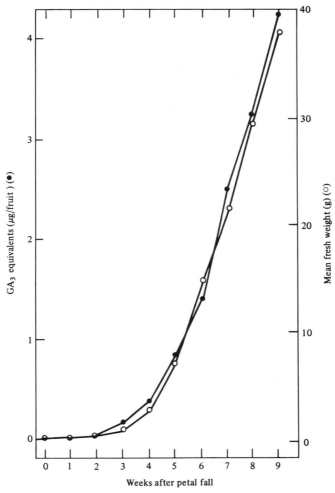

FIGURE 3.25. Correlation between concentration of GA-like substances and growth of navel orange (*Citrus sinensis* Osbeck) fruits in a 9-week period following petal fall. Concentrations of GA-like substances are expressed in terms of GA_3 equivalents determined by a fluorometric assay. (Redrawn, with permission, from Wiltbank and Krezdorn, 1969.)

development of seeds ceases at an early stage. The authors stated that the earlier decrease of GA in the seedless Tokay berries probably is associated with the abortion of the embryo and cessation of seed development.

Are GAs essential to the development of seeds? Ironically, an unequivocal affirmative answer may not yet be fully justified, although on the basis of available evidence it seems reasonable to conclude that GAs are indeed essential for the development of seeds, just as for the sporophytes that potentially develop from them. Seeds apparently accumulate GAs in forms from which some free, active GAs potentially can be mobilized to the seedling shoot–root axis during germination. However, it is certain that far more GA accumulates in seeds than is required for their own maturation.

In summary, whether GAs produced in seeds contribute to the growth of fleshy fruits in which they are contained is not clear from presently available evidence. Some data suggest that each tissue of a fleshy fruit has a relatively independent GA metabolism, while other evidence indicates that seeds in fleshy fruits do contribute GAs or GA precursors to the surrounding tissues. In general, very good, direct correlations have been found between GA content and growth of whole fleshy fruits and the constituent tissues thereof. A decrease in endogenous GAs appears to be associated with the ripening of fleshy fruits.

Sites of GA Biosynthesis in Seed Plants

J. A. Lockhart in 1957 provided the first evidence on the sites of GA biosynthesis using etiolated pea seedlings. His method of establishing the organ of production was to remove individual parts of the plant suspected of being the source and then attempt to reproduce the response normally elicited by the organ by application of pure hormone. His rationale was that, if the added substance can completely or to a large extent replace the organ in restoring normal growth, then the presumption is strong that the organ removed normally supplies a comparable substance to the plant. It may be noted that the converse conclusion is not valid. He showed that applied GA_3 could almost completely replace the stem tip in promoting the elongation of the subapical region of the pea stem and concluded that the natural GA of the pea plant is produced in the stem tip.

R. L. Jones and I. D. J. Phillips studied the sites of GA synthesis in light-grown sunflower plants in 1966 and 1967 using techniques of organ excision and application of GA, extraction, diffusion, and treatment with inhibitors of GA biosynthesis. They found that GA synthesis occurred in the young leaves (not the apical meristem per se) of the apical bud. Older but not fully grown leaves supply GA to the stem although in lower quantity. Root tips also were shown to be sites of GA synthesis, this ability being confined to the apical 3 to 4 mm of the root. The findings that the amounts of GAs obtained over a period of time exceeded the amounts of GAs obtained by extraction, and that CCC reduced the amounts of GAs obtainable over time by the diffusion technique, provided strong corroborative evidence for GA biosynthesis in the excised organs.

D. Sitton, A. Richmond, and Y. Vaadia provided additional evidence in 1967 for potential GA biosynthesis in sunflower root tips. They extracted GA-like substances from the roots of decapitated plants, and reported further that incubation of the root tips in [2-^{14}C]mevalonate yielded kaurenol, an intermediate in GA biosynthesis.

As indicated in the discussion of the biosynthetic pathway, developing seeds are rich sources of GAs and are comparatively intensely active in GA biosynthesis. Not only do developing seeds contain relatively large amounts of

GAs, inhibitors of GA biosynthesis also can be demonstrated to inhibit accumulation of GA in seeds *in situ.*

Direct evidence on sites of localization of enzymes involved in GA biosynthesis in developing seeds is fragmentary and incomplete and is derived mainly from studies of immature seeds of two species, pea (*Pisum sativum* L.) and wild cucumber (*Marah macrocarpus*). R. C. Coolbaugh and T. C. Moore in 1971 performed comparative assays of kaurene accumulation in cell-free enzyme extracts of whole pea (cv. Alaska) seeds, isolated seed coats, cotyledons, and embryonic shoot–root axes. Kaurene, and presumably GA, biosynthesis from mevalonic acid was localized exclusively in the cotyledons.

These results differ rather strikingly from comparable data for wild cucumber seeds in which kaurene synthesis is localized in the liquid endosperm. This difference becomes intriguing when it is realized that the early morphological development of pea seeds also includes the formation of liquid endosperm. During early embryogenesis, the pea embryo occupies a small part of the embryo sac, the remainder being filled with liquid (free-nuclear) endosperm. By the time the seed is 11 to 12 days old, the endosperm is entirely digested, and the seed consists of the embryo and the seed coat. The pea seeds utilized in the investigations of R. C. Coolbaugh and T. C. Moore were 12 to 16 days of age.

Interesting changes in kaurene-synthesizing capacity occur during the development of both pea and wild cucumber seeds. In extracts of *Marah macrocarpus* the highest incorporation of mevalonic acid into a lipid fraction (containing kaurene) occurs in seeds when the cotyledons are not yet visible and the seed cavity is mostly filled with endosperm, and the activity declines as the seeds mature and the cotyledons form. Cell-free extracts of pea seeds exhibit a relatively low capacity for kaurene biosynthesis during early morphological development when liquid endosperm is present. In fact, direct assays for kaurene biosynthesis in the liquid endosperm of pea seeds yielded negative results. The activity in pea seeds increases to a maximum when the seeds are approximately half-maximum fresh weight (see Fig. 3.22) and the cotyledons nearly fill the seed cavity and then decreases as the seeds mature. Thus, so far as can be determined on the basis of studies to date, GA biosynthesis in developing seeds of *Pisum sativum* and *Marah macrocarpus* occurs in anatomically quite dissimilar parts of the seeds, but parts that, in both kinds of seeds, are most commonly associated with a nutritive function. Of course the incorporation of mevalonic acid into kaurene in cell-free enzyme extracts does not necessarily mean that these intermediates are converted to GAs in these tissues. However, the available data on extractable GA content in legume and *Marah macrocarpus* seeds correlate well with the published results concerning localization of kaurene-synthesizing enzymes.

Very little information is available on sites of GA synthesis in germinating seeds. Margaret Radley showed by extraction and bioassay in 1967 that GA-like substances are produced in the scutellum of barley embryos during the first 2 days of germination. On the third day, the activity of the scutellum ceases and the embryonic axis begins to produce GA.

Transport of GA

Most of the data concerning transport of GA in seed plants pertain to movement of applied (exogenous) GA. Information regarding movement of endogenous GA is largely indirect. A. J. McComb was among the first to investigate movement of exogenous GA. He applied [^{14}C]GA$_3$ to pea seedlings and reported in 1964 that translocation of applied GA was devoid of any unique features, GA transport apparently occurring in the phloem by the same mechanisms and according to similar patterns for other organic metabolites. Additional, more direct evidence for the transport of GAs in phloem comes from the findings that GAs occur in the sieve tube sap of plants.

T. Y. Chin and J. A. Lockhart concurred in 1965 with McComb's conclusions in their investigations of translocation of applied GA$_3$ in pinto bean seedlings. They concluded, as have other authors, that the movement of applied GA is related to carbohydrate transport within the plant.

Other, more recent investigations have shown that GAs are transported both in the phloem and the xylem, and there is experimental evidence from studies using [^{14}C]GA that interchange between phloem and xylem occurs. Thus, there appears to be a ready transport of GAs in the entire conductive (vascular) system of the plant, proceeding passively with the flow of assimilates or of water plus salts and some other organic compounds.

Most investigations of GA transport have employed excised sections of coleoptiles, stems, or petioles, exogenous radioactive GA, and the donor–receptor agar block technique that was described in the chapter on auxins. In general, movement has appeared to be nonpolar (in contrast to transport of endogenous auxin) and to occur at comparative rates of 1.4 to 2.0 mm per hour. However, there have been reports of polar GA transport as well. When reported, polar transport has been described as being in the basipetal vector in stem, coleoptile, and subapical root sections. Perhaps the apparent polar movement, when observed, really is a movement from a source to a growth center (in accord with the concept of source-to-sink movement of metabolites in phloem generally), and really is not true polar movement.

Anatomical and Biophysical Basis of GA-Induced Growth

The most dramatic and most thoroughly studied effect of GAs is promotion of stem elongation, particularly in many genetic dwarf plants (Figs. 3.10 and 3.11) and the rosette stages of biennials and other plants that require vernalization and/or photoperiodic induction by long days to flower (Fig. 8.24). Indeed, as previously indicated, it was the capacity of GAs to enhance elongation of rice stems that led to their discovery.

Since the early days of investigation of the effects of GAs on growth, it has been evident that these hormones can stimulate both cell division and cell enlargement. In the case of bean (*Phaseolus vulgaris*) seedling stems *in situ,* both parameters were stimulated by exogenous GA, but the largest effect was thought to be on cell elongation. Enhancement of cell enlargement has also been reported to be the major or exclusive response to exogenous GAs by *Tradescantia* stamen hairs, mesocarp parenchyma of parthenocarpic peaches, and intercalary meristems of excised internodal segments of *Avena sativa* stems.

In other cases, the major or primary effect of GAs is stimulation of cell division, as in subapical regions of the apical meristem of rosette plants such as *Hyoscyamus niger* and *Samolus parviflorus* (Fig. 3.26) and dwarf peas. Clearly the response of a given cell or tissue to divide or elongate depends upon age and stage of development. Younger cells generally can respond by dividing while older cells may respond only by enlargement.

In general, the stem elongation response manifested by whole plants to treatment with exogenous GAs is correlated with effects of the hormones on both cell division and cell enlargement, the former effect of course preceding the latter in time (Fig. 3.26). As noted in Chapter 1, cell division alone does not

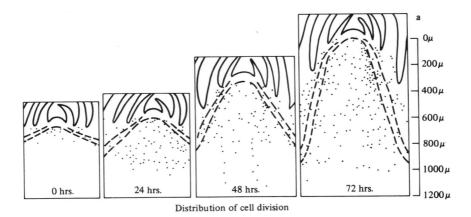

Distribution of cell division

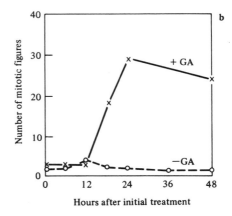

FIGURE 3.26. GA applications to rosette plants produce bolting by increasing transverse cell divisions in the tissues immediately below the apical meristem. Longitudinal sections through the shoot apex of *Hyoscyamus niger* illustrate where division increases following application of GA. **a.** (Each dot represents one mitotic figure in a 64-μm-thick section.) The numbers of mitoses in the subapical tissues of shoot apices of *Hyoscyamus niger*, with and without GA, are shown in **b.** (Redrawn, with permission, from Sachs et al., 1959.)

result in growth, and so stimulation of cell division must be coupled with a normal or enhanced rate of cell elongation for there to be a growth response to exogenous GAs.

Interestingly, it appears that GAs are not essential for growth of roots. It has not been shown convincingly that GAs either generally stimulate or inhibit root growth or that GAs affect cell division or cell elongation in roots.

Although both auxins and GAs promote cell enlargement, it is not yet clear whether they do so in the same manner. Whereas the major effect of auxins is to increase cell wall plasticity (see Chapter 2), the effects of GAs are less well understood. Cleland and associates (1968) compared the effect of auxin and GAs on the cell wall extensibility of cucumber hypocotyls, the growth of which is promoted by both hormones. Their results showed a consistently strong promotion of plasticity by IAA, but GA caused a variable plasticity response with values only slightly above control levels. They concluded, therefore, that GA had little or no effect on cell wall plasticity. However, in later investigations with *Avena* stem segments, Adams et al. (1975) found that GA markedly increased cell wall plasticity, in a manner resembling the action of auxin (Fig. 3.27). The plasticity values of GA-treated segments were, on the average, three

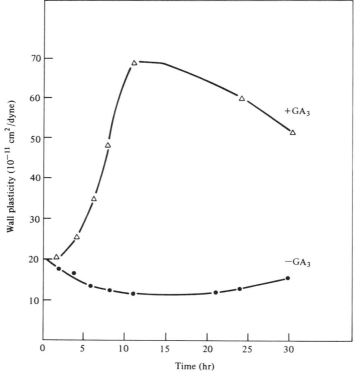

FIGURE 3.27. Effect of GA on cell wall plasticity in *Avena* stem segments. Segments were incubated in 0.1 *M* sucrose plus or minus 30 μM GA. (Redrawn with modifications, by permission, from Adams et al., 1975.)

times greater than those of control *Avena* stem segments. These authors were led to speculate that "plant cell elongation might occur by a similar mechanism regardless of where it occurs and which factors control its occurrence."

Mechanism of Action of GA

It is well known that germinating seeds produce hydrolytic enzymes that digest the fat, carbohydrate, and protein reserves of the storage tissues. Investigations of the tissue source of these hydrolases and the control of their production and release have provided the most definitive information on the mechanism of action of GA. In particular, investigations of germinating barley grains (*Hordeum vulgare*) have been fruitful.

Early History

Haberlandt noted in 1890 that the aleurone layers of rye produce a substance (or substances) that causes liquefaction of the starchy portion of the endosperm and dissolution of the starch grains. Brown and Escombe confirmed this observation in 1898, using barley grains.

In 1960, L. G. Paleg, in Australia, and H. Yomo, in Japan, showed independently that when gibberellic acid (GA_3) is added to endosperm portions of barley grains, it initiates the apparent production of certain amylolytic enzymes, including α-amylase, and the release of sugars as well as stimulating germination. D. E. Briggs extended this observation in 1963 to include several other hydrolytic enzymes (protease, phosphatase, and β-glucanase). Beginning in 1964, J. E. Varner and several associates at the MSU/DOE Plant Research Laboratory at Michigan State University, East Lansing, became leaders in this field.

Description of the Barley Experimental System

In the course of natural germination of a barley grain, the embryo is the source of endogenous GA. Margaret Radley showed in 1967 that GA-like substances are produced by the scutellum during the first 2 days of germination and thereafter by the embryo axis (Fig 3.28). The GAs liberated by the embryo diffuse across the endosperm to the aleurone cells, in which the hydrolytic enzymes are produced. Besides the embryo, the only other living cells in the grain are those of the aleurone, which is composed of several layers of cells constituting the outermost layer of endosperm. Hence the aleurone cells are a homogeneous collection of respiring, non-dividing cells that have the highly specialized function of producing and releasing some of those enzymes that are required to digest the starchy endosperm in preparation for its utilization by the growing embryo.

By 1961, investigators were utilizing embryoless half-grains of barley, with which it could be shown that the increase in α-amylase activity is absolutely dependent on added GA. Within a few more years, many investigators turned to

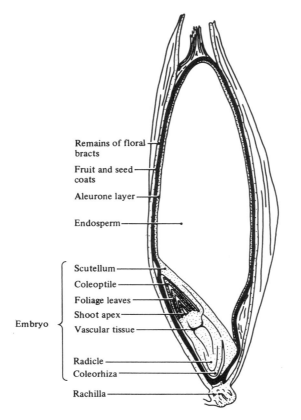

FIGURE 3.28. Longitudinal section of a barley grain. (Redrawn from *Control Mechanisms in Plant Development* by A. W. Galston and P. J. Davies, p. 108. Copyright © 1970 by Prentice-Hall, Inc. By permission of Prentice-Hall, Inc.)

Remains of floral bracts
Fruit and seed coats
Aleurone layer
Endosperm
Scutellum
Coleoptile
Foliage leaves
Shoot apex
Vascular tissue
Embryo
Radicle
Coleorhiza
Rachilla

investigations of isolated aleurone layers, which are separated from the starchy portion of the endosperm after the embryoless half seeds have been imbibed for 3 days.

When isolated aleurone layers of barley are incubated in a solution containing GA, they produce and secrete several hydrolytic enzymes, GA-dependent *de novo* synthesis has been demonstrated for α-amylase, protease, β-1,3-glucanase, and ribonuclease and more recently 1,3:1,4-β-glucanase, acid phosphatase, and DNase (Brown and Ho, 1986). In addition, a GA-dependent release of ribonuclease and β-1,3-glucanase has been demonstrated. The increase in activity of at least one hydrolase, β-amylase, in the presence of GA is due to release of preformed enzyme and not to *de novo* synthesis. Also, carboxypeptidase activity that develops also in the aleurone layers during imbibition arises with or without GA, although its release is enhanced by GA.

Evidence that the increase in activity of several hydrolases induced by GA actually occurs as a result of *de novo* synthesis, rather than by activation of preformed enzyme, has been obtained in various ways. However, the most unequivocal proof comes from density labeling experiments, first performed by P. Filner and J. E. Varner in 1967. The procedure was to incubate barley aleurone layers in a solution containing GA_3 and either normal water or water containing the heavy isotope of oxygen ($H_2{}^{18}O$). Hydrolysis of stored proteins in the aleurone cells liberates amino acids for new protein synthesis. Hence, hydrolysis of stored proteins in the presence of $H_2{}^{18}O$ results in the liberation of amino acids labeled with the heavy isotope of oxygen:

Protein (n amino acid residues) $+ n\ H_2{}^{18}O$

$$\rightarrow n\ RCH\text{–}NH_2\text{–}C\,{}^{18}OOH \text{ and } RCH\text{–}NH_2\text{–}CO{}^{18}OH$$

Any protein synthesized from the labeled amino acids will be heavier than proteins formed from amino acids released in the presence of ordinary $H_2{}^{16}O$; that is, the former will be "density-labeled" (Fig 3.29). On conducting

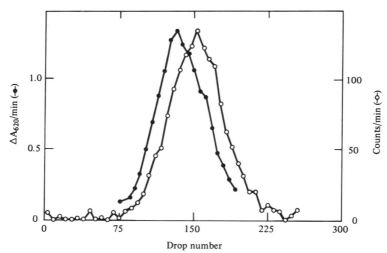

FIGURE 3.29. Demonstration of the stimulation by GA_3 of *de novo* synthesis of α-amylase in barley aleurone layers by the density-labeling technique of Filner and Varner (1967). Isolated barley aleurone layers were incubated for 36 hours in the presence of GA_3 and either $H_2{}^{18}O$ or ordinary $H_2{}^{16}O$. The endogenous reserve proteins are hydrolyzed to amino acids that then are available for reuse in protein synthesis. If this protein turnover occurs in the presence of $H_2{}^{18}O$, the carbonyl oxygens of the peptide chain contain one-half of the atom% ^{18}O present in the $H_2{}^{18}O$:

$$\text{Reserve protein} + H_2{}^{18}O \rightarrow (RCHNH_2CO{}^{18}OH \leftrightarrow RCHNH_2C{}^{18}OOH)$$

In the experiment results of which are illustrated above, marker α-amylase labeled with ^3H and nonradioactive but ^{18}O-labeled α-amylase were simultaneously subjected to isopycnic equilibrium centrifugation on a CsCl gradient. The ^3H-labeled marker enzyme was located by radioactivity, and ^{18}O-labeled enzyme was located by enzyme assay. The distribution of ^{18}O-labeled enzyme was displaced 15 drops in the direction of higher density, which represents a 1.1% increase in mass of the newly synthesized enzyme. Calculations showed that essentially all of the α-amylase formed in aleurone layers incubated in $H_2{}^{18}O$ and GA_3 arose by *de novo* synthesis from free amino acids derived from the preexisting aleurone proteins. (Redrawn, with permission, from Filner and Varner, 1967.)

isopycnic equilibrium centrifugation (e.g., 40,000 rpm for 65 to 68 hours) in a cesium chloride gradient, it could be shown that the α-amylase appearing in GA₃-treated aleurone cells incubated with $H_2{}^{18}O$ was of approximately the theoretical 1% greater buoyant density than α-amylase from $H_2{}^{16}O$-incubated aleurone cells. These results proved that all the induced α-amylase activity was due to *de novo* synthesis of the enzyme.

In 1967 J. V. Jacobsen and J. E. Varner proved by the same procedures that protease also is synthesized *de novo* in response to GA. And P. A. Bennett and M. J. Chrispeels (1972) proved GA-induced *de novo* synthesis of ribonuclease and β-1,3-glucanase in barley aleurone cells. Incidentally, they used D_2O instead of $H_2{}^{18}O$.

Characteristics of the GA-Induced *de Novo* Synthesis of Hydrolases in Barley Aleurone Cells

Induction of *de novo* synthesis of α-amylase by GA in isolated aleurone layers is evident after a lag period of approximately 8 hours following administration of the hormone (Fig.3.30). α-Amylase appears to be released from the aleurone cells as it is synthesized. But ribonuclease (Fig. 3.31) and β-1,3-glucanase first accumulate in the aleurone cells and then are released or secreted rapidly. In keeping with hormone responses generally, GA must be present continuously if the *de novo* synthesis of hydrolases is to be sustained (Fig. 3.32).

Synthesis of new RNA is essential to the GA-induction of *de novo* synthesis of hydrolases. Actinomycin D, an inhibitor of RNA synthesis, inhibits the

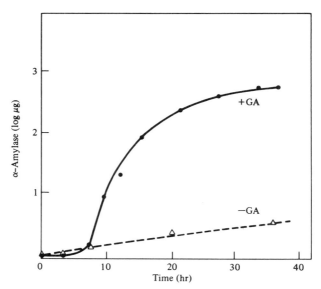

FIGURE 3.30. Effect of GA on α-amylase production by barley endosperm. Ten endosperm halves were incubated in buffer with 10^{-6} M GA₃ where indicated. (Redrawn, with permission, from Varner et al., 1965.)

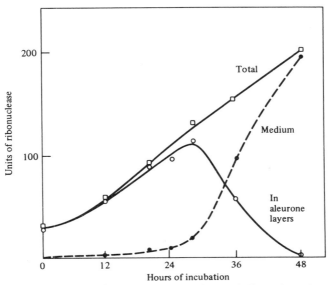

FIGURE 3.31. Time-course of production and release of ribonuclease by 10 aleurone layers incubated with 5 n*M* GA. (Redrawn, with permission, from Chrispeels and Varner, 1967a.)

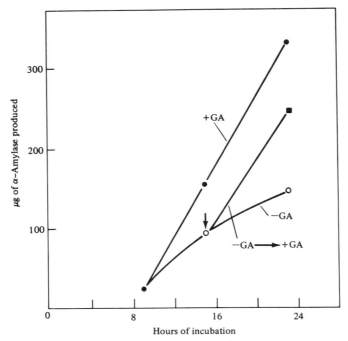

FIGURE 3.32. Effect of removing GA and of adding it back after being withheld for 6 hours on *de novo* synthesis of α-amylase in isolated aleurone layers of barley grains. (Redrawn, with permission, from Chrispeels and Varner, 1967b.)

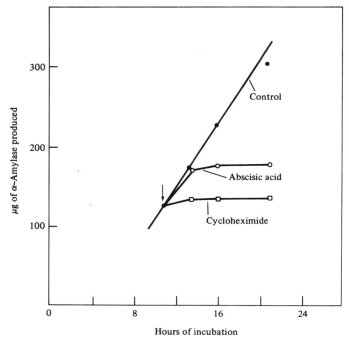

FIGURE 3.33. Inhibition of α-amylase synthesis by abscisic acid and cycloheximide. Aleurone layers were incubated in 0.1 μM GA for 11 hours. At this time abscisic acid (5 μM) or cycloheximide (10 μg/ml) was added and α-amylase synthesis was measured 2.5, 5, and 10 hours later. (Redrawn, with permission, from Chrispeels and Varner, 1967b.)

synthesis and release of α-amylase if the inhibitor is presented during the first 7 to 8 hours after treatment. Thereafter the process is insensitive to actinomycin D. This is interpreted as indicating that all the necessary messenger RNA (mRNA) for α-amylase synthesis is formed during the initial (lag) period. Inhibitors of protein synthesis, such as cycloheximide, also inhibit GA-induction of hydrolases, as expected. Interestingly, abscisic acid, a growth-inhibiting hormone (see Chapter 5), also inhibits GA-induced α-amylase synthesis (Fig. 3.33).

Yet the simple concept of GA as a promoter and ABA as an inhibitor is no longer tenable, because, in fact, both GA and ABA each *promotes and inhibits* the synthesis of particular proteins in the barley aleurone system. Tuan-hua David Ho and associates have demonstrated that the synthesis of several mRNA species is suppressed by GA in barley aleurone. They have showed that three cDNA clones of several independent suppressible messages hybridize to messages of approximately 3, 2.5, and 0.7 kb, respectively. All the clones were maintained at high levels in excised aleurone layers until GA^3 (10^{-6} M) was added to the incubation medium. On addition of GA$_3$, steady-state levels fell to trace quantities within 24 hours.

Ho and associates have also reported on the induction by ABA of some proteins in the barley aleurone system. They have reported that ABA, in fact, induces the synthesis of at least 16 polypeptides, including interestingly an α-amylase inhibitor. ABA-inducible cDNA clones were isolated by differential screening of a λgt 10 cDNA library with mRNA isolated from tissue treated with or without ABA. Northern analysis showed that one of the cDNAs hybridized to a 1.2-kb mRNA in ABA-treated tissues. The induction of this mRNA appeared as early as 40 minutes after the addition of ABA. An ABA concentration as low as 10^{-9} M induced this mRNA. Hybrid-select translation followed by two-dimensional electrophoresis indicated that the mRNA encoded a basic 27-KDa polypeptide. The physiological significance of the ABA-induced proteins, except for the α-amylase inhibitor, remains to be determined.

Site and Mechanism of Action of GA in the Aleurone System

It is clear that GA causes derepression of the genes for the synthesis of α-amylase and some other hydrolases in barley aleurone layers. However, it is not clear from the evidence presented thus far whether the hormone operates directly at the transcriptional level (during mRNA synthesis), at the posttranscriptional level, or at the translational level (during protein synthesis). A close look at the events that occur during the lag period in GA induction of *de novo* synthesis of new enzymes has provided some important clues, however.

W. H. Evins and J. E. Varner in 1972 reported on studies of the biochemical sequence of events during the lag period in the GA-induced synthesis of α-amylase in isolated aleurone layers. They found an increase in polyribosome formation and an increased synthesis of ribosomes and endoplasmic reticulum membranes. All these effects begin within 2 to 4 hours after application of GA. They showed that the polyribosomes responsible for new enzyme synthesis were those polyribosomes produced in the presence of the hormone, and that those polyribosomes were attached to the endoplasmic reticulum. These observations led Evins and Varner to conclude that the GA-stimulated increases in the number of monoribosomes and the percentage of polyribosomes probably are prerequisite for the hormone induction of protein synthesis. Incidentally, abscisic acid (Chapter 5), which already has been shown to counteract the effect of GA in the barley aleurone system, inhibits RNA synthesis, which in turn results in inhibition of polyribosome formation. The results of Evins and Varner are consistent with the notion that GA evokes selective mRNA synthesis, which is correlated with enhanced polyribosome formation, and *de novo* synthesis of particular hydrolases in the aleurone system. Regarding the effect of GA on membrane proliferation before the increase of hydrolase activity, there is much evidence for the participation of rough endoplasmic reticulum in the synthesis and exportation of secreted proteins in eukaryotic cells.

K. D. Johnson and H. Kende reported in 1971 on some early biochemical effects that occur in barley aleurone layers during the lag period in the response to exogenous GA. They found that the earliest detectable changes after GA treatment were in the activities of two enzymes involved in lecithin (a common component of membranes) biosynthesis, namely phosphorylcholine glyceride transferase and phosphorycholine cytidyl transferase. Both of these enzymes increase in activity about threefold during the first 2 hours after treatment of aleurone layers with GA. Accompanying, perhaps even preceding, changes in the two enzyme activities, they observed an increased incorporation of ^{32}P into cytidyl triphosphate (CTP), and several hours later an increased incorporation of labeled choline into microsomal lipid and of ^{32}P into phospholipid. The latter changes all were detectable within 4 hours after treatment. On the basis of these results, Johnson and Kende proposed a posttranscriptional model for GA action, according to which synthesis of endoplasmic reticulum is initiated within the first few hours after treatment, completion of that activity occurs some 4 to 8 hours later, and finally there is secretion of hydrolytic enzymes the synthesis of which occurs on mRNA templates already present in the aleurone before GA treatment. The latter point is not supported by direct evidence, and indeed conflicts obviously with much data signifying the dependence of GA action on new RNA synthesis.

J. A. Zwar and J. V. Jacobsen (1972) showed that GA stimulates selective incorporation of labeled precursor into polydisperse (5 to 14 S) RNA. More recent reports that GA increases the synthesis of poly(A)-containing RNA provides further support for the view that the hormone acts at the transcriptional level, since poly(A) segments are characteristic of eukaryotic mRNAs. The enhancement of poly(A)-RNA synthesis begins within 3 to 4 hours after presentation of GA and attains a maximum at 10 to 12 hours, which is just about the time α-amylase first becomes detectable.

Somewhat later, in 1976, T. J. V. Higgins and associates provided an even more direct link between GA-stimulated *de novo* enzyme synthesis and appearance of the complementary mRNA. They isolated total RNA and poly(A)-containing RNA from aleurone layers and assayed for α-amylase in a cell-free system from wheat germ, and analyzed the products of translation by immunoprecipitation and polyacrylamide gel electrophoresis. They demonstrated convincingly that the level of translatable α-amylase mRNA increased in GA-treated tissue in parallel with the increased rate of enzyme synthesis (Fig. 3.34). These results provide still more evidence that GA acts to induce selective mRNA and *de novo* enzyme synthesis in aleurone cells.

Action of GAs in Other Systems

The barley aleurone system is an excellent one for investigating the site and mechanism of action of GAs, and much has been learned with that system. But it must be considered to what extent analogous action is observed in other systems if we are to eventually fully understand the role of GAs in the regulation of growth and development.

One very interesting investigation with another system was conducted by M. M. Johri and J. E. Varner in 1968. They studied enhancement of RNA synthesis

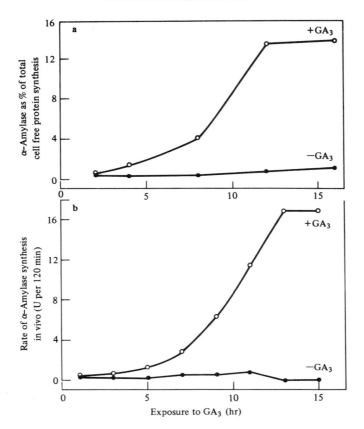

FIGURE 3.34. The increase with time in the level of translatable mRNA for α-amylase and the increase in the rate of synthesis of the enzyme *in vivo* in response to GA₃ treatment. **a** α-Amylase synthesized by the cell-free system was estimated from densitometric scans of autoradiographs of gels that had been in contact with X-ray film for different times. The gels were scanned with a microdensitometer and the contribution made by the α-amylase peak to total protein synthesized was calculated by weight. **b** The rate of α-amylase synthesis *in vivo* was calculated as the increment in enzyme level (units of α-amylase) during successive 2-hour intervals following hormone treatment. Total α-amylase activity was assayed, using replicates of 10 aleurone layers. U, Units of α-amylase. (Redrawn, with permission, from Higgins et al., 1976.)

in isolated nuclei of dwarf pea seedlings by GA. They detected marked increases in RNA synthesis when GA₃ was present throughout the isolation procedures (Fig.3.35). GA₃ is also active in promoting stem elongation in dwarf peas. Hence it was of interest to see what effect GA₈, which does not promote growth of dwarf peas, would have on the system. The result was as predicted: GA₈ did not evoke enhanced RNA synthesis by the isolated dwarf pea nuclei.

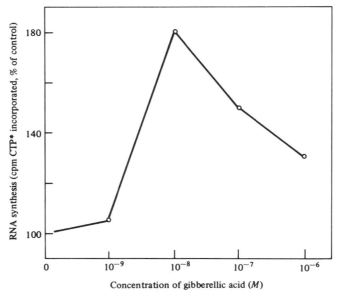

FIGURE 3.35. The application of GA$_3$ during the isolation of nuclei from pea seedlings results in an increased RNA synthesis in the nuclei, as indicated by the incorporation of [^{14}C]cytidine triphosphate, (CTP). (Drawn from data of Johri and Varner, 1968.)

Recently, Chory et al. (1987) reported on GA-induced changes in the populations of translatable mRNAs and accumulated polypeptides in dwarfs of maize and pea. They used the d_5 mutant of maize and the Progress No. 9 cultivar of pea, which lack GA$_1$ but which become phenotypically normal when treated with exogenous GA$_1$. Using two-dimensional gel electrophoresis, they found both increases and decreases in the levels of specific polypeptides in both maize and pea, and these changes were observed within 30 minutes of treatment with GA. They concluded that GA induces changes in the expression of a subset of gene products within elongating dwarfs, but noted that this might be due to changes in transcription rate, mRNA stability, or increased efficiency of translation of certain mRNAs.

Gibberellin Receptor

Cereal aleurone would appear to be an ideal tissue for GA receptor studies. However, a group of scientists at the University of Bristol found that conventional binding assays did not reveal a candidate GA receptor in aleurone tissue of *Avena fatua*. They have resorted to the use of aleurone protoplasts of *Avena fatua* instead of intact tissue and have exploited these using novel GA derivatives as affinity probes. A [^{3}H]aryl azide derivative of GA$_4$ has been used to photoaffinity label aleurone proteins *in vivo*. Identification of GA receptors by this approach may be feasible. Likewise, use of immobilized GAs to probe subcellular location of the receptor and their potential in affinity chromatography may be profitable.

Interaction of GAs with Polyamines

There is some evidence that a group of naturally occurring diamines, triamines, and tetraamines—collectively known as polyamines (PAs)—interacts with GAs, and other hormones as well, in the control of growth and development in higher plants. Arthur W. Galston and associates at Yale University have emphasized repeatedly the physiological importance of PAs in higher plants (see, e.g., Galston and Kaur-Sawhney, 1987). They point out that all cells contain the diamine putrescine and the triamine spermidine, and that eukaryotic cells contain the tetraamine spermine as well (Fig. 3.36).

Evidently mutants of higher plants lacking the ability to produce PAs do not grow and develop normally. The titers of the PAs are responsive to physiological controls such as light, hormones, and stress, and exogenous PAs can affect physiological processes. Moreover, inhibitors of PA biosynthesis have noteworthy effects. The PAs are most commonly associated with rapidly growing tissues, and their metabolism and accumulation are sometimes stimulated by plant hormones. Among their several physiological effects are inhibition of senescence of plant tissues, e.g., excised oat leaves, stimulation of cell division in some plant tissue cultures, and others.

Dai et al. (1982) investigated the role of PAs in GA-induced internode growth in light-grown dwarf peas. They found upon treatment of the seedlings with GA_3 the activity of arginine decarboxylase, a PA-forming enzyme, increased markedly, and that the titers of putrescine and spermidine also increased, paralleling the effect of GA_3 on internode growth. GA_3 also reversed the red light-induced inhibition of arginine decarboxylase activity in etiolated Alaska (normal) pea epicotyls. This was interpreted as additional evidence for GA-light interaction in the control of PA biosynthesis. Smith et al. (1985) also investigated the role of polyamines in GA-induced internode growth in peas. Exogenous polyamines were not effective in promoting growth unless intracellular polyamines were partially depleted. They concluded that their results

$$H_2N-(CH_2)_4-NH_2$$

Putrescine

$$H_2N-(CH_2)_3-\overset{\overset{\displaystyle H}{\displaystyle |}}{N}-(CH_2)_4-NH_2$$

Spermidine

$$H_2N-(CH_2)_3-\overset{\overset{\displaystyle H}{\displaystyle |}}{N}-(CH_2)_4-\overset{\overset{\displaystyle H}{\displaystyle |}}{N}-(CH_2)_3-NH_2$$

Spermine

FIGURE 3.36. Polyamines that occur in eukaryotic cells.

suggested that polyamines do not have a role in cell elongation, but that they may be required to support cell proliferation.

Galston and Kaur-Sawhney (1987) concluded about the PAs that, "Although they cannot yet be considered as hormones, because of scant evidence about their translocatability, they are present in all cells, and essential to normal growth and development." It will be interesting to see what future research reveals about the physiological importance of the PAs.

Conclusions Regarding GA Action

On the basis of collective data now available, it may confidently be concluded that in the barley aleurone system GA evokes selectively the synthesis of particular molecular species of mRNA that in turn lead to *de novo* synthesis of certain hydrolytic enzymes. Left unanswered unequivocally at this stage of our knowledge are two important questions. One is whether the action of GA just described alone accounts exclusively for the fundamental mechanism of action of GA in the aleurone system. The other question is whether the action of GA in stimulating *de novo* synthesis of particular enzymes as described for the barley aleurone system is universal. That is, is the mechanism of GA action in the barley aleurone system directly indicative of the mechanism by which GA hormones participate generally in the regulation of the growth and development of higher plants? Obviously, the answers to these questions must await additional research.

References

Adams, P. A., M. J. Montague, M. Tepfer, D. L. Rayle, H. Ikuma, and P. B. Kaufman. 1975. Effect of gibberellic acid on the plasticity and elasticity of *Avena* stem segments. *Plant Physiol.* **56**: 757–760.

Altman, A., R. Kaur-Sawhney, and A. W. Galston. 1977. Stabilization of oat leaf protoplasts through polyamine-mediated inhibition of senescence. *Plant Physiol.* **60**: 570–574.

Atzorn, R. and E. W. Weiler. 1983. The role of endogenous gibberellins in the formation of α-amylase by aleurone layers of germinating barley caryopses. *Planta* **159**: 287–299.

Barendse, G. W. M. 1971. Formation of bound gibberellins in *Pharbitis nil*. *Planta* **99**: 290–301.

Barendse, G. W. M. 1975. Biosynthesis, metabolism, transport and distribution of gibberellins. In: Krishnamoorthy, H. N., ed. *Gibberellins and Plant Growth*. Wiley Eastern Limited, New Delhi. Pp. 65–89.

Barendse, G. W. M., H. Kende, and A. Lang. 1968. Fate of radioactive gibberellin A_1 in maturing and germinating seeds of peas and Japanese morning glory. *Plant Physiol.* **43**: 815–822.

Barendse, G. W. M. and A. Lang. 1972. Comparison of endogenous gibberellins and of the fate of applied radioactive gibberellin A_1 in a normal and a dwarf strain of Japanese morning glory. *Plant Physiol.* **49**: 836–841.

Bennett, P. A. and M. J. Chrispeels. 1972. *De novo* synthesis of ribonuclease and β-1,3-

glucanase by aleurone cells of barley. *Plant Physiol.* **49**: 445–447.

Biswas, A. K. and M. A. Choudhuri. 1980. Mechanism of monocarpic senescence in rice. *Plant Physiol.* **65**: 340–345.

Brian, P. W., G. W. Elson, H. G. Hemming, and M. Radley. 1954. The plant-growth promoting properties of gibberellic acid, a metabolic product of the fungus *Gibberella fujikuroi, J. Sci. Food. Agr.* **5**: 602–612.

Brown, P. H. and T.-H. D. Ho. 1986. Barley aleurone layers secrete a nuclease in response to gibberellic acid. Purfication and partial characterization of the associated ribonuclease, deoxyribonuclease, and 3′-nucleotidase activities. *Plant Physiol.* **82**: 801–806.

Callis, J. and T.-H. D. Ho. 1983. Multiple molecular forms of the gibberellin-induced α-amylase from the aleurone layers of barley seeds. *Arch. Biochem. Biophys.* **224**: 224–234.

Chory, J., D. F. Voytas, N. E. Olszewski, and F. M. Ausubel. 1987. Gibberellin-induced changes in the populations of translatable mRNAs and accumulated polypeptides in dwarfs of maize and pea. *Plant Physiol.* **83**: 15–23.

Chrispeels, M. J. and J. E. Varner, 1967a. Gibberellic acid-enhanced synthesis and release of α-amylase and ribonuclease by isolated barley aleurone layers. *Plant Physiol.* **42**: 398–406.

Chrispeels, M. J. and J. E. Varner. 1967b. Hormonal control of enzyme synthesis: On the mode of action of gibberellic acid and abscisin in aleurone layers of barley. *Plant Physiol.* **42**: 1008–1016.

Chung, C. H. and R. C. Coolbaugh. 1986. *Ent*-kaurene biosynthesis in cell-free extracts of excised parts of tall and dwarf pea seedlings. *Plant Physiol.* **80**: 544–548.

Cleland, R., M. L. Thompson, D. L. Rayle, and W. K. Purves. 1968. Differences in effects of gibberellins and auxins on wall extensibility of cucumber hypocotyls. *Nature (London)* **219**: 510–511.

Coolbaugh, R. C. 1985. Sites of gibberellin biosynthesis in pea seedlings. *Plant Physiol.* **78**: 655–657.

Coolbaugh, R. C., D. R. Heil, and C. A. West. 1982. Comparative effects of substituted pyrimidines on growth and gibberellin biosynthesis in *Gibberella fujikuroi. Plant Physiol.* **69**: 712–716.

Coolbaugh, R. C. and R. Hamilton. 1976. Inhibition of *ent*-kaurene oxidation and growth by α-cyclopropyl-α-(*p*-methoxyphenyl)-5-pyrimidine methyl alcohol. *Plant Physiol.* **57**: 245–248.

Coolbaugh, R. C. and T. C. Moore. 1969. Apparent changes in rate of kaurene biosynthesis during the development of pea seeds. *Plant Physiol.* **44**: 1364–1367.

Coolbaugh, R. C. and T. C. Moore. 1971. Localization of enzymes catalysing kaurene biosynthesis in immature pea seeds. *Phytochemistry* **10**: 2395–2400.

Coolbaugh, R. C., T. C. Moore, S. A. Barlow, and P. R. Ecklund. 1973. Biosynthesis of *ent*-kaurene in cell-free extracts of *Pisum sativum* shoot tips. *Phytochemistry* **12**: 1613–1618.

Coolbaugh, R. C., D. I. Swanson, and C. A. West. 1982. Comparative effects of ancymidol and its analogs on growth of peas and *ent*-kaurene oxidation in cell-free extracts of immature *Marah macrocarpus* endosperm. *Plant Physiol.* **69**: 707–711.

Dai, Y.-R., R. Kaur-Sawhney, and A. W. Galston. 1982. Promotion by gibberellic acid of polyamine biosynthesis in internodes of light-grown dwarf peas. *Plant Physiol.* **69**: 103–105.

Deikman, J. and R. L. Jones. 1985. Control of α-amylase mRNA accumulation by gibberellic acid and calcium in barley aleurone layers. *Plant Physiol.* **78**: 192–198.

Deikman, J. and R. L. Jones. 1986. Regulation of the accumulation of mRNA for α-amylase isoenzymes in barley aleurone. *Plant Physiol.* **80**: 672–675.

Duncan, J. D. and C. A. West. 1981. Properties of kaurene synthetase from *Marah macrocarpus* endosperm: Evidence for the participation of separate but interacting enzymes. *Plant Physiol.* **68**: 1128–1134.

Ecklund, P. R. and T. C. Moore. 1968. Quantitative changes in gibberellin and RNA correlated with senescence of the shoot apex in the 'Alaska' pea. *Am. J. Bot.* **55**: 494–503.

Ecklund, P. R. and T. C. Moore. 1974. Correlations of growth rate and de-etiolation with rate of *ent*-kaurene biosynthesis in pea (*Pisum sativum* L.). *Plant Physiol.* **53**: 5–10.

Evins, W. H. and J. E. Varner. 1972. Hormonal control of polyribosome formation in barley aleurone layers. *Plant Physiol.* **49**: 348–352.

Filner, P. and J. E. Varner. 1967. A test for *de novo* synthesis of enzymes: Density labeling with $H_2{}^{18}O$ of barley α-amylase induced by gibberellic acid. *Proc. Natl. Acad. Sci. U.S.A.* **58**: 1520–1526.

Fry, S. C. and H. E. Street. 1980. Gibberellin-sensitive suspension cultures. *Plant Physiol.* **65**: 472–477.

Gafni, Y. and I. Shechter. 1981. Isolation of a kaurene synthetase inhibitor from castor bean seedlings and cell suspension cultures. *Plant Physiol.* **67**: 1169–1173.

Galston, A. W. 1961. *The Life of the Green Plant.* Prentice-Hall, Englewood Cliffs, New Jersey.

Galston, A. W. 1983. Polyamines as modulators of plant development. *Bioscience* **33**: 382–388.

Galston, A. W. and P. J. Davies. 1970. *Control Mechanisms in Plant Development.* Prentice-Hall, Englewood Cliffs, New Jersey.

Galston, A. W. and R. Kaur-Sawhney. 1987. Polyamines as endogenous growth regulators. In: Davies, P. J., ed. *Plant Hormones and Their Role in Plant Growth and Development.* Martinus Nijhoff Publishers, Dordrecht, The Netherlands. Pp. 280–295.

González, E. and M. A. Delsol. 1981. Induction of glyconeogenic enzymes by gibberellin A_3 in endosperm of castor bean seedlings. *Plant Physiol.* **67**: 550–554.

Graebe, J. E. 1968. Biosynthesis of kaurene, squalene and phytoene from mevalonate-2-^{14}C in a cell-free system from pea fruits. *Phytochemistry* **7**: 2003–2020.

Graebe, J. E. 1986. Gibberellin biosynthesis from gibberellin A_{12}-aldehyde. In: Bopp, M., ed. *Plant Growth Substances 1985.* Springer-Verlag, Berlin, Heidelberg, New York, Tokyo. Pp. 74–82.

Graebe, J. E., D. H. Bowen, and J. MacMillan. 1972. The conversion of mevalonic acid into gibberellin A_{12}-aldehyde in a cell-free system from *Cucurbita pepo. Planta.* **102**: 261–271.

Graebe, J. E., D. T. Dennis, C. D. Upper and C. A. West. 1965. Biosynthesis of gibberellins. I. The biosynthesis of (−)-kaurene, (−)-kaurene-19-ol, and *trans*-geranylgeraniol in endosperm nucellus of *Echinocystis macrocarpa* Greene. *J. Biol. Chem.* **240**: 1847–1854.

Hammerton, R. W. and T.-H. D. Ho. 1986. Hormonal regulation of the development of protease and carboxypeptidase activities in barley aleurone layers. *Plant Physiol.* **80**: 692–697.

Hasson, E. P. and C. A. West, 1976a. Properties of the system for the mixed function oxidation of kaurene and kaurene derivatives in microsomes of the immature seed of *Marah macrocarpus.* Cofactor requirements. *Plant Physiol.* **58**: 473–478.

Hasson, E. P. and C. A. West. 1976b. Properties of the system for the mixed function oxidation of kaurene and kaurene derivatives in microsomes of the immature seed of *Marah macrocarpus*. Election transfer components. *Plant Physiol.* **58**: 479–484.

Hedden, P. and J. E. Graebe. 1982. Cofactor requirements for the soluble oxidases in the metabolism of the C_{20}-gibberellins. *J. Plant Growth Regul.* **1**: 105–116.

Hedden, P., J. MacMillan, and B. O. Phinney. 1978. The metabolism of the gibberellins. *Annu. Rev. Plant Physiol.* **29**: 149–192.

Hedden, P. and B. O. Phinney. 1976. The dwarf-5 mutant of *Zea mays*: A genetic lesion controlling the cyclization step (B activity) in kaurene biosynthesis. In: Pilet, P.-E., ed. *Collected Abstracts of the Paper-Demonstrations of the 9th International Conference on Plant Growth Substances.* P. 136 (Abstract).

Hedden, P. and B. O. Phinney. 1979. Comparison of *ent*-kaurene and *ent*-isokaurene synthesis in cell-free systems for etiolated shoots of normal and dwarf-5 maize seedlings. *Phytochemistry* **18**: 1475–1479.

Higgins, T. J. V., J. A. Zwar, and J. V. Jacobsen. 1976. Gibberellic acid enhances the level of translatable mRNA for α-amylase in barley aleurone layers. *Nature (London)* **260**: 166–169.

Ho, T.-H. D., R. C. Nolan, and D. E. Shute. 1981. Characterization of a gibberellin-insensitive dwarf wheat, D6899. Evidence for a regulatory step common to many diverse responses to gibberellins. *Plant Physiol.* **67**: 1026–1031.

Iwahori, S., R. H. Weaver, and R. M. Pool. 1968. Gibberellin-like activity in berries of seeded and seedless Tokay grapes. *Plant Physiol.* **43**: 333–337.

Jackson, D. I. 1968. Gibberellin and the growth of peach and apricot fruits. *Austral. J. Biol. Sci.* **21**: 209–215.

Jackson, D. I. and B. G. Coombe. 1966. Gibberellin-like substances in the developing apricot fruit. *Science* **154**: 277–278.

Jacobsen, J. V. 1977. Regulation of ribonucleic acid metabolism by plant hormones. *Annu. Rev. Plant Physiol.* **28**: 537–564.

Jacobsen, J. V. and L. R. Beach. 1985. Control of transcription of α-amylase and rRNA genes in barley aleurone protoplasts by gibberellin and abscisic acid. *Nature (London)* **316**: 275–277.

Jacobsen, J. V. and P. M. Chandler. 1987. Gibberellin and abscisic acid in germinating cereals. In: Davies, P. J., ed. *Plant Hormones and Their Role in Plant Growth and Development.* Martinus Nijhoff Publishers, Dordrecht, The Netherlands. Pp. 164–193.

Jacobsen, J. V. and J. E. Varner. 1967. Gibberellic acid-induced synthesis of protease by isolated aleurone layers of barley. *Plant Physiol.* **42**: 1596–1600.

Johnson, K. D. and H. Kende. 1971. Hormonal control of lecithin synthesis in barley aleurone cells: Regulation of the CDP-choline pathway by gibberellin. *Proc. Natl. Acad. Sci. U. S. A.* **68**: 2674–2677.

Johri, M. M. and J. E. Varner. 1968. Enhancement of RNA synthesis of isolated pea nuclei by gibberellic acid. *Proc. Natl. Acad. Sci. U.S.A.* **59**: 269–276.

Jones. R. L. 1973. Gibberellins: their physiological role. *Annu. Rev. Plant Physiol.* **24**: 571–598.

Jones, R. L. and A. Lang. 1968. Extractable and diffusible gibberellins from light- and dark-grown pea seedlings. *Plant Physiol.* **43**: 629–634.

Jones, R. L. and J. Carbonell. 1984. Regulation of the synthesis of barley aleurone α-amylase by gibberellic acid and calcium ions. *Plant Physiol.* **76**: 213–218.

Jones, R. L. and I. D. J. Phillips. 1966. Organs of gibberellin synthesis in light-grown sunflower plants. *Plant Physiol.* **41**: 1381–1386.

Kamiya, Y. and J. E. Graebe. 1983. The biosynthesis of all major pea gibberellins in a cell-free system from *Pisum sativum*. *Phytochemistry* **22**: 681–690.

Kende, H. 1967. Preparation of radioactive gibberellin A₁ and its metabolism in dwarf peas. *Plant Physiol.* **42**: 1612–1618.

Kende, H. and A. Lang. 1964. Gibberellins and light inhibition of stem growth in peas. *Plant Physiol.* **39**: 435–440.

Knotz, J., R. C. Coolbaugh, and C. A. West. 1977. Regulation of the biosynthesis of *ent*-kaurene from mevalonate in the endosperm of immature *Marah macrocarpus* seeds by adenylate energy charge. *Plant Physiol.* **60**: 81–85.

Koehler, S. and T.-H. D. Ho. 1988. Purification and characterization of gibberellic acid-induced cysteine endoproteases in barley aleurone layers. *Plant Physiol.* **87**: 95–103.

Köhler, D., 1966. Die Abhangigkeit der Gibberellinproduktion von Normalerbsen vom Phytochrom-system. *Planta* **69**: 27–33.

Krishnamoorthy, H. N., ed. 1975. *Gibberellins and Plant Growth*. Wiley Eastern Limited, New Delhi.

Lang, A. 1970. Gibberellins: Structure and metabolism. *Annu. Rev. Plant Physiol.* **21**: 537–570.

Leopold, A. C. and P. E. Kriedemann. 1975. *Plant Growth and Development*. 2nd ed. McGraw-Hill Book Company, New York.

Liu, P. B. W. and J. B. Loy. 1976. Action of gibberellic acid on cell proliferation in the subapical shoot meristem of watermelon seedlings. *Am. J. Bot.* **63**: 700–704.

Lockhart, J. A. 1957. Studies on the organ of production of the natural gibberellin factor in higher plants. *Plant Physiol.* **32**: 204–207.

MacMillan. J. 1971. Diterpenes—the gibberellins. In: Goodwin, T. W., ed. *Aspects of Terpenoid Chemistry and Biochemistry*. Academic Press, London. Pp. 153–180.

MacMillan, J. 1977. Some aspects of gibberellin metabolism in higher plants. In: Pilet, P. E., ed. *Plant Growth Regulation*. Springer-Verlag, New York. Pp. 129–138.

MacMillan, J. and N. Takahashi. 1968. Proposed procedure for the allocation of trivial names to the gibberellins. *Nature (London)* **217**: 170–171.

McComb, A. J. 1964. The stability and movement of gibberellic acid in pea seedlings. *Ann. Bot. (N.S.)* **28**: 669–687.

Métraux, J.-P. 1987. Gibberellins and plant cell elongation. In: Davies, P. J., ed. *Plant Hormones and Their Role in Plant Growth and Development*. Martinus Nijhoff Publishers, Dordrecht, The Netherlands. Pp. 296–317.

Metzger, J. D. 1985. Role of gibberellins in the enviromental control of stem growth in *Thlaspi arvense* L. *Plant Physiol.* **78**: 8–13.

Metzger, J. D. 1987. Hormones and reproductive development. In: Davies, P. J., ed. *Plant Hormones and Their Role in Plant Growth and Development*. Martinus Nijhoff Publishers, Dordrecht, The Netherlands. Pp. 431–462.

Milborrow, B. V. 1974. Biosynthesis of abscisic acid by a cell-free system. *Phytochemistry* **13**: 131–136.

Mitchell, J. W., D. P. Skaggs, and W. P. Anderson. 1951. Plant growth-stimulating hormones in immature bean seeds. *Science* **114**: 159–161.

Moore, T. C. 1967. Gibberellin relationships in the 'Alaska' pea *(Pisum sativum)*. *Am. J. Bot.* **54**: 262–269.

Moore, T. C., S. A. Barlow, and R. C. Coolbaugh. 1972. Participation of noncatalytic 'carrier' protein in the metabolism of kaurene in cell-free extracts of pea seeds. *Phytochemistry* **11**: 3225–3233.

Moore, T. C. and R. C. Coolbaugh. 1976. Conversion of geranylgeranyl pyrophosphate to *ent*-kaurene in enzyme extracts of sonicated chloroplasts. *Phytochemistry.* **15**: 1241–1247.

Moore, T. C. and P. R. Ecklund. 1974. Biosynthesis of *ent*-kaurene in cell-free extracts of pea shoots. In: *Plant Growth Substances 1973.* Hirokawa Publishing Company, Tokyo. Pp. 252–259.

Moore, T. C. and P. R. Ecklund. 1975. Role of gibberellins in the development of fruits and seeds. In: Krishnamoorthy, H. N., ed. *Gibberellins and Plant Growth.* Wiley Eastern Limited, New Delhi. Pp. 145–182.

Moore, T. C., H. Yamane, N. Murofushi, and N. Takahashi. 1988. Concentrations of *ent*-kaurene and squalene in vegetative rice shoots. *J. Plant Growth Regul.* **7**: 145–151.

Murakami, Y. 1961. Paper-chromatographic studies on change in gibberellins during seed development and germination in *Pharbitis nil. Bot. Mag. (Tokyo)* **74**: 241–247.

Murakami, Y. 1972. Dwarfing genes in rice and their relation to gibberellin biosynthesis. In: Carr, D. J., ed. *Plant Growth Substances 1970.* Springer-Verlag, Berlin. Pp. 166–174.

Murphy, P. J. and C. A. West. 1969. The role of mixed function oxidases in kaurene metabolism in *Echinocystis macrocarpa* Greene endosperm. *Arch. Biochem. Biophys.* **133**: 395–407.

Ogawa, Y. 1964. Changes in the amount of gibberellin-like substances in the seedling of *Pharbitis nil* with special reference to expansion of cotyledon. *Plant Cell Physiol.* **5**: 11–20.

O'Neill, S. D., B. Keith, and L. Rappaport. 1986. Transport of gibberellin A_1 in cowpea membrane vesicles. *Plant Physiol.* **80**: 812–817.

Paleg, L. G. 1965. Physiological effects of gibberellins. *Annu. Rev. Plant Physiol.* **16**: 291–322.

Phillips. I. D. J. 1971. *Introduction to the Biochemistry and Physiology of Plant Growth Hormones.* McGraw-Hill Book Company, New York.

Phinney, B. O., M. Freeling, D. S. Robertson, C. R. Spray, and J. Silverthorne. 1986. Dwarf mutants in maize—the gibberellin biosynthetic pathway and its molecular future. In: Bopp, M., ed. *Plant Growth Substances 1985.* Springer-Verlag, Berlin, Heidelberg, New York, Tokyo. Pp. 55–64.

Phinney, B. O. and C. Spray. 1982. Chemical genetics and the gibberellin pathway in *Zea mays* L. In: Wareing, P. F., ed. *Plant Growth Substances 1982.* Academic Press, London. Pp. 101–110.

Phinney, B. O., C. A. West, M. Ritzel, and P. M. Neely. 1957. Evidence for "gibberellin-like" substances from flowering plants. *Proc. Natl. Acad. Sci. U.S.A.* **43**: 398–404.

Radley, M. 1956. Occurrence of substances similar to gibberellic acid in higher plants. *Nature (London)* **178**: 1070–1071.

Radley, M. 1958. The distribution of substances similar to gibberellic acid in higher plants. *Ann. Bot. (N.S.)* **22**: 297–307.

Railton, I. D., B. Fellows, and C. A. West. 1984. *Ent*-kaurene synthesis in chloroplasts from higher plants. *Phytochemistry* **23**: 1261–1267.

Reeve, D. R. and A. Crozier. 1974. An assessment of gibberellin structure-activity relationships. *J. Exp. Bot.* **25**: 431–445.

Reid, D. M., M. S. Tuing, R. C. Durley, and I. D. Railton. 1972. Red-light-enhanced conversion of tritiated gibberellin A_9 into other gibberellin-like substances in homogenates of etiolated barley leaves. *Planta* **108**: 67–75.

Reid, J. B. 1987. The genetic control of growth via hormones. In: Davies, P. J., ed. *Plant Hormones and Their Role in Plant Growth and Development.* Martinus Nijhoff Publishers, Dordrecht, The Netherlands. Pp. 318–340.

Robinson, D. R. and C. A. West. 1970. Biosynthesis of cyclic diterpenes in extracts from seedlings of *Ricinus communis* L. I. Identification of diterpene hydrocarbons formed from mevalonate. *Biochemistry* **9**: 70–79.

Rood, S. B., F. D. Beall, and R. P. Pharis. 1986. Photocontrol of gibberellin metabolism *in situ* in maize. *Plant Physiol.* **80**: 448–453.

Runeckles, V. C., E. Sondheimer, and D. C. Walton, eds. 1974. *The Chemistry and Biochemistry of Plant Hormones.* Vol. 7. *Recent Advances in Phytochemistry.* Academic Press, New York.

Sachs, R. M. 1965. Stem elongation. *Annu. Rev. Plant Physiol.* **16**: 73–96.

Sachs, R. M., C. F. Bretz, and A. Lang. 1959. Shoot histogenesis: the early effects of gibberellin upon stem elongation in two rosette plants. *Annu. Am. J. Bot.* **46**: 376–384.

Sembdner, G., J. Weiland, O. Aurich, and K. Schreiber. 1968. Isolation, structure and metabolism of a gibberellin glucoside. *S. C. I. Monograph No. 31.* 70–86.

Shen-Miller, J. and C. A. West. 1982. *Ent*-kaurene biosynthesis in extracts of *Helianthus annuus* L. seedlings. *Plant Physiol.* **69**: 637–641.

Simcox, P. D., D. T. Dennis, and C. A. West. 1975. Kaurene synthetase from plastids of developing plant tissues. *Biochem. Biophys. Res. Commun.* **66**: 166–172.

Sitton, D., A. Richmond, and Y. Vaadia. 1967. On the synthesis of gibberellins in roots. *Phytochemistry* **6**: 1101–1105.

Smith, M. A. and P. J. Davies. 1985. Effect of photoperiod on polyamine metabolism in apical buds of G2 peas in relation to the induction of apical senescence. *Plant Physiol.* **79**: 400–405.

Smith, M. A., P. J. Davies, and J. B. Reid. 1985. Role of polyamines in gibberellin-induced internode growth in peas. *Plant Physiol.* **78**: 92–99.

Smith, T. A. 1985. Polyamines. *Annu. Rev. Plant Physiol.* **36**: 117–143.

Stodola, F. H., K. B. Raper, D. I. Fennell, H. F. Conway, V. E. Sohns, C. T. Langford, and R. W. Jackson. 1955. The microbiological production of gibberellins A and X. *Arch. Biochem. Biophys.* **54**: 240–245.

Sponsel, V. M. 1985. Gibberellins in *Pisum sativum*—their nature, distribution and involvement in growth and development of the plant. *Plant Physiol.* **65**: 533–538.

Sponsel, V. M. 1987. Gibberellin biosynthesis and metabolism. In: Davies, P. J., ed. *Plant Hormones and Their Role in Plant Growth and Development.* Martinus Nijhoff Publishers, Dordrecht, The Netherlands. Pp. 43–75.

Stowe, B. B. and T. Yamaki. 1957. The history and physiological action of the gibberellins. *Annu. Rev. Plant Physiol.* **8**: 181–216.

Stuart, D. A. and R. L. Jones. 1977. Roles of extensibility and turgor in gibberellin- and dark-stimulated growth. *Plant Physiol.* **59**: 61–68.

Takahashi, N., I. Yamaguchi, and H. Yamane. 1986. Chapter 3. Gibberellins. In: Takahashi, N., ed. *Chemistry of Plant Hormones.* CRC Press, Boca Raton, Florida. Pp. 57–151.

Thimann, K. V. 1974. Fifty years of plant hormone research. *Plant Physiol.* **54**: 450–453.

Thimann, K. V. 1977. *Hormone Action in the Whole Life of Plants.* University of Massachusetts Press, Amherst.

Varner, J. E. and D. T. Ho. 1976. Hormones. In: Bonner, J. and J. E. Varner, eds. *Plant Biochemistry*. 3rd ed. Academic Press, New York. Pp. 713–770.

Varner, J. E., G. Ram Chandra, and M. J. Chrispeels. 1965. Gibberellic acid-controlled synthesis of α-amylase in barley endosperm. *J. Cell. Physiol.* **66**: Suppl. 1: 55–68.

Wang, S. Y., T. Sun, and M. Faust. 1986. Translocation of paclobutrazol, a gibberellin biosynthesis inhibitor, in apple seedlings. *Plant Physiol.* **82**: 11–14.

Wareing, P. F. and A. K. Seth. 1967. Ageing and senescence in the whole plant. *Symp. Soc. Exp. Biol.* **21**: 543–558.

West, C. A. 1973. Biosynthesis of gibberellins. In: Milborrow, B. V., ed. *Biosynthesis and Its Control in Plants*. Academic Press, London. Pp. 143–169.

West, C. A. and B. O. Phinney. 1956. Properties of gibberellin-like factors from extracts of higher plants. *Plant Physiol.* **31**: (Suppl.) XX (Abstract).

Wiltbank, W. J. and A. H. Krezdorn. 1969. Determination of gibberellins in ovaries and young fruits of navel oranges and their correlation with fruit growth. *J. Am. Soc. Hort. Sci.* **94**: 195–201.

Yokota, T., N. Murofushi, N. Takahashi, and M. Katsumi. 1971. Biological activities of gibberellins and their glucosides in *Pharbitis nil*. *Phytochemistry* **10**: 2943–2949.

Zwar, J. A. and R. Hooley. 1986. Hormonal regulation of α-amylase gene transcription in wild oat (*Avena fatua* L.) aleurone protoplasts. *Plant Physiol.* **80**: 459–463.

Zwar, J. A. and J. V. Jacobsen. 1972. A correlation between a ribonucleic acid fraction selectively labeled in the presence of gibberellic acid and amylase synthesis in barley aleurone layers. *Plant Physiol.* **49**: 1000–1006.

CHAPTER 4

Cytokinins

History of Discovery

The definitive discovery of the cytokinins occurred in 1955 when C. O. Miller, Folke Skoog, M. H. von Saltza, and F. M. Strong, working in the laboratories of Skoog and Strong at the University of Wisconsin, isolated a substance called "kinetin" (6-furfurylaminopurine, $C_{10}H_9N_5O$) (Fig. 4.1) from an autoclaved sample of herring sperm DNA and demonstrated it to be very active in promoting mitosis and cell division in tobacco callus tissue *in vitro*.

The idea of specific substances required for cell division occurring in plants actually dates back to the German physiologist J. Wiesner who, in 1892, proposed that initiation of cell division is evoked by endogenous factors, indeed a proper balance among endogenous factors. Somewhat later, Haberlandt reported in 1913 that diffusates from phloem tissue could induce cell division in parenchyma tissue of potato tubers. Repeating and extending some of Haberlandt's experiments, Jablonski and Skoog (1954) showed that a piece of vascular tissue cultured on top of tobacco pith tissue could cause division of the pith cells, which would not otherwise divide. These observations led to a search for pure substances that could induce cell division in a manner similar to that of the unknown substance or substances in vascular tissue.

Hence the ultimate discovery of the cytokinins was a direct outcome of tissue culture investigations that were carried on since circa 1941 in the laboratory of Folke Skoog at the University of Wisconsin. Skoog and his associates have utilized mostly asceptically excised portions of the nondividing mature pith from stems of well-developed tobacco (*Nicotiana tabacum* cv. Wisconsin No. 38) plants.

Some important features of this tobacco pith tissue are as follows: (1) the tissue is mature and the parenchyma cells comprising it normally do not divide if efforts are made to culture the tissue *in vitro* on a simple culture medium containing no hormones; (2) both exogenous auxin and cytokinin are required for growth of the pith explants, and at appropriate concentrations of the two hormones the tissue grows into an amorphous, relatively undifferentiated callus; (3) if only auxin (and not cytokinin) is supplied, the pith tissue exhibits pronounced cell enlargement entirely unaccompanied by cell division; and (4) cytokinin is inactive in promoting cell division in the absence of auxin.

In passing it should be noted that the behavior of tobacco pith in tissue culture is not exhibited by all tissues. In fact a great many tissues are like tobacco pith in that they do require a cytokinin or auxin or both for growth on a tissue culture medium. But there are some examples of tissues that so far have not been grown

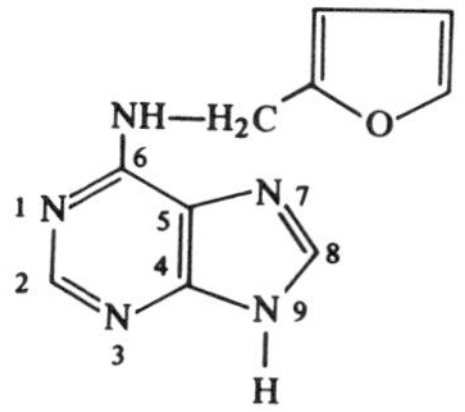

FIGURE 4.1. Structure of kinetin 6-furfurylaminopurine, the first cytokinin to be discovered.

on defined media successfully at all. And, at the other extreme, there are many examples of tissues, mostly meristematic tissues, which will grow very well on a relatively simple tissue culture medium containing only a few inorganic salts, a carbon-energy source, and certain vitamins—and no exogenous auxin or cytokinin. Meristematic and tumor tissues—such as crown gall tumor induced by *Agrobacterium tumefaciens*—exhibit this behavior. Evidently the cytokinin-synthesizing system is operative in the latter types of tissues, whereas it is turned off or repressed during the normal course of differentiation into a nondividing tissue such as pith.

Terminology

The name "kinetin" was assigned to 6-furfurylaminopurine by Miller et al. (1955) when they first reported on the isolation of the substance from DNA. The name of the substance obviously was based on its property of causing cell division or "cytokinesis" in tobacco pith tissue. Later, in 1956, the Wisconsin group proposed the generic term "kinin" for the group of synthetic and naturally occurring substances that resemble kinetin in inducing cell division in certain excised tissues in the presence of exogenous auxin.

Rather quickly, however, confusion began to develop over terminology, for animal physiologists at about the same time proposed the term "kinin" to designate a class of polypeptides (compounds of quite different nature than the plant hormones) that stimulates smooth muscle contraction and nerve endings in the manner of various stings and venoms. Hence, in efforts to avoid confusion, plant physiologists coined new terms such as "phytokinin," "cytonin," and "cytokinin" to replace "kinin" in the plant literature. Finally, after consultation with other investigators in the field, Skoog and his associates proposed that "cytokinin" be used to designate all compounds that promote cell division and exert other growth regulatory functions in the same manner as kinetin, and this term now is generally adopted. The term "kinetin," of course, is reserved for one specific cytokinin, namely 6-furfurylaminopurine.

Isolation of Kinetin and the Search for Other Naturally Occurring Cytokinins

Kinetin was first isolated from an autoclaved sample of DNA from herring sperm. Lest this be regarded incorrectly as a mere fortuitous discovery, it should be noted that Skoog and his associates had previously found that adenine had slight cell division-promoting activity in tobacco pith and they speculated that nucleic acids might be involved in the biological activity of the then unknown stimulus. In fact, a variety of purine-containing substances was tested, and among them an old sample of herring sperm DNA was found to be particularly effective in promoting growth. One can only imagine the surprise when Skoog et al. repeated the experiments that resulted in the discovery of kinetin with fresh samples of DNA and found that it was inactive.

It turns out that kinetin is an artifact, at least in terms of the high concentrations present in the old DNA from which it first was isolated. Only old or heated DNA preparations yielded substantial quantities of kinetin. Kinetin has the empirical formula of a dehydrated deoxyadenosine. It arises spontaneously in DNA preparations (as during autoclaving), presumably from deoxyadenosine. Apparently, during the transformation, the pentose derivative (i.e., the furfuryl constituent) migrates from the 9-position to the 6-position of the adenine ring, as illustrated below:

Hall and deRopp (1955) demonstrated that kinetin could be produced simply by autoclaving a mixture of adenine and furfuryl alcohol. Most investigators now agree that kinetin is thus an artifact, although there has been some speculation that kinetin may also be formed *in vivo*.

Artifact or not, still the biological activity of kinetin was real, and the substance had activity such as had been expected of the elusive cell division-promoting hormone. Moreover, investigations of the biological activity of dozens of compounds resembling kinetin chemically showed that kinetin-like biological activity was fairly well restricted to N^6-substituted adenine derivatives, suggesting that perhaps some compound or compounds closely resembling kinetin was present naturally in plants. Indeed, already by 1970, it was estimated that at least 100 potent cytokinins, all N^6-substituted adenine compounds, had been synthesized (Fig. 4.2). Thus the search for naturally occurring kinetin-like compounds (i.e., cytokinins) intensified.

Discovery of Natural Cytokinins

The first isolation of a naturally occurring cytokinin should be credited to both D. S. Letham of the Fruit Research Division, DSIR, Auckland, New Zealand, and C. O. Miller of the University of Indiana, circa 1963. In 1963, D. S. Letham presented a report at a symposium in which he announced the isolation in crystalline form of a natural cytokinin from immature corn (*Zea mays*) kernels, which he called "zeatin." In a longer subsequent paper Letham et al. (1964)

FIGURE 4.2. Structures of some synthetic cyto-
kinins. (Redrawn from *Plant Growth and Develop-
ment,* 2nd ed., by A. C. Leopold and P. E.
Kriedemann. Copyright © 1975 by McGraw-Hill
Book Company. Used with permission of McGraw-
Hill Book Company.)

Kinetin

Benzyladenine (BA)

Tetrahydropyranylbenzyladenine (PBA)

Ethoxyethyladenine

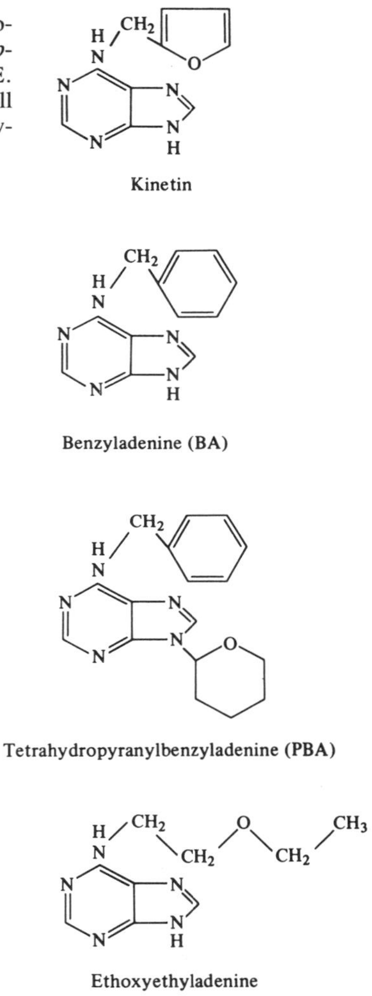

reported on many of the chemical properties of zeatin and postulated as the most
probable structure 6-(4-hydroxy-3-methyl-*trans*-2-butenylamino)purine (Fig.
4.3), a structure promptly confirmed by Shaw and Wilson (1964). Interestingly,
C. O. Miller (1961b) earlier had detected a kinetin-like substance in immature
corn kernels, although he did not fully characterize the active compound. Later
reports from Miller's laboratory and from the collaboration of Miller and
Letham (1965) showed conclusively that the substance isolated by Miller also
was zeatin. Zeatin and its riboside and ribonucleotide have since been reported
from many sources. As a matter of fact, Letham (1968) identified zeatin riboside
as the major active cell division-promoting substance in coconut milk, which for
some 27 years had been known to have growth-promoting properties. In the

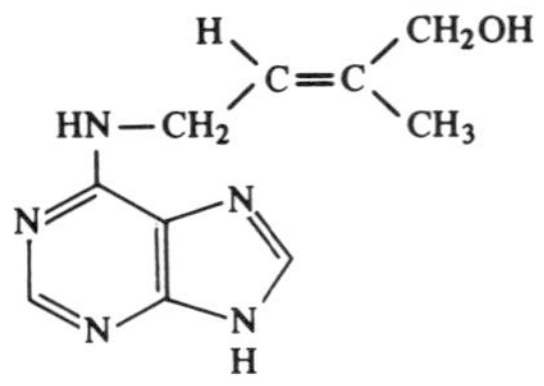

Zeatin (io^{6}Ade)

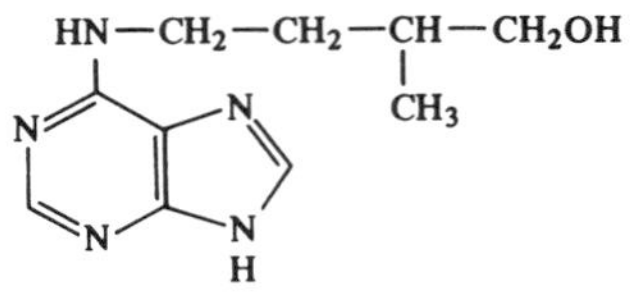

Dihydrozeatin

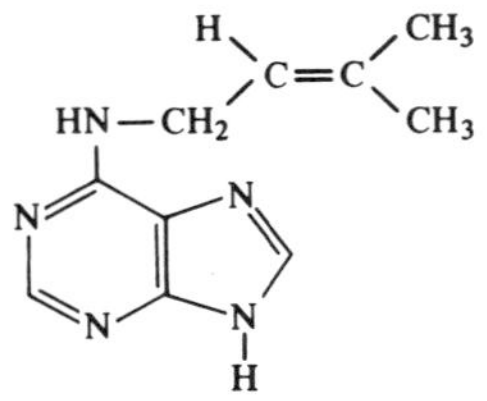

N^6-(Δ^2-Isopentenyl) adenine (i^{6}Ade)

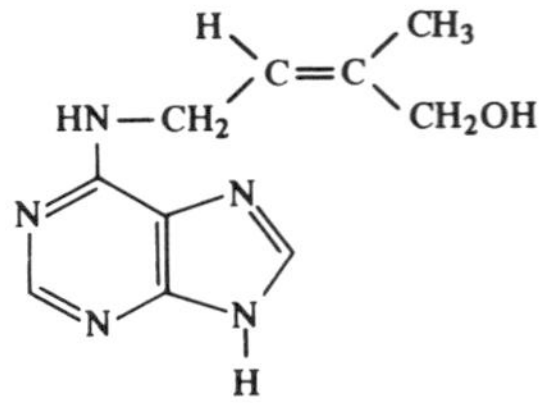

cis-Zeatin (cis-io^{6}Ade)

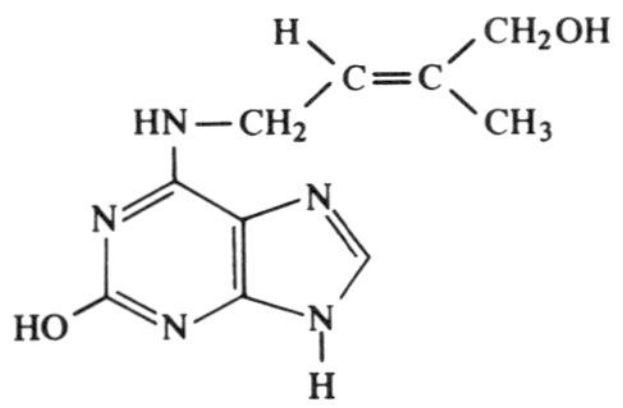

2-Hydroxyzeatin

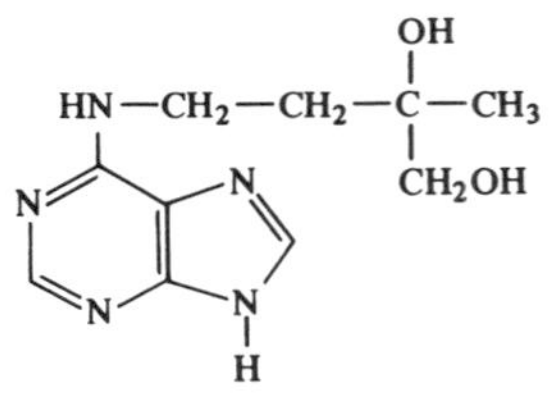

6-(3, 4-Dihydroxy-3-methylbutylamino) purine

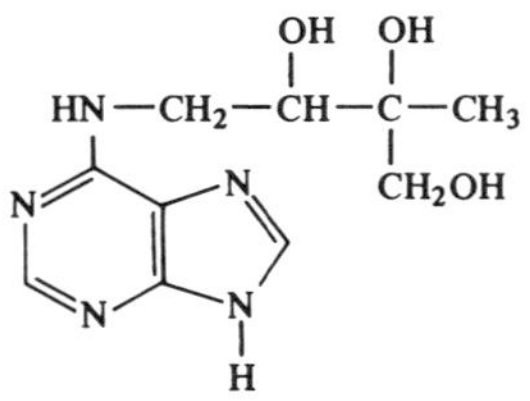

6-(2, 3, 4-Trihydroxy-3-methylbutylamino) purine

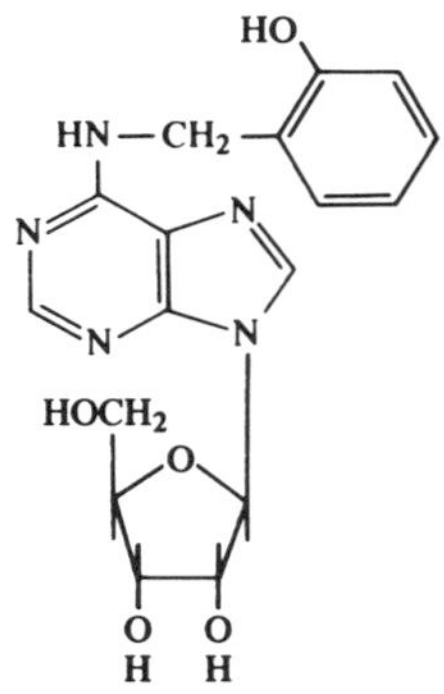

N^6-(2-Hydroxybenzyl) adenosine

FIGURE 4.3. Structures of some naturally occurring cytokinins. Illustrations include some free bases, ribosides, and ribotides.

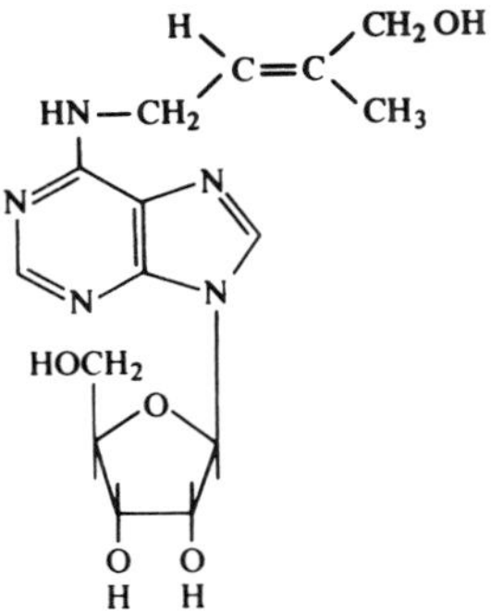

Ribosylzeatin (io⁶A)

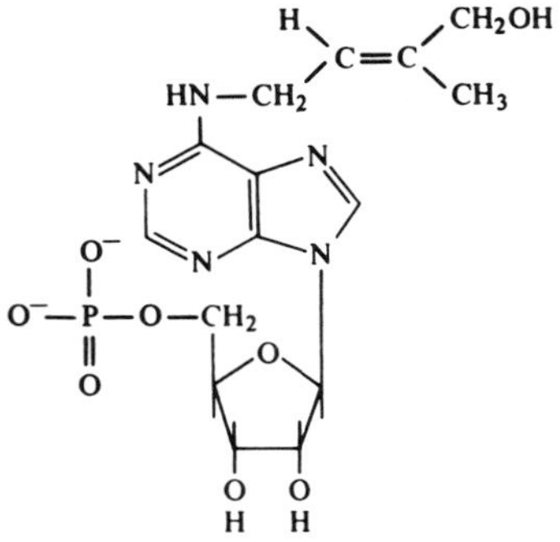

Ribosylzeatin-5'-monophosphate

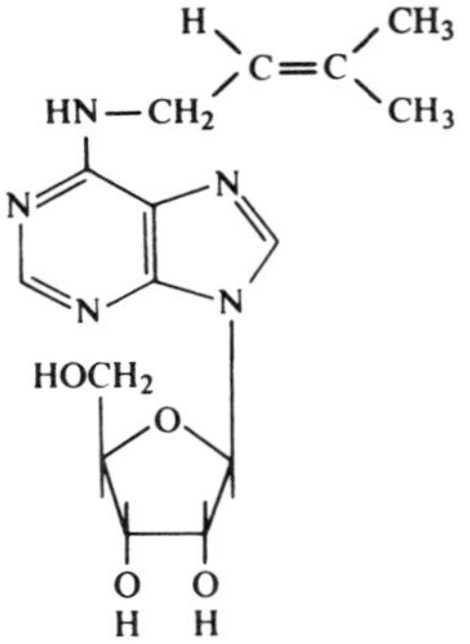

N^6-(Δ^2-Isopentenyl) adenosine (i⁶A)

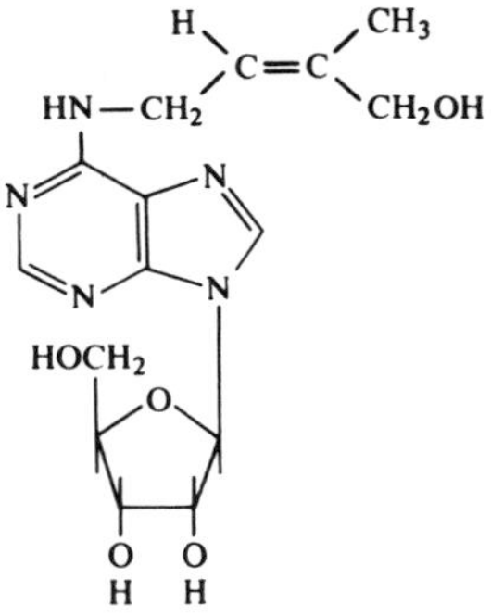

cis–Ribosylzeatin (io⁶A)

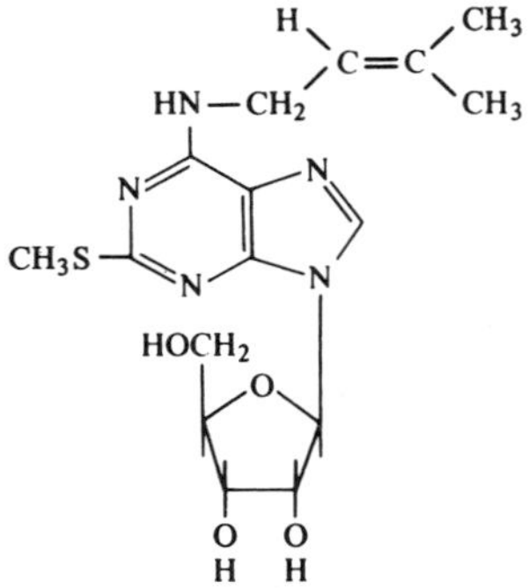

2-Methylthio–N^6-(Δ^2-isopentenyl) adenosine (ms²i⁶A)

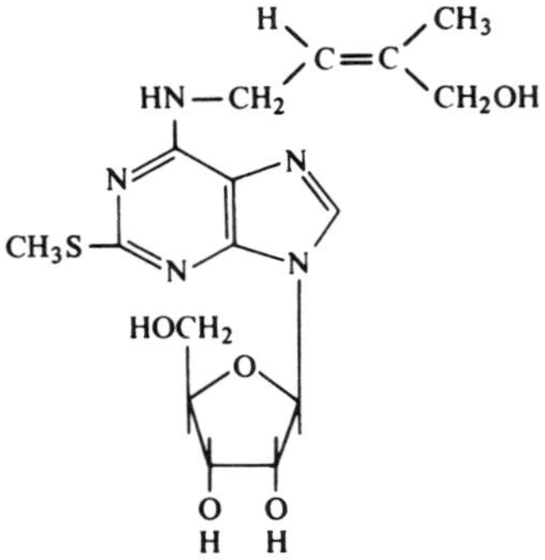

2-Methylthio–cis–ribosylzeatin (ms²io⁶A)

FIGURE 4.3. *Continued.*

TABLE 4.1. Free bases of natural cytokinins.[a]

R₁	R₂	Chemical Name	Common name or abbreviation
—CH₂—CH=C(CH₂OH)(CH₃)	H	6-(4-Hydroxy-3-methyl-*trans*-2-butenyl)aminopurine	*trans*-Zeatin
—CH₂—CH=C(CH₃)(CH₂OH)	H	6-(4-Hydroxy-3-methyl-*cis*-2-butenyl)aminopurine	*cis*-Zeatin
—CH₂—CH₂—CH(CH₂OH)(CH₃)	H	6-(4-Hydroxy-3-methylbutyl)aminopurine	Dihydrozeatin

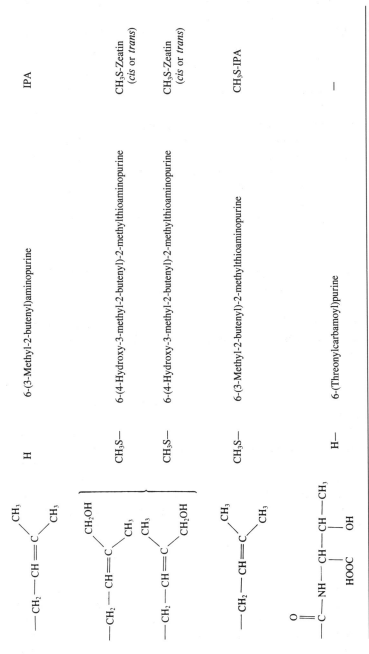

$-CH_2-CH=C\begin{smallmatrix}CH_3\\CH_3\end{smallmatrix}$	H	6-(3-Methyl-2-butenyl)aminopurine
		IPA
$-CH_2-CH=C\begin{smallmatrix}CH_2OH\\CH_3\end{smallmatrix}$	CH_3S-	6-(4-Hydroxy-3-methyl-2-butenyl)-2-methylthioaminopurine
		CH_3S-Zeatin (cis or trans)
$-CH_2-CH=C\begin{smallmatrix}CH_3\\CH_2OH\end{smallmatrix}$	CH_3S-	6-(4-Hydroxy-3-methyl-2-butenyl)-2-methylthioaminopurine
		CH_3S-Zeatin (cis or trans)
$-CH_2-CH=C\begin{smallmatrix}CH_3\\CH_3\end{smallmatrix}$	CH_3S-	6-(3-Methyl-2-butenyl)-2-methylthioaminopurine
		CH_3S-IPA
$\underset{O}{\overset{\parallel}{-C}}-NH-\underset{HOOC}{\overset{}{CH}}-\underset{OH}{\overset{}{CH}}-CH_3$	H—	6-(Threonylcarbamoyl)purine
		—

years following the discovery of zeatin, several other cytokinins were isolated from various sources and were characterized chemically (Fig. 4.3). All the naturally occurring cytokinins are considered to be isopentenyl adenine derivatives. N^6-(Δ^2-isopentenyl)adenosine (i^{6}A) occurs in all tRNA preparations that have been examined. In fact, i^{6}A is the only cytokinin that has been found in tRNA from animals. Four cytokinin-active ribonucleosides—i^{6}A, io^{6}A, and their 2-methylthio derivatives, ms^{2}i^{6}A and ms^{2}io^{6}A (Fig. 4.3)—have been identified from plant tRNA, although ms^{2}i^{6}A typically is present in only trace amounts if at all. Aside from glycosylation and methylthiolation of the purine ring, the important differences obviously are in the side chains. The free bases, illustrating these differences in side chain, are tabulated in Table 4.1.

By 1967, cytokinin-containing extracts had been prepared from approximately 40 species of higher plants, and it is now confidently assumed that the cytokinins are ubiquitous among seed plants and perhaps throughout the plant kingdom. Indeed cytokinins have been reported to occur also, specifically in the tRNA, of numerous animals and microorganisms, indicating a high probability of occurrence, in fact, in all organisms.

Effects of Cytokinins and Other Hormones in Organisms Other Than Seed Plants

Not only do cytokinins occur in microbial and animal cells, but some interesting effects of exogenous cytokinins on some of these organisms have been reported. For example, cytokinins have been reported to promote the growth (i.e., increase the rate of cell division) of some bacteria, fungi, protozoans, and even cultured mammalian cells. However, it is not clear whether the reported effects of cytokinins in animal cells are directly related to the hormonal effects of cytokinins in plants.

The reported effects of cytokinins in some animal systems point up a fact of more general importance about plant hormones. The mere presence of a substance in an organism does not mean that the substance plays any essential role in the metabolism of that organism. This principle most certainly applies to the substances that we regard as hormones of higher plants. All the major types of hormones, except perhaps abscisic acid, occur in plants of all phyletic groups, but the substances evidently do not have hormone-like activity in all plants. Indeed the ''hormone concept'' is not even applicable to primitive organisms such as the bacteria. The indoleacetic acid produced in comparative abundance by many bacteria and fungi is not known to be essential to the metabolism of those organisms. GAs produced in relatively large amounts by *Fusarium moniliforme* apparently are unnecessary for the growth of the fungus, although they are important to the host–parasite relationship and pathogenesis into which the fungus enters together with an angiosperm host. There are uncertainties still as to whether auxins and GAs are even essential to the growth of all green plants, such as the algae, for example.

In general, it appears that the hormones of higher plants evolved quite early in comparatively less complex life forms in which they do not have a role directly essential to metabolism—although they are essential to certain types of symbiotic relationships with plant organisms that are phyletically more advanced. The major symptom of the bakanae disease of rice caused by the fungus *Fusarium moniliforme*, of course, is abnormal elongation of stems and leaf sheaths evoked by the GAs that the pathogen produces and secretes into the host tissue. Fasciation disease, a common disease or pathological overgrowth that attacks a variety of dicotyledonous species, is caused by the cytokinins liberated by the pathogenic bacterium *Corynebacterium tumefaciens* into the host tissue. Extreme cases of this type of pathological condition are called witches brooms, which attack many species and which typically are the result of dozens of lateral buds growing very close together. There are many other cases of symbiotic and host–pathogen relationships between microorganisms and higher plants in which the microbial symbiont or pathogen either liberates hormone into the higher plant partner, or alters the hormone metabolism of the higher plant, or both, causing an abnormal growth response in the higher plant. Examples are the formation of crown gall tumors, caused by *Agrobacterium tumefaciens*, and nodulation of legume roots by *Rhizobium* species. The formation of certain galls, caused by particular species of insects, is still another case in which the hormone metabolism of the higher plants is affected by an invading organism.

Structure/Activity Relationships of the Cytokinins

Previously the cytokinins were characterized as derivatives of adenine, a purine, having a substituent attached to the N^6 position of the adenine ring. Early work indicated that high cytokinin activity was evidently restricted to these N^6-substituted adenines. But an important question that was asked soon after the initial isolation of kinetin was: Just how specific, in fact, are the chemical structural requirements for cytokinin-like biological activity? This is an important question about any type of hormone for at least two reasons: (1) knowing the essential structural features is critical to ultimate elucidation of the specific kind of molecular interaction it undergoes as it exerts hormonal action; and (2) knowing the essential structural features is necessary to efforts to synthesize biologically active analogs and competitive inhibitors of the hormone.

In an extensive study, Skoog et al. (1967) investigated structure/activity relationships of several dozen (69) cytokinin-related substances, most of which were purely synthetic. They found that only N^6-substituted adenine compounds definitely possess cytokinin activity (Fig. 4.4); however, the activity of these compounds varies considerably with the length, degree of unsaturation, and substitutions on the side chain. Modification of the adenine moiety, for example substitution of N for C at the 8-position, reduces activity as measured in the tobacco callus assay by as much as 95%. Even minor substitutions on the adenine nucleus usually reduce activity. A second substituent at the 1- or 3-

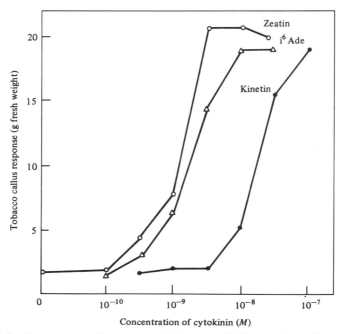

FIGURE 4.4. Comparative effectiveness of three cytokinins in stimulating growth of tobacco callus *in vitro*. The compound abbreviated i⁶Ade is N^6-(Δ^2-isopentenyl) adenine. (Redrawn with modifications, by permission, from Leonard et al., 1968.)

position of the purine ring nearly eliminates cytokinin activity. However, a second substituent at the N^6, N^7, or N^9 position has much less of a depressing effect. N^9 substituents are metabolically labile, and metabolic conversions of both 3- and 9-substituted derivatives to their corresponding N^6 isomers have been demonstrated. N^6-substituted adenosines possess high cytokinin activity; in some cases they are only slightly less active than their corresponding adenines. It remains to be fully ascertained, however, whether ribosides and ribotides are active per se or whether cytokinin activity is acquired only on the conversion to free bases.

Making use of the information on structure/activity relationships, Skoog and associates (1973) synthesized a compound, 3-methyl-7-(3-methylbutylamino)-pyrazolo-[4,3-d]pyrimidine (Fig. 4.5), which acts as an active cytokinin antagonist, or a competitive inhibitor of N^6-adenine cytokinins (Fig. 4.6). In Fig. 4.5 three steps are shown that were found to be necessary to convert the cytokinin 6-(3-methyl-2-butenylamino)purine (2iP or i⁶Ade) to the pyrazolo pyrimidine antagonist, namely (1) saturation of the side chain, (2) exchange of the C-9 and N-9 atoms, and (3) addition of a methyl group onto the pyrimidine nucleus. Since the development of the pyrazolo pyrimidine cytokinin antagonist, several other compounds with similar action have been synthesized.

There have been reports that certain nonpurine compounds have cytokinin or cytokinin-like activity, and some workers have suggested that they be considered

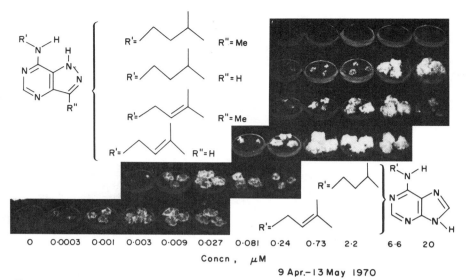

9 Apr.–13 May 1970

FIGURE 4.5. Three steps that convert the active cytokinin 6-(3-methyl-2-butenyl-amino)purine (2iP or i^{6}Ade) (lower right) to a cytokinin antagonist, 3-methyl-7-(3-methylbutylamino)pyrazolo[4,3-d]pyrimidine (upper left). Also growth of tobacco callus on medium containing serial concentrations of each compound. (Reproduced, with permission, from Skoog et al., 1973; photo courtesy of F. Skoog, 1978.)

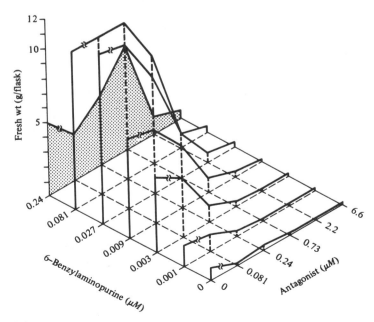

FIGURE 4.6. Interaction of a synthetic cytokinin antagonist (see structure of pyrazolo pyrimidine compound in Fig. 4.5) and N^6-benzylaminopurine in the tobacco callus bioassay. (Redrawn, with permission, from Skoog et al., 1973.)

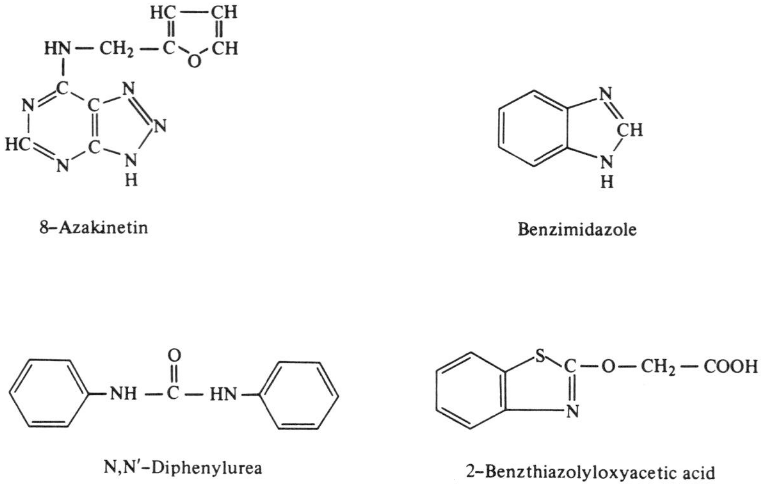

8–Azakinetin

Benzimidazole

N,N'–Diphenylurea

2–Benzthiazolyloxyacetic acid

FIGURE 4.7. Examples of nonpurine compounds with cytokinin-like activity. [Copyright © 1969 McGraw-Hill Book Co. (UK) Ltd. From The cytokinins by J. Eugene Fox in M. B. Wilkins: *Physiology of Plant Growth and Development.* Reproduced by permission.] See also Thidiazuron, p. 173.

true cytokinins, although not all physiologists agree. Four such compounds are illustrated in Fig. 4.7. Of these, only N,N'-diphenylurea is naturally occurring; the other three are purely synthetic. Diphenylurea occurs in coconut milk, and at one time the cell division-promoting activity of coconut milk was attributed to that compound. However, as previously noted, Miller (1968) showed that most of the cytokinin activity of coconut milk is attributable to zeatin and its ribonucleoside. Perhaps the so-called "urea cytokinins" are not active as such but instead serve as inducers of or precursors—i.e., side chain donors—to the N^6-substituted adenine-type cytokinins. Doubt has been expressed about whether 2-benzthiazolyloxyacetic acid really has cytokinin activity; instead it may have auxin activity. 8-Azakinetin and benzimidazole may in fact have cytokinin activity, although the former is only less than 10% as active as kinetin. At the present state of our knowledge, it still seems a valid generalization that the cytokinins are N^6-substituted adenine types of compounds and close analogs.

Biosynthesis and Metabolism

Cytokinin biosynthesis in seed plants evidently occurs generally in tissues and loci that are either meristematic or still retain growth potential. Of particular interest is the fact that cytokinins apparently are synthesized in roots and translocated acropetally in shoots, since cytokinins very frequently have been detected in bleeding sap (xylem sap) of such plants as sunflowers, grapevines, and tobacco plants. This acropetal transport of cytokinins in the vascular tissue appears to be involved in certain so-called "correlative" growth phenomena, such as apical dominance. For example, cytokinin moving acropetally past buds at the nodes evidently opposes the effect of the basipetally moving endogenous auxin. Whereas the auxin inhibits differentiation of the vascular tissues in the bud traces connecting bud and stele, cytokinin promotes differentiation of the bud traces.

More recently, Chong-maw Chen et al. (1985) at the University of Wisconsin at Parkside, Kenosha investigated the biosynthesis of cytokinins in pea (*Pisum sativum* L.) plant organs and carrot (*Daucus carota* L.) root tissues. They found that pea roots, stems, and leaves all synthesized cytokinins when these organs were grown separately for 3 weeks on a culture medium containing [8-^{14}C]adenine without an exogenous supply of cytokinin and auxin. Incubation of carrot root cambium and noncambium tissues for 3 days in a liquid culture medium containing [8-^{14}C]adenine without cytokinin showed that radioactive cytokinins were synthesized in the cambium but not in the noncambium tissues. On the basis of the amounts of cytokinin synthesized per gram fresh tissues, their results indicated that the root indeed is the major site, but not the only site, of cytokinin biosynthesis. They speculated that, furthermore, cambium and possibly all actively dividing tissues are responsible for the synthesis of cytokinins (Chen et al., 1985).

Our understanding of the biochemistry of cytokinin production is quite incomplete, and information that is available is comparatively very complicated. This complexity derives chiefly from the fact that cytokinins occur not only as free bases and the ribonucleosides and ribonucleotides but also as constituents of particular molecular species of tRNA. Most investigations of cytokinin biogenesis in fact have been concerned with the mechanism by which cytokinins come to be present in particular tRNA fractions, and it is this topic to which we will give major consideration.

As regards biosynthesis of the free bases, the cytokinins evidently originate by a condensation of adenine with an appropriate donor of the N^6-substituent. A cell-free enzyme system from the slime mold *Dictyostelium* was reported to form a cytokinin from adenosine monophosphate and Δ^2-isopentenyl pyrophosphate as substrates. The general isoprenoid nature of the N^6-substituents of natural cytokinins indicates that the biosynthesis of the cytokinins parallels, to some extent, the biosynthesis of GAs and abscisic acid.

The metabolic reactions that have been described for the cytokinins include (1) interconversions, (2) formation of ribosides and ribotides from free bases, (3) glucosylation, (4) methylthiolation, and (5) cleavage of the side chain and breakdown of the purine ring. Examples of interconversions are the formation of N^6-(3-methyl-3-hydroxybutyl)aminopurine from N^6-(3-methyl-2-butenyl)-aminopurine (i^6Ade or IPA) and the conversion of the IPA (or i^6Ade) riboside to the zeatin riboside. Formation of the ribosides and ribotides has been demonstrated for zeatin and dihydrozeatin and for the synthetic cytokinins kinetin and benzyladenine. Zeatin also forms glucosides, 7-glucosylzeatin (also called "raphanatin," from *Raphanus sativus*, radish), and 9-glucosylzeatin (see numbering system in Fig. 4.1). In 9-glucosylzeatin the glucose moiety is attached to the 9-position of the purine ring, at which ribose is attached in zeatin nucleoside. In other cases, cytokinin glucosides are formed by the addition of a glucose moiety to the hydroxyl group on the side chain of zeatin. Presently it is not known whether cytokinin ribosides, ribotides and glucosides are biologically active per se or whether, as seems more likely, they are active only on liberation of the free bases. If the latter, then some conjugated derivatives—e.g., glucosides—may be important storage forms of cytokinins. The glucosides probably are important also as transport forms in the vascular system. Another type of reaction that cytokinins can undergo is addition of a methylthio group (CH_3S—) at position 2 of the purine ring (Fig. 4.3). Finally, the N^6-substituent (side chain) may be cleaved, and a residual purine ring may be successively oxidized by xanthine oxidase to yield uric acid and then urea, or carbon 8 may be oxidized away leaving a triaminopurine. Undoubtedly, some or all of these types of metabolic reactions, together with turnover of tRNA containing cytokinin moieties and regulation of biosynthesis, are important in the regulation of levels of active cytokinins in tissues.

Chen and Leisner (1984) reported on modification of cytokinin molecules by microsomal cytochrome P-450 enzymes from cauliflower (*Brassica oleracea* L.), particularly the hydroxylation of i^6Ado to io^6Ado and i^6Ade to io^6Ade. They showed also that cytokinins were dealkylated by microsomal enzymes and formed adenine from cytokinin base and adenosine from cytokinin nucleoside. They suggested that plant cytochrome P-450 is involved in the conversion of one type of cytokinin to another, and in the modification of cytokinin molecules.

Earlier Chen and Kristopeit (1981a,b) described a dephosphorylation of AMP and i^6Ado-5'-P by a 5'-ribonucleotide phosphorylase (5'-nucleotidase) from wheat germ cytosol, and the deribosylation of i^6Ado by an adenosine nucleosidase (adenosine ribohydrolase) from wheat germ cells. They hypothesized (Chen and Kristopeit, 1981b) that an adequate level of ''active cytokinin'' in plant cells may be provided through the deribosylation of cytokinin riboside in concert with other cytokinin metabolic enzymes.

Lee et al. (1985) investigated the metabolism of *trans*-[8-^{14}C]zeatin in embryos of *Phaseolus vulgaris* cv Great Northern and *P. lunatus* cv Kingston. Five major metabolites were recovered from *P. vulgaris* embryo extracts: ribosylzeatin, ribosylzeatin 5'-monophosphate, and *O*-glucoside of ribosylzeatin and two novel, unidentified metabolites. In embryos of *P. lunatus* were ribosylzeatin, ribosylzeatin 5'-monophosphate, and the *O*-glucosyl derivatives of zeatin and ribosylzeatin. In a later paper from the Moks' laboratory (Mok and Mok, 1987), the metabolism of *trans*-[8-^{14}C]zeatin was examined in embryos of *P. acutifolius* and *P. coccineus*. In both of these species zeatin was converted to ribosylzeatin, ribosylzeatin 5'-monophosphate, *O*-glucosyl-9-ribosylzeatin, and *O*-xylosyl derivatives of zeatin and ribosylzeatin. Two new metabolites, identified as *O*-xylosyldihydrozeatin and its ribonucleoside, were recovered from *P. coccineus* embryos.

The catabolism of cytokinins in plant tissues appears to be due, in large measure, to the activity of a specific enzyme, cytokinin oxidase. The properties of cytokinin oxidase have been the subject of several investigations. The enzyme

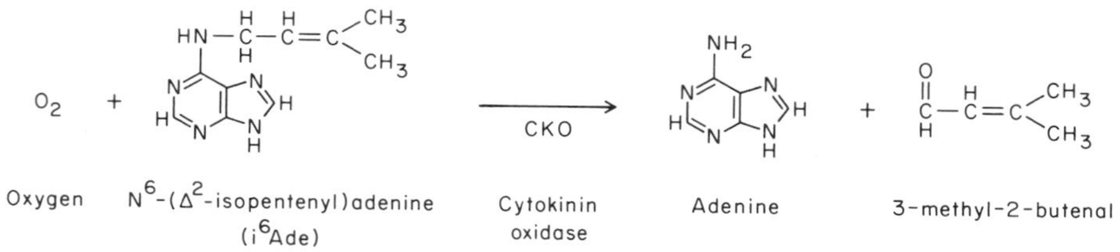

FIGURE 4.8. Generalized reaction catalyzed by cytokinin oxidase.

was first reported by Paces et al. (1971) in crude homogenates of cultured tobacco cells. Subsequently it was isolated and partially purified from immature corn kernels (Whitty and Hall, 1974) and from *Vinca rosea* crown gall tumor tissue (McGaw and Horgan, 1983). In 1988 Roger Horgan and associates purified the cytokinin oxidase to homogeneity. The enzyme requires molecular oxygen and catalyzes the oxidative cleavage of the N^6-side chains of i^6Ade and zeatin (io^6Ade) and their ribonucleosides (i^6Ado and io^6Ado). The products of the reaction are adenine (or adenosine) and a side chain fragment, which was identified as 3-methyl-2-butenal in the case of the oxidation of i^6Ade. Cytokinins with saturated isoprenoid side chains (e.g., N^6-isopentenyladenosine and dihydrozeatin) and some synthetic cytokinins (e.g., benzyladenine and kinetin) are reported to be resistant to cytokinin oxidase (see Chatfield and Armstrong, 1986).

The regulation of cytokinin oxidase activity in callus tissues of *Phaseolus vulgaris* L. cv Great Northern has been investigated recently by Chatfield and Armstrong (1986, 1987, 1988). Their assay was based on the oxidation of N^6-[8-^{14}C](Δ^2-isopentenyl)adenine, [8-^{14}C](i^6Ade), to adenine (Fig. 4.8). Solutions of exogenous cytokinins applied directly to the surface of the callus tissues induced relatively rapid increases in cytokinin oxidase activity. The increase in activity was detectable after 1 hour and continued for about 8 hours, reaching levels two to three times higher than the controls. The cytokinin-induced increase in cytokinin oxidase activity was inhibited in tissues pretreated with cordycepin or cycloheximide, suggesting that RNA and protein synthesis are required for the response. All cytokinin-active compounds tested, including both substrates and nonsubstrates of cytokinin oxidase, were effective in inducing levels of the enzyme. Thidiazuron (Fig. 4.9), a cytokinin-active urea derivative, was as active as any adenine derivative in inducing the response, although it was not determined whether Thidiazuron was a substrate for cytokinin oxidase (Chatfield and Armstrong, 1986).

Later it was reported (Chatfield and Armstrong, 1987) that the addition of cupric ions to reaction mixtures containing imidazole buffer enhanced cytokinin oxidase activity more than 20-fold. The effect was enzyme dependent, specific

FIGURE 4.9. Structural formula of Thidiazuron (N-phenyl-N′-1,2,3-thidiazole-5-yl urea).

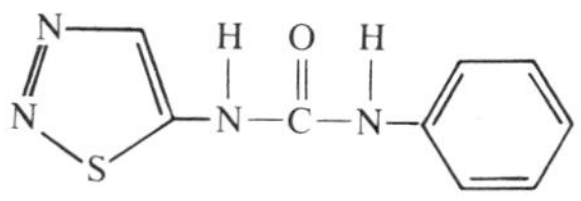

for copper, and observed only in the presence of imidazole. The mechanism by which copper and imidazole enhanced cytokinin oxidase activity was not determined for certain. However, it was suggested that copper–imidazole complexes were substituting for oxygen in the reaction. Quite recently Chatfield and Armstrong (1988) postulated that cytokinin oxidase of *Phaseolus vulgaris* callus tissues is a glycoprotein, based on its affinity for the lectin concanavalin A. Later in 1988 Armstrong and Kaminek reported on a comparison of cytokinin oxidase activities in *P. vulgaris* cv Great Northern and *P. lunatus* cv Kingston callus cultures. The substrate specificities of the two activities appeared to be identical. However, the two activities differed in other ways, such as pH optimum, affinity for concanavalin A, and behavior on DEAE-cellulose. The enzyme from *P. lunatus* did not bind to concanavalin A.

Auxin and Cytokinin Biosynthesis in Crown Gall Tumors

Infection of dicotyledonous plants and many gymnosperms with *Agrobacterium tumefaciens* causes production of tumorous growths called crown galls. Cells from such galls can be transferred without contamination by *Agrobacteria* to simple culture media for plant tissues where, unlike most normal plant tissues, they continue to grow and divide without an exogenous supply of auxins and cytokinins. The crown gall tissues are autonomous for the production of auxin and cytokinin.

Hormone independence in crown gall tissues is the result of expression of genes for hormone production from the T-DNA, that is, the part of Ti plasmids that is transferred from *Agrobacteria* to plant cells during the infection. Evidently there are at least four genes coding for enzymes of hormone biosynthesis. No other T-DNA gene is required for hormone independence. Two of the genes' code for enzymes for a two-step biosynthetic pathway from tryptophan through indole-3-acetamide to indole-3-acetic acid. A third gene codes for an isopentenyltransferase active in cytokinin biosynthesis. The fourth Ti plasmid hormone gene is contained in the Vir-region, the part of the Ti plasmid responsible for transfer of the T-DNA into plant cells. The T-region and the Vir-region (vir from virulence) are physically separated parts of the Ti plasmid, and they have different functions. The T-region is transferred to and stably maintained in the plant cells. It is responsible for tumorous growth, but it contains no genes that are necessary for its own transfer. These functions are encoded in the Vir-region. The functional importance of the fourth (Vir-region) gene is unclear. It leads to *trans*-zeatin production *in vivo*, and it may qualify as a "virulence gene" if increased hormone production by *Agrobacteria* during infection participates in conditioning the plant cells for gene transfer (Schröder et al., 1986; Morris et al., 1986).

Mechanisms of Origin in tRNA

Now to a consideration of the more extensive investigations of the mechanism of origin of cytokinins in tRNA. Cytokinins are unique among plant hormones in that adenine compounds of identical structure occur in nucleic acids. That is, specifically, particular cytokinin-active ribonucleosides occur as components of certain molecular species of tRNA. Several laboratories with quite different directions of research came on this discovery. Actually, as early as 1960, Thimann and Lalorya first proposed that cytokinins might be rapidly incorporated into macromolecules such as RNA on the basis of their findings that cytokinins promoted RNA and protein synthesis and that exogenous cytokinins often have highly localized effects when applied to whole plants and plant organs. Let us now examine briefly what we know about the occurrence of natural cytokinins in tRNA.

H. G. Zachau et al. (1966), during their determination of the base sequences of serine tRNA in yeast, first reported an "odd" base immediately adjacent to the 3′ end of the anticodon (Fig. 4.10). In collaboration with K. Biemann et al. (1966), this "odd" base was identified as the natural cytokinin isopentenyladenosine [(IPA) or 6-(γ,γ-dimethylallylamino)purine], which is one of the most

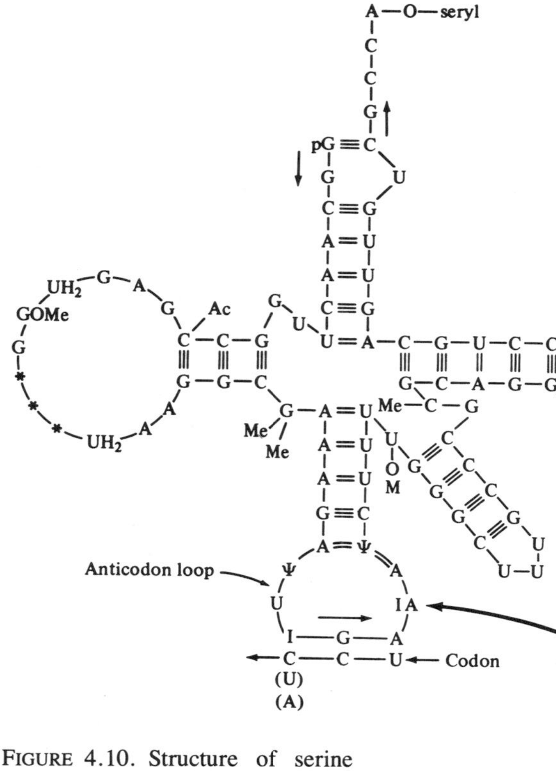

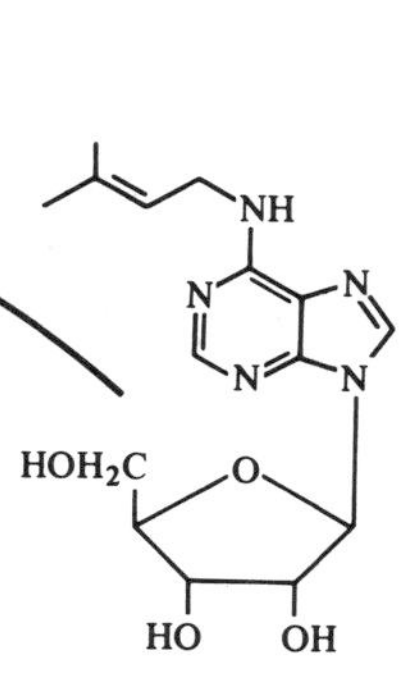

FIGURE 4.10. Structure of serine tRNA from Brewers yeast showing the position of isopentenyladenosine adjacent to the 3′ end of the anticodon. (Redrawn, with permission, from Zachau et al., 1966b.)

Isopentenyladenosine

1st Letter	2nd Letter				3rd Letter
	U	C	A	G	
U	PHE	SER	TYR	CYS	U
	PHE	SER	TYR	CYS	C
	LEU	SER	C.T.	C.T.	A
	LEU	SER	C.T.	TRY	G
C	LEU	PRO	HIS	ARG	U
	LEU	PRO	HIS	ARG	C
	LEU	PRO	GLN	ARG	A
	LEU	PRO	GLN	ARG	G
A	ILEU	THR	ASN	SER	U
	ILEU	THR	ASN	SER	C
	ILEU	THR	LYS	ARG	A
	MET (C.I.)	THR	LYS	ARG	G
G	VAL	ALA	ASP	GLY	U
	VAL	ALA	ASP	GLY	C
	VAL	ALA	GLU	GLY	A
	VAL (C.I.)	ALA	GLU	GLY	G

FIGURE 4.11. Distribution of cytokinin-containing tRNA species in *E. coli*. Striped codons correspond to cytokinin-containing tRNA species. Active constituent identified in most cases as 6-(3-methyl-2-butenylamino)-9-β-D-ribofuranosylpurine or derivative. Boxed-in codons correspond to tRNA species reported to contain threoninecarbamoyl-purine derivatives. (Reproduced, with permission, from the *Annual Review of Plant Physiology*, Volume 21. © 1970 by Annual Reviews Inc. From Skoog and Armstrong, 1970.)

highly active cytokinins known. At about the same time, R. H. Hall et al. (1966) reported on the isolation of IPA from crude tRNA preparations. As it was later determined, a cytokinin occurs next to the anticodon actually in all four of the tRNA fractions specific for serine (Fig. 4.11). A short time later, tRNA for tyrosine (tRNA^{Tyr}) was shown by Harada et al. (1968) to contain 2-methylthio-N^6-(Δ^2-isopententyl)adenosine (ms^{2}i^{6}A) adjacent to the 3′ end of the anticodon.

As this research progressed, it came to be known that cytokinins occur in tRNA (certain molecular species only) of a large number of—indeed it is safe to assume now all—organisms. Moreover, in all cases where the specific location of a cytokinin moiety has been determined, it invariably is immediately adjacent to the 3′ end of the anticodon. And, interestingly, all tRNA species that are known to contain a cytokinin recognize codons with the initial base as uracil (U) (Fig. 4.11), although not all tRNA species recognizing codons beginning with U contain a cytokinin. The cytokinin-containing tRNA species are tRNA^{Ser}, tRNA^{Phe}, tRNA^{Cys}, tRNA^{Trp}, tRNA^{Leu}, and tRNA^{Tyr}, which are specific for the amino acids serine, phenylalanine, cysteine, tryptophan, leucine, and tyrosine, respectively. In tRNA from microbial sources, cytokinin-active ribonucleosides appear to be present in most, if not all, of the tRNA species that recognize codons beginning with U. But in higher organisms the distribution appears to be more restricted. In the cases of wheat germ and etiolated bean (*Phaseolus vulgaris*) seedlings, cytokinin-active ribonucleosides are limited to only tRNA^{Ser} and tRNA^{Leu}.

Before discussing further the localization of natural cytokinins in tRNA, it should be noted that the natural occurrence of cytokinins is not restricted exclusively to tRNA. Recently, Taller et al. (1987) reported on the occurrence of cytokinin-active ribosides also in ribosomal RNA (rRNA) from two higher plant sources. Pea epicotyl rRNA contained ribosylzeatin, isopentenyladenosine, and 2-methylthioribosylzeatin. Wheat germ rRNA contained *cis*- and *trans*-ribosylzeatin and 2-methylthioribosylzeatin. Nothing can yet be concluded about the physiological significance of cytokinin occurrence in rRNA. Also, as previously noted, cytokinins occur also as the free bases, ribonucleosides and ribonucleotides. Unfortunately there is little information on the relative amounts of cytokinin that occur in these forms, compared to being incorporated into tRNA, for any biological material. It is important to bear in mind that cytokinins occur in multiple chemical forms as the mechanism of hormonal action of the cytokinins is contemplated.

Acknowledging that natural cytokinins do occur in particular tRNAs, several important questions immediately arise: (1) Are the cytokinins precursors in the formation of tRNA (i.e., is the whole cytokinin molecule incorporated into a molecule of tRNA as the polynucleotide chain is synthesized), or does the presence of a cytokinin moiety in a molecule of tRNA result from the addition of the N^6-substituent to an adenosine residue of preformed tRNA? (2) What, if any, are the metabolic consequences of the presence of cytokinin-active bases in tRNA? (3) Is the observed presence of cytokinin-active bases in tRNA related to the hormonal activity of the free cytokinins? (4) What is the molecular mechanism or mechanisms of action of the cytokinins? Each of these questions will be addressed, but as we do so it will be well to bear in mind that the presence of cytokinins in tRNA does have consequences for the metabolic activity of those tRNAs in protein synthesis, and that, even so, tissues that contain potent cytokinins in their tRNA (e.g., tobacco callus) still require exogenous cytokinin for growth *in vitro*. Should these facts necessarily lead us to speculate that free cytokinins have a biological role or mechanism of action quite distinct from the cytokinin moieties that occur in tRNA? Later discussion will aid in trying to answer this as well as the preceding questions.

Now, as to the first question posed above. Are the cytokinins incorporated intact into tRNA, in a manner like all other nucleoside triphosphates that are polymerized in the synthesis of tRNA, or does the presence of the cytokinins result from alkylation (transfer of isoprenoid side chain) of existing adenosine moieties already present in preformed tRNA? In fact, experimental evidence has been reported for both processes. In 1966 Eugene Fox of the University of Kansas reported that ^{14}C label from the synthetic cytokinin N^6-benzyladenine (whether labeled in the 8-position of the adenine moiety or the methylene carbon of the benzyl side chain) was incorporated into RNA in soybean and tobacco tissue cultures, both of which have an absolute cytokinin requirement. Further work by Fox and associates and workers in Skoog's group at Wisconsin has extended that early report considerably. Evidence was presented that two synthetic cytokinins—kinetin (N^6-furfuryladenine) and benzyladenine—are incorporated into RNA of cytokinin-dependent tobacco callus tissue. Moreover, it was shown that N^6-benzyladenine is incorporated as the intact molecule into both tRNA and ribosomal RNA (rRNA). Indeed, both synthetic cytokinins were recovered as the corresponding ribonucleosides to about three times greater extent per absorbance unit at 260 nm from the rRNA than from the tRNA preparations. But the amount of incorporation was very small, and it remains uncertain yet whether the observed incorporation of exogenous cytokinins into RNA is the result of specific events involved in the mechanism of cytokinin action or merely the result of transcriptional errors.

Meanwhile, the evidence for cytokinin moieties in tRNA arising by alkylation of specific adenosine residues in preformed tRNA is conclusive. For example, in the bacterium *Escherichia coli* N^6-(Δ^2-isopentenyl)adenosine (i^6A) and 6-(3-methyl-2-butenylamino)-2-methylthio-9-β-D-ribofuranosylpurine (ms^2i^6A) have been identified as cytokinin-active ribonucleosides in tRNA, and both cytokinin moieties are located next to the 3′ end of the anticodon of tRNA species recognizing codons beginning with uridine. In this system, it is known definitely that the cytokinin i^6A is synthesized by the transfer of the isopentenyl moiety from Δ^2-isopentenyl pyrophosphate to the specific adenosine residues adjacent to the 3′ end of the anticodon in preformed tRNA molecules. Moreover, as will be discussed later, there is good evidence from the *E. coli* system that the cytokinin-active components of tRNA function to ensure proper codon–anticodon interaction of the mRNA–aminoacyl tRNA complex on ribosomes. Also in tobaccco callus—indeed for all organisms examined thus far—this process of transfer of an isopentenyl group to specific adenosine residues in preformed tRNA molecules is a conclusively demonstrated means by which cytokinin-active moieties come to be present in specific tRNA molecules. At this stage of our knowledge, therefore, we reasonably can conclude that the latter process accounts for the presence of cytokinins in certain tRNAs, but we cannot say whether it is the only way.

Metabolic Consequences of the Presence of Cytokinins in tRNA

Regarding another question posed previously, it does indeed seem that there are significant metabolic consequences to the occurrence of cytokinins in tRNA. That is to say, there is good evidence that the cytokinin moieties in tRNA are functionally significant and that they do affect the behavior of those tRNA molecules in the process of protein synthesis. One of the earliest and most conclusive investigations from which this important fact emerged was conducted by M. L. Gefter and R. L. Russell (1969) with *Escherichia coli*.

Gefter and Russell (1969) infected *E. coli* cells with a ''defective transducing bacteriophage'' carrying the tyrosine tRNA gene. This caused an amplification of the production by *E. coli* of actually three molecular species of tRNA for tyrosine (tRNA^{Tyr}). The three forms of tRNA^{Tyr} had the same nucleotide sequence and differed only in the degree of modification of the adenosine residue adjacent to the 3′ end of the anticodon: (1) unmodified adenosine (A), (2) N^6-(Δ^2-isopentenyl)adenosine (i^{6}A), and (3) 6-(3-methyl-2-butenylamino)-2-methylthio-9-β-D-ribofuranosylpurine (ms^{2}i^{6}A). All three forms of the tRNA were then tested for tyrosine acceptor activity and for binding of the tyrosyl-tRNA to an mRNA–ribosome complex. The results were very interesting. No significant differences were found among the three forms of tRNA^{Tyr} as regards amino acid acceptor capacity. However, the tRNA^{Tyr} containing an unmodified adenosine residue adjacent to the anticodon was markedly less effective, in the *in vitro* experiments, in the binding of the amino acyl–tRNA^{Tyr} to the mRNA–ribosome complex than tRNA^{Tyr} containing i^{6}A or ms^{2}i^{6}A. Thus, in this very important case, a cytokinin moiety evidently must be present adjacent to the anticodon of tRNA^{Tyr} if this tRNA species is to function effectively in protein synthesis.

Somewhat earlier, Fittler and Hall (1966) also found that a chemical change in a cytokinin moiety located in tRNA modified the activity of that tRNA in protein synthesis. Specifically, they observed that iodination (treatment with aqueous iodine) of the tRNA for serine of yeast decreased the binding affinity of the seryl-tRNA for the mRNA–ribosome complex without affecting the acceptor activity for serine.

In contrast, other investigators produced both normal and i^{6}A-deficient tRNAs from *Lactobacillus acidophilus* by limiting the mevalonate uptake in the latter to that amount that was just sufficient for maximum growth of the bacteria. Under these conditions, the i^{6}A content of the tRNA was reduced to about half the normal amount. Yet both the normal and undermodified tRNAs were about equally effective in amino acid acceptor activity and protein synthesis *in vitro*.

In summary, clearly the cytokinin moieties in particular species of tRNA are functionally significant. It appears that the cytokinin moieties have an important regulatory function for tRNA. It is not known by what mechanism that regulation is achieved. Perhaps it is by preventing a wrong set of three nucleotides from being recognized by a codon, but this is only one of several speculative ideas. By whatever mechanism the regulation is achieved, however, it appears that cytokinin-active bases in certain types of tRNA may have a modulating effect on protein synthesis at the translational level. However, the important question remains as to whether the reported action of cytokinin moieties in tRNA is related, directly or indirectly, to the hormonal action of free cytokinins.

Hormonal Activity of Free Cytokinins

There are good reasons presently for believing that free cytokinins have important biological activity independently of any direct association with tRNA: (1) there are some investigations, to be described shortly, that have been interpreted as showing directly biological and biochemical activity of exogenous cytokinin independently of incorporation into tRNA; (2) ethanolic extracts of corn (*Zea mays*) kernels contain the *trans*-isomer of zeatin, while the tRNA hydrolysates of corn kernels contain the *cis*-isomer, suggesting that zeatin is not a precursor in the synthesis of tRNA in that material; (3) likewise, dihydrozeatin is a major free cytokinin in beans (*Phaseolus vulgaris*), but it apparently does not occur in bean tRNA; (4) even tissues that contain potent cytokinins in their RNA still require exogenous cytokinin for growth *in vitro*; and (5) the data on direct incorporation of exogenous cytokinins into tRNA do not show conclusively that this process is anything more than transcriptional error.

A prevalent idea about mechanisms of action of plant hormones, cytokinin free bases and others, is that the hormone first binds—by weak hydrogen or ionic bonds—to some receptor. The receptor commonly is envisaged to be an allosteric protein that, as a consequence of binding the hormone, can, as the altered receptor or as a receptor–hormone complex, evoke hormonal action. This idea, well established in animal physiology, has found experimental support in plant physiology only comparatively quite recently (see Venis, 1985, cited in Chapter 2).

Cytokinin Binding Protein

Since about 1970, some progress regarding binding of cytokinins to proteins has been made by J. Eugene Fox and associates. A basic observation that Fox and Erion (1975) made was that cytokinins would bind, with rather high specificity, to ribosomes isolated from higher plants. Extending that observation, Fox and Erion (1977) described actually three cytokinin-binding proteins isolated from ribosomal preparations from wheat germ. They found a fraction of medium molecular weight of 93,000 (CBF-1) with high binding affinity, a low-molecular-weight fraction (30,000) with a high affinity for cytokinins (CBF-2), and a large-molecular-weight fraction of more than 250,000 that had a low affinity for cytokinins (CBF-3). The CBF-1 fraction was isolated specifically from ribosomes, although it was found to occur also in the high-speed postribosomal supernatant together with CBF-2 and CBF-3. A large number of cytokinins, cytokinin nucleotides, analogs, and related purines, and other plant growth substances were tested for their ability to compete with 6-benzylamino-purine (or N^6-benzyladenine) for binding sites on CBF-1. Apparently wheat germ ribosomes contain one copy of CBF-1 per ribosome. The competition studies showed a high degree of specificity for cytokinin-active moieties. The biological role of these cytokinin-binding proteins presently is unknown, but it will be very interesting to follow further research in this area. The binding of cytokinins to ribosomes conceivably could have profound consequences for the regulation of protein synthesis.

J. Eugene Fox and his associates have continued to characterize extensively one cytokinin-binding protein from wheat embryos. The cytokinin-binding factor-1 (CBF-1) is an embryo-specific protein from wheat that binds in a noncovalent manner but with relatively high affinity N^6-substituted purines having cytokinin activity. The native protein was purified by affinity chromatography, and was found to consist of three 54-KDa subunits. The protein lost its affinity for cytokinins on treatment with denaturing agents, proteolytic enzymes, and heating. Interestingly, this protein is synthesized in large amounts during embryogenesis and disappears rapidly on seed germination. Immunologically similar proteins have been detected in rye, barley, oats, triticale, rice, and maize, although the molecular weights of the subunits may differ from CBF-1.

Probable full length cDNA clones for CBF-1 between 2 and 3 kb have been selected and a substantial part of the coding region has been sequenced. Independently the protein was being sequenced using triple quadrapole mass spectrometry. The radiolabeled photoaffinity ligand [^{14}C]2-azido-N^6-benzyladenine was covalently cross-linked to CBF-1 and a single labeled peptide was subsequently obtained. Analysis of 10 pmol of the peptide by laser photodissociation Fourier transform mass spectrometry enabled Fox and Brinegar to determine the amino acid sequence of the cytokinin-modified peptide as follows: Ala-Phe-Leu-Gln-Pro-Ser-His-His (azido-benzyladenine)-Asp-Ala-Asp-Glu.

Recently H. Huang, S. Y. Yang, and Y. W. Tang of the Shanghai Institute of Plant Physiology reported on a cytokinin-binding protein located on the surface of the chloroplast membrane system. This protein can be solubilized from the membrane fraction with 0.5 to 2 M sodium chloride. The protein in the soluble state retains the ability to bind benzyladenine. Benzyladenine together with the binding protein reportedly enhanced photophosphorylation in chloroplasts *in vitro*.

Effects on Moss Protonemata

Some of the most intriguing data on biological effects of exogenous cytokinins have come from investigations of "bud" (actually gametophore) production by protonemata of mosses (Figs. 4.12, 4.13, and 4.14). Evidently "bud" formation in these cryptogams is naturally regulated by an endogenous adenine-type of cytokinin. The haploid moss spore germinates to form a green algal-like filament called a chloronema, which develops into a caulonema, and it is from subapical cells of the caulonema that the "buds" that give rise to gametophores are formed. The chloronema and caulonema are but sequential stages in the development of the green, branching, septate protonema.

H. Brandes and Hans Kende of the MSU/DOE Plant Research Laboratory expanded and quantified some of the early observations obtained with the moss system (Figs. 4.12, 4.13, and 4.14). Observations were made that were interpreted as showing that the synthetic cytokinin when supplied exogenously to protonemata of *Funaria hygrometrica* was active independently of any covalent bonding of the hormone. Using ^{14}C-labeled N^6-benzyladenine, they found that during early stages of their development moss "buds" revert to protonemal filaments if protonemata were washed in an effort to remove the free cytokinin.

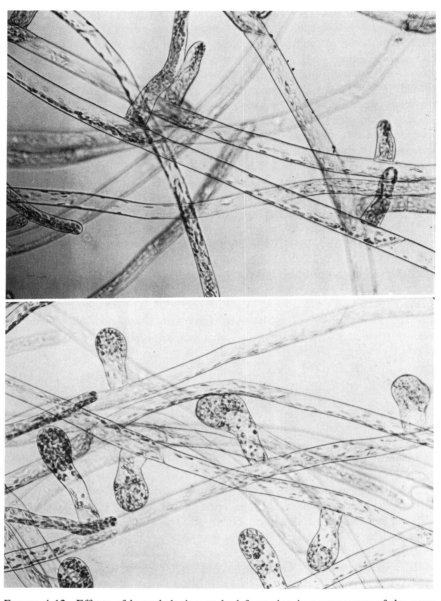

FIGURE 4.12. Effects of benzyladenine on bud formation in protonemata of the moss *Funaria hygrometrica*. **a** Filaments of a protonema grown on basal medium for 15 days. **b** Bud formation on filaments of a protonema that was cultured on basal medium for 14 days and then transferred to basal medium containing 1 μM benzyladenine for 24 hours. (With permission, from Brandes and Kende, 1968; photos courtesy of H. Kende, 1978.)

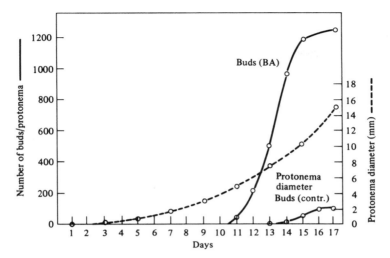

FIGURE 4.13. Growth curve of the protonemata of the moss *Funaria hygrometrica in vitro* and spontaneous and benzyladenine-induced (1 μM) bud formation. (Redrawn, with permission, from Brandes and Kende, 1968.)

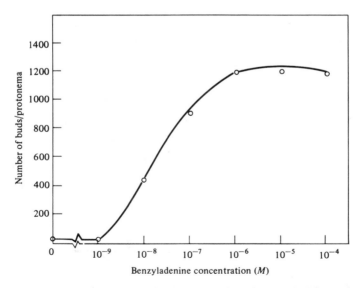

FIGURE 4.14. Dose–response curve for benzyladenine effect on bud formation in moss protonemata. (Redrawn, with permission, from Brandes and Kende, 1968.)

These results were interpreted further as indicating that the exogenous cytokinin was not acting as a "trigger" but that, like hormone responses generally, the cytokinin had to be present during a critical period of time until "bud" differentiation was stabilized.

Some Physiological Effects on Seed Plants

Like auxins, gibberellins, and all other plant hormones, the cytokinins have many diverse physiological effects on seed plants, some of which have been mentioned previously. This diversity of physiological effects complicates efforts to elucidate molecular mechanisms of hormones. However, such diversity of effects is only natural and logically to be expected when it is realized that, in fact, the regulation of growth and development in plants is achieved by several types of hormones acting simultaneously. Hence it really is not surprising to learn that cytokinins have been demonstrated to be involved in regulation of seed plants literally throughout their ontogeny. Of these multiple effects only three of the major ones will be discussed in this section.

Effects of Auxin and Cytokinin on Morphogenesis in Tobacco Callus Tissue

In a classic paper published in 1957, Skoog and Miller described a unique example of hormonal regulation of morphogenesis in tobacco pith explants (Fig. 4.15). This important research beautifully documented a delicate and quantita-

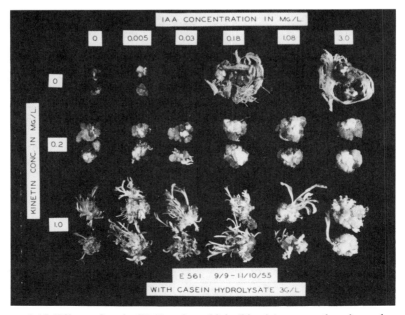

FIGURE 4.15 Effects of auxin (IAA) and cytokinin (kinetin) on growth and morphogenesis in tobacco callus. (With permission, from Skoog and Miller, 1957; photo courtesy of F. Skoog, 1978.)

tive interaction between auxin and cytokinin in the control of bud and root formation from pith explants. They showed that with a particular combination of concentrations of IAA and kinetin, the pith tissues grew as relatively undifferentiated callus. By varying the ratio of IAA:kinetin, however, they could successfully cause the explants to give rise to buds or to roots. High cytokinin: auxin ratios favored bud formation, and low cytokinin: auxin ratios favored root formation. Thus by varying the amounts of the two types of growth substances in the culture medium, morphogenesis in tobacco pith explants can be controlled to a remarkable degree.

Effects of Cytokinins on Leaf Growth and Senescence

Senescence is one of the most poorly defined and least understood—yet very important—phenomena in biology. By a generally acceptable connotation, senescence refers to the collective progressive and deteriorative process that naturally terminate in the death of an organ or organism. In this sense "senescence" is synonymous with "aging."

There are many familiar manifestations of senescence in higher plants. Whole plant senescence is obvious in annual plants in which the life cycle is terminated at the end of one growing season. In herbaceous perennials, the aerial portion of the shoot senesces and dies at the end of a growing season, but the underground system remains viable. Annual senescence and abscission of leaves are exhibited by many woody perennials. Progressive senescence and death of leaves from the basal portion acropetally on a shoot are common. Finally, the ripening and subsequent deterioration of fleshy fruits are examples of senescence.

The biochemical changes that occur in leaves as they grow old have been characterized extensively. However, the endogenous factors that regulate these changes and the reasons why the leaf eventually dies are incompletely understood. The most prominent changes that occur in the senescent leaf are declines in the proteins and nucleic acids and an irreversible yellowing due to the loss of chlorophyll. Catabolism exceeds anabolism, and there is massive export of a variety of soluble metabolites out of the senescent leaf to other parts of the plant. Photosynthesis declines, and generally the respiration rate decreases also. Cytological changes include structural, as well as functional, disorganization of organelles, increased membrane permeability, destruction of cytoplasmic polyribosomes, and eventual disintegration of nuclei.

Leaf senescence and abscission (in those species in which abscission occurs) actually are conservative biological phenomena. Lower (older) leaves on a shoot axis often become quite inefficient organs of photosynthesis because of the shading by the upper portion of the shoot. Thus death and possible elimination of the organ, preceded by export of soluble metabolites, are conservative processes that contribute to the overall welfare of the plant.

When a mature leaf is excised from the plant and the petiole (or basal part) is submerged basally in water, senescence occurs rapidly, provided the petiole does not form adventitious roots. If adventitious roots form on the petiole of an excised leaf rapid senescence, typical of the excised mature leaf, does not occur, because the roots export endogenous cytokinin to the leaf in the xylem sap. The protein and chlorophyll content of the nonrooted leaf declines to less than half the original value in a matter of just a few days. The same process can be observed with excised sections or disks of leaves floated on water.

In 1957, A. E. Richmond and A. Lang discovered that kinetin retarded the senescence of detached leaves of cocklebur (*Xanthium*). This discovery was followed rapidly by numerous reports showing that kinetin and benzyladenine, another synthetic cytokinin, retard senescence of whole excised leaves and excised sections of leaves of numerous species. If whole excised leaves are quite young when excised, cytokinin stimulates growth appreciably.

This area of research was advanced dramatically in 1959 and the years following by K. Mothes and associates in Germany. They found by spraying solutions of kinetin directly onto leaves that the effect of the applied cytokinin was quite localized. Only those areas to which the chemical was applied remained green. Furthermore, the treated areas of yellowing leaves actually became greener. In an effort to explain these intriguing observations, Mothes and associates applied radioactive amino acids and nonradioactive kinetin to leaves. They found that radioactive amino acids migrated to and accumulated in the kinetin-treated parts of the leaves. Thus it came to be understood that cytokinin causes mobilization of metabolites from an untreated to a treated portion of a leaf—i.e., creates new source-sink relationships—and that the chemical somehow acts to sustain nucleic acid and protein synthesis.

A. C. Leopold and M. Kawase (1964) advanced this research a step further using shoot cuttings of bean (*Phaseolus vulgaris*) seedlings. They observed another interesting effect of benzyladenine in addition to stimulation of growth of intact whole young leaves and retardation of senescence of localized treated parts of mature leaves. Treatment of one leaf with cytokinin actually caused inhibition of growth and hastened senescence of other, untreated leaves on the cuttings (Fig. 4.16).

Roles of Cytokinins and Auxin in Apical Dominance

"Apical dominance" refers to the inhibition of growth of subtending lateral (axillary) buds by a growing shoot apex. Some degree of apical dominance is found in all seed plants, and the phenomenon obviously is of profound significance in determining growth form. In some herbaceous plants [e.g., pea (*Pisum sativum*) and sunflower (*Helianthus annuus*)] apical dominance is quite strong, the lateral buds commonly are suppressed, and the shoot thus develops as a single main axis. In other species [e.g., tomato (*Lycopersicum esculentum*) and potato (*Solanum tuberosum*)] apical dominance is much weaker and the shoot is extensively branched. Sometimes apical dominance is only partial, in which case some of the lateral buds develop while others remain fully inhibited.

In certain species of vascular plants—e.g., *Ginkgo biloba*, the Maidenhair tree, and *Cercidiphyllum japonicum*, the Katsura tree of Japan—there is a

FIGURE 4.16. Senescence of a trifoliate leaf of a bean (*Phaseolus vulgaris*) shoot cutting caused by treating the subtending pair of primary leaves with benzyladenine. (With permission, from Leopold and Kawase, 1964; photo courtesy of A. C. Leopold, 1978.)

special and very interesting case of lateral buds that open but that do not elongate appreciably. Instead they produce "short shoots" as opposed to the more usual "long shoots." When a vigorous terminal bud develops first it establishes dominance over the laterals below. This inhibition is transmitted by the auxin produced in the elongating terminal ("long") shoot. In these plants a lateral "short shoot" behaves fundamentally like an inhibited bud of an herbaceous plant, differing in that it opens while an inhibited lateral bud does not. Therefore the occurrence of "short shoots" and "long shoots" on a plant such as *Ginkgo* is regarded as a special instance of the phenomenon of apical dominance.

The two cases just described indicate that apical dominance is very important in determining the growth habits of woody perennials. However, some confusion has existed in the past because of attempts to explain the adult forms of trees and shrubs by too strict and sometimes erroneous analogy with apical dominance in herbaceous plants. The relationship between bud inhibition and form in woody plants is much more complex than bud inhibition in herbaceous plants because of the time sequence in the formation and release of lateral buds. The life cycles of woody perennials are characterized by periods of dormancy. A dormant apical bud does not exert an inhibitory influence on the subtending lateral buds, and more than one bud may develop when dormancy is broken and growth resumes. This may partially explain why so much of the branching of woody perennials occurs at the beginning of a period of active growth. Apical dominance certainly is directly involved in determining the pattern of growth of individual twigs and branches. However, it probably would be well, as suggested by C. L. Brown et al. (1967), to restrict the term to the pattern of bud inhibition in currently elongating twigs, and to denote the complex of physiological conditions giving rise to excurrent and decurrent growth forms of whole trees and shrubs as "apical control."

While the mechanism of apical dominance is incompletely understood at present, K. V. Thimann and T. Sachs have showed that the phenomenon involves an antagonism between cytokinins and auxins. Auxin, produced at the shoot tip and transported basipetally, is inhibitory to the release of buds from apical dominance, and this action of auxin is antagonized by cytokinin, some of which is synthesized in the suppressed lateral bud itself, and some of which is transported past the bud in the xylem sap. Lateral buds typically are relieved of inhibition in a growing shoot experimentally by removal of the shoot apex. Inhibition of the lateral buds is reinstated if auxin is applied to the cut surface of the decapitated stem. Local application of cytokinin directly to inhibited lateral buds causes a release from inhibition. However, buds thus released from apical dominance do not elongate as much as uninhibited buds released by decapitation of the shoot apex, unless they are treated with auxin following release with cytokinin. Thus, in the intact plant, the inhibited lateral bud may—as paradoxical as it might at first appear—fail to grow because of a deficiency in both cytokinin and auxin.

Correlated directly with the effects of auxin and cytokinin on lateral bud inhibition and release therefrom are well-documented effects of these two kinds of hormones on differentiation of vascular tissues between axillary buds and the primary stele of the main shoot axis. Histological studies have revealed that inhibited lateral buds of some plants lack well-developed vascular connections with the vascular system of the main stem. This observation suggests, of course, that inhibited buds receive a deficient supply of metabolites, which evidence shows actually to be the case. Removal of the apical bud from a growing shoot leads to rapid development of vascular connections between the released bud and the vascular system of the main shoot axis. As would be predicted, auxin applied to a decapitated shoot tip inhibits the formation of vascular connections, and cytokinin applied locally to the lateral buds promotes the differentiation of the vascular traces.

Translocation

Somewhat of a paradox exists with regard to translocation of the cytokinins. On the one hand, there are abundant data that reveal that exogenous cytokinins, when applied to leaves, stems, and buds, exhibit markedly little if any movement from the site of application. This relative immobility is an essential feature of the so-called "mobilization effect" of cytokinins, according to which a localized application of exogenous cytokinin to a leaf or part of a leaf delays senescence in that localized portion and creates a sink to which metabolites are transported from other parts of the leaf or even other leaves (see Fig. 4.16).

On the other hand, however, it is well documented that xylem sap and phloem sap as well contain cytokinins. The ability of cytokinins to move upward in the xylem sap has been confirmed by detecting cytokinins in the exudate from decapitated root systems, and by injecting exogenous cytokinins into stems and measuring the resulting systemic effects. Cytokinins, especially as glucosides, also occur naturally in phloem sap, as indicated, for example, by their occurrence in aphid honeydew.

Possibly these apparently conflicting observations really are not irreconcilable. Cytokinins present in the vascular tissues move passively with other solutes in strict accordance with source-to-sink relationships. The major sinks for cytokinins are parts of the plant that are meristematic or that otherwise have growth potential, such as young leaves, buds, young internodes, developing seeds, and fruits—parts that also are the primary centers of cytokinin production together, evidently, with root tips. Moreover, when physiologists apply exogenous cytokinins to intact whole plants or plants organs, they are most often applied to the same sink areas or areas of cytokinin synthesis, because those are the parts that are most responsive.

References

Alvim, R., E. W. Hewett, and P. F. Saunders. 1976. Seasonal variation in the hormone content of willow. I. Changes in abscisic acid content and cytokinin activity in the xylem sap. *Plant Physiol.* **57**: 474–476.

Armstrong, D. J., N. Murai, B. J. Taller, and F. Skoog. 1976. Incorporation of the cytokinin N^6-benzyladenine into tobacco callus transfer ribonucleic acid and ribosomal ribonucleic acid preparations. *Plant Physiol.* **57**: 15–22.

Beimann, K., S. Tsunakawa, J. Sonnenbichler, H. Feldmann, D. Dütting, and H. G. Zachau. 1966. Structure of an odd nucleoside from serine-specific transfer ribonucleic acid. *Angew. Chem. Int. Edit.* **5**: 590–591.

Black, M. K. and D. J. Osborne. 1965. Polarity of transport of benzyladenine, adenine and indole-3-acetic acid in petiole segments of *Phaseolus vulgaris. Plant Physiol.* **40**: 676–680.

Brandes, H. and H. Kende. 1968. Studies on cytokinin-controlled bud formation in moss protonemata. *Plant Physiol.* **43**: 827–837.

Brown, C. L., R. G. McAlpine, and P. P. Kormanik. 1967. Apical dominance and form in woody plants: a reappraisal. *Am. J. Bot.* **54**: 153–162.

Burrows, W. J. 1975. Mechanism of action of cytokinins. *Current Adv. Plant Sci.* **7**: 837–847.

Chatfield, J. M. and D. J. Armstrong. 1986. Regulation of cytokinin oxidase activity in callus tissues of *Phaseolus vulgaris* L. cv Great Northern. *Plant Physiol.* **80**: 493–499.

Chatfield, J. M and D. J. Armstrong. 1987. Cytokinin oxidase from *Phaseolus vulgaris* callus tissues. Enhanced *in vitro* activity of the enzyme in the presence of copper-imidazole complexes. *Plant Physiol.* **84**: 726–731.

Chatfield, J. M. and D. J. Armstrong. 1988. Cytokinin oxidase from *Phaseolus vulgaris* callus cultures. Affinity for concanavalin A. *Plant Physiol.* **88**: 245–247.

Chen, C.-M., J. R. Ertl, S. M. Leisner, and C.-C. Chang. 1985. Localization of cytokinin biosynthetic sites in pea plants and carrot roots. *Plant Physiol.* **78**: 510—513.

Chen, C.-M. and R. H. Hall. 1969. Biosynthesis of N^6-(Δ^2-isopentenyl)adenosine in the transfer ribonucleic acid of cultured tobacco pith tissue. *Phytochemistry* **8**: 1687–1695.

Chen, C.-M. and S. M. Kristopeit. 1981a. Metabolism of cytokinin. Dephosphorylation of cytokinin ribonucleotide by 5'-nucleotidases from wheat germ cytosol. *Plant Physiol.* **67**: 494–498.

Chen, C.-M. and S. M. Kristopeit. 1981b. Metabolism of cytokinin: Deribosylation of cytokinin ribonucleoside by adenosine nucleosidase from wheat germ cells. *Plant Physiol.* **68**: 1020–1023.

Chen, C.-M. and S. M. Leisner. 1984. Modification of cytokinins by cauliflower microsomal enzymes. *Plant Physiol.* **75**: 442–446.

Chen, C.-M. and S. M. Leisner. 1985. Cytokinin-modulated gene expression in excised pumpkin cotyledons. *Plant Physiol.* **77**: 99–103.

Crafts, C. B, and C. O. Miller. 1974. Detection and identification of cytokinins produced by mycorrhizal fungi. *Plant Physiol.* **54**: 586–588.

Deleuze, G. G., J. D. McChesney, and J. E. Fox. 1972. Identification of a stable cytokinin metabolite. *Biochem. Biophys. Res. Commun.* **48**: 1426–1432.

Doree, M. and J. Guern. 1973. Short-time metabolism of some exogenous cytokinins in *Acer pseudoplatanus* cells. *Biochim. Biophys. Acta* **304**: 611–622.

Dyson, W. H., J. E. Fox, and J. D. McChesney. 1972. Short term metabolism of urea and purine cytokinins. *Plant Physiol.* **49**: 506–513.

Einset, J. W. and F. Skoog. 1973. Biosynthesis of cytokinins in cytokinin-autotrophic tobacco callus. *Pro. Natl. Acad. Sci. U.S.A.* **70**: 658–660.

Erion, J. L. and J. E. Fox. 1981. Purification and properties of a protein which binds cytokinin-active 6-substituted purines. *Plant Physiol.* **67**: 156–162.

Fittler, F. and R. H. Hall. 1966. Selective modification of yeast seryl-tRNA and its effect on the acceptance and binding functions. *Biochem. Biophys. Res. Commun.* **25**: 441–446.

Fittler, F., L. K. Kline, and R. H. Hall. 1968. Biosynthesis of N^6-(Δ^2-isopentenyl) adenosine. The precursor relationship of acetate and mevalonate to the Δ^2-isopentenyl group of the transfer ribonucleic acid of microoganisms. *Biochemistry* **7**: 940–944.

Fox, J. E. 1966. Incorporation of a kinin, N,6-benzyladenine into soluble RNA. *Plant Physiol.* **41**: 75–82.

Fox, J. E. 1969. The cytokinins. In: Wilkins, M.B., ed. *The Physiology of Plant Growth and Development.* McGraw-Hill Book Publishing Company Limited, London. Pp. 85–123.

Fox, J. E. and C.-M. Chen. 1967. Characterization of labeled ribonucleic acid from tissue grown on ^{14}C containing cytokinins. *J. Biol. Chem.* **242**: 4490–4494.

Fox, J. E., J. Cornette, G. Deleuze, W. Dyson, C. Giersak, P. Niu, J. Zapata, and J. McChesney. 1973. The formation, isolation, and biological activity of a cytokinin 7-glucoside. *Plant Physiol.* **52**: 627–632.

Fox, J. E. and J. L. Erion. 1975. A cytokinin binding protein from higher plant ribosomes. *Biochem. Biophys. Res. Commun.* **64**: 694–700.

Fox, J. E. and J. Erion. 1977. Cytokinin-binding proteins in higher plants. In: Pilet, P.E. ed. *Plant Growth Regulation.* Springer-Verlag, New York. Pp. 139–146.

Fox, J. E., C. K. Sood, B. Buckwalter, and J. D. McChesney. 1971. The metabolism and biological activity of a 9-substituted cytokinin. *Plant Physiol.* **47**: 275–281.

Gefter, M. L. and R. L. Russell. 1969. Role of modifications in tyrosine transfer RNA: A modified base affecting ribosome binding. *J. Mol. Biol.* **39**: 145–157.

Goldthwaite, J. J. 1987. Hormones in plant senescence. In: Davies, P. J., ed. *Plant Hormones and Their Role in Plant Growth and Development.* Martinus Nijhoff Publishers, Dordrecht, The Netherlands. Pp. 553–573.

Grayburn, W. S., P. B. Green, and G. Steucek. 1982. Bud induction with cytokinin. A local response to local application. *Plant Physiol.* **69**: 682–686.

Hall, R. H. 1970. N^6-(Δ^2-isopentenyl)adenosine: Chemical reactions, biosynthesis, metabolism, and significance to the structure and function of tRNA. *Prog. Nucl. Acid Res. Mol. Biol.* **10**: 57–86.

Hall, R. H. 1973. Cytokinins as a probe of development processes. *Annu. Rev. Plant Physiol.* **24**: 415–444.

Hall, R. H., L. Csonka, H. David, and B. McLennan. 1976. Cytokinins in the soluble RNA of plant tissues. *Science* **156**: 69–71.

Hall, R. H. and G. Mintsioulis. 1973. Enzymatic activity that catalyzes degradation of N^6-(Δ^2-isopentenyl)adenosine. *J. Biochem.* **73**: 739–748.

Hall, R. H., M. J. Robins, L. Stasiuk, and R. Thedford. 1966. Isolation of ^{6}N-(γ, γ-dimethylallyl)adenosine from soluble ribonucleic acid. *J. Am. Chem. Soc.* **88**: 2614–2615.

Harada, F., H. J. Gross, F. Kimura, S. H. Chang, S. Nishimura, and U. L. Raj Bhandary. 1968. 2-Methylthio-N^6-(Δ^2-isopentenyl)adenosine: a component of *E. coli* tyrosine transfer RNA. *Biochem. Biophys. Res. Commun.* **33**: 299–306.

Harvey, B. M. R., B. C. Lu, and R. A. Fletcher. 1974. Benzyladenine accelerates chloroplast differentiation and stimulates photosynthetic enzyme activity in cucumber cotyledons. *Can. J. Bot.* **52**: 2581–2586.

Helgeson, J. P. and N. J. Leonard. 1966. Cytokinins: Identification of compounds isolated from *Corynebacterium fascians. Proc. Natl. Acad. Sci. U. S. A.* **56**: 60–63.

Horgan, R., E. W. Hewett, J. G. Purse, J. M. Horgan, and P. F. Wareing. 1973. Identification of a cytokinin in sycamore sap by gas chromatography-mass spectrometry. *Plant Sci. Lett.* **1**: 321–324.

Ihara, M., Y. Taya, S. Nichimura, and Y. Tanaka. 1984. Purification and some properties of Δ^2-isopentenylpyrophosphate: 5′ Amp Δ^2-isopentenyltransferase from the cellular slime mold *Dictyostelium discoideum. Arch. Biochem. Biophys.* **230**: 652–660.

Itai, C. and Y. Vaadia. 1965. Kinetin-like activity in root exudate of water-stressed sunflower plants. *Physiol. Plantarum* **18**: 941–944.

Jablonski, J. R. and F. Skoog. 1954. Cell enlargement and cell division in excised tobacco pith tissue. *Physiol. Plantarum* **7**: 16–24.

Kende, H. 1964. Preservation of chlorophyll in leaf sections by substances obtained from root exudate. *Science* **145**: 1066–1067.

Kende, H. 1965. Kinetin like factors in the root exudate of sunflowers. *Proc. Natl. Acad. Sci. U.S.A.* **53**: 1302–1307.

Kende, H. and G. Gardner. 1976. Hormone binding in plants. *Annu. Rev. Plant Physiol.* **27**: 267–290.

Klambt, D., G. Thies, and F. Skoog. 1966. Isolation of cytokinins from *Corynebacterium fascians. Proc. Natl. Acad. Sci. U. S. A.* **56**: 52–59.

Krikorian, A. D., K. Kelly, and D. L. Smith. 1987. Hormones in tissue culture and micropropagation. In: Davies, P. J., ed. *Plant Hormones and Their Role in Plant Growth and Development*. Martinus Nijhoff Publishers, Dordrecht, The Netherlands. Pp. 593–613.

Lagerstedt, H. B. and R. G. Langston. 1967. Translocation of radioactive kinetin. *Plant Physiol.* **42**: 611–622.

Lee, Y.-H., M. C. Mok, D. W. S. Mok, D. A. Griffin, and G. Shaw. 1985. Cytokinin metabolism in *Phaseolus* embryos. Genetic difference and the occurrence of novel zeatin metabolites. *Plant Physiol.* **77**: 635–641.

Leonard, N. J. 1974. Chemistry of the cytokinins. In: Runeckles, V. C., E. Sondheimer, and D. C. Walton, eds. *The Chemistry and Biochemistry of Plant Hormones.* Vol. 7. *Recent Advances in Phytochemistry.* Academic Press, New York. Pp. 21–56.

Leonard, N. J., S. M. Hecht, F. Skoog, and R. Y. Schmitz. 1968. Cytokinins: Synthesis of 6(3-methyl-3-butenylamino)9-β-D-ribofuranosylpurine (3iPA), and the effect of side-chain unsaturation on the biological activity of isopentylaminopurines and their ribosides. *Proc. Natl. Acad. Sci. U.S.A.* **59**: 15–21.

Leopold, A. C. and M. Kawase. 1964. Benzyladenine effects on bean leaf growth and senescence. *Am. J. Bot.* **51**: 294–298.

Leopold, A. C. and P. E. Kriedemann. 1975. *Plant Growth and Development.* 2nd ed. McGraw-Hill Book Company, New York.

Letham, D. S. 1963. Zeatin, a factor inducing cell division isolated from *Zea mays. Life Sci.* **2**: 569–573.

Letham, D. S. 1967. Chemistry and physiology of kinetin-like compounds. *Annu. Rev. Plant Physiol.* **18**: 349–364.

Letham, D. S. 1973. Transfer RNA and cytokinins. In: Stewart, P. R. and D. S. Letham, eds. *The Ribonucleic Acids.* Springer-Verlag, New York. Pp. 81–106.

Letham, D. S. and C. O. Miller. 1965. Identity of kinetin-like factors from *Zea mays. Plant Cell Physiol.* **6**: 355–359.

Letham, D. S. and R. K. Ralph. 1967. A cytokinin in soluble RNA from a higher plant. *Life Sci.* **6**: 387–394.

Letham, D. S., J. S. Shannon, and I. R. McDonald. 1964. The structure of Zeatin, a factor inducing cell division. *Proc. Chem. Soc.* 230–231.

Matsubara, S., D. J. Armstrong, and F. Skoog. 1968. Cytokinins in tRNA of *Corynebacterium fascians. Plant Physiol.* **43**: 451–453.

McGaw, B. A. 1987. Cytokinin biosynthesis and metabolism. In: Davies, P. J., ed. *Plant Hormones and Their Role in Plant Growth and Development.* Martinus Nijhoff Publishers, Dordrecht, The Netherlands. Pp. 76–93.

McGaw, B. A. and R. Horgan. 1983. Cytokinin oxidase from *Zea mays* kernels and *Vinca rosea* crown-gall tissue. *Planta* **159**: 30–37.

Miller, C. O. 1961a. Kinetin and related compounds in plant growth. *Annu. Rev. Plant Physiol.* **12**: 395–408.

Miller, C. O. 1961b. A kinetin-like compound in maize. *Proc. Natl. Acad. Sci. U.S.A.* **47**: 170–174.

Miller, C. O. 1974. Ribosyl-*trans*-zeatin, a major cytokinin produced by crown gall tumor tissue. *Proc. Natl. Acad. Sci. U.S.A.* **71**: 334–338.

Miller, C. O. 1985. Possible regulatory roles of cytokinins. NADH oxidation by peroxidase and a copper interaction. *Plant Physiol.* **79**: 908–910.

Miller, C. O., F. Skoog, F. S. Okumura, M. H. von Saltza, and F. M. Strong. 1955. Structure and synthesis of kinetin. *J. Am. Chem. Soc.* **77**: 2662–2663.

Miller, C. O., F. Skoog, M. H. von Saltza, and F. M. Strong. 1955. Kinetin, a cell division factor from deoxyribonucleic acid. *J. Am. Chem. Soc.* **77**: 1392.

Miura, G. A. and C. O. Miller. 1969. 6-(γ, γ-dimethylallylamino)purine as a precursor of zeatin. *Plant Physiol.* **44**: 372–376.

Mok, D. W. S. and M. C. Mok. 1987. Metabolism of ^{14}C-zeatin in *Phaseolus* embryos. Occurrence of *O*-xylosyldihydrozeatin and its ribonucleoside. *Plant Physiol.* **84**: 596–599.

Mok, M. C., D. W. S. Mok, D. J. Armstrong, K. Shudo, Y. Isogai, and T. Okamoto. 1982. Cytokinin activity of N-phenyl-N'-1,2,3-thiadiazol-5-yl urea (Thidiazuron). *Phytochemistry* **21**: 1509–1511.

Morris, R. O. 1987. Genes specifying auxin and cytokinin biosynthesis in prokaryotes. In: Davies, P. J., ed. *Plant Hormones and Their Role in Plant Growth and Development.* Martinus Nijhoff Publishers, Dordrecht, The Netherlands. Pp. 636–655.

Morris, R. O., G. K. Powell, J. S. Beaty, R. C. Durley, N. G. Hommes, L. Lica, and E. M. MacDonald. 1986. Cytokinin biosynthetic genes and enzymes from *Agrobacterium tumefaciens* and other plant-associated prokaryotes. In: Bopp, M., ed. *Plant Growth Substances 1985.* Springer-Verlag, Berlin, Heidelberg, New York, Tokyo. Pp. 185–196.

Mothes, K. and L. Engelbrecht. 1961. Kinetin-induced directed transport of substances in excised leaves in the dark. *Phytochemistry* **1**: 58–62.

Paces, V., E. Werstiuk, and R. H. Hall. 1971. Conversion of N^6-(Δ^2-isopentenyl)adenosine to adenosine by enzyme activity in tobacco tissue. *Plant Physiol.* **48**: 775–778.

Parker, C. W., M. M. Wilson, and D. S. Letham. 1973. The glucosylation of cytokinins. *Biochem. Biophys. Res. Commun.* **55**: 1370–1376.

Person, C., D. J. Samborski, and F. R. Forsyth. 1957. Effect of benzimidazole on detached wheat leaves. *Nature (London)* **180**: 1294–1295.

Peterkofsky, A. 1968. The incorporation of mevalonic acid into the N^6-(Δ-isopentenyl)-adenosine of transfer ribonucleic acid in *Lactobacillus acidophilus. Biochemistry* **7**: 472–482.

Phillips, I. D. J. 1969. Apical dominance. In: Wilkins, M. B., ed. *The Physiology of Plant Growth and Development.* McGraw-Hill Book Company, New York. Pp. 165–202.

Phillips, I. D. J. 1971. *Introduction to the Biochemistry and Physiology of Plant Growth Hormones.* McGraw-Hill Book Company, New York.

Richmond, A. E. and A. Lang. 1957. Effect of kinetin on protein content and survival of detached Xanthium leaves. *Science* **125**: 650–651.

Sachs, T. and K. V. Thimann. 1964. Release of lateral buds from apical dominance. *Nature (London)* **201**: 939–940.

Sachs, T. and K. V. Thimann. 1967. The role of auxins and cytokinins in the release of buds from dominance. *Am. J. Bot.* **54**: 136–144.

Scarbrough, E., D. J. Armstrong, F. Skoog, C. R. Frihart, and N. J. Leonard. 1973. Isolation of *cis*-zeatin from *Corynebacterium fascians* cultures. *Proc. Natl. Acad. Sci. U.S.A.* **70**: 3825–3829.

Schröder, J., I. Buchmann, and G. Schröder. 1986. Enzymes of auxin and cytokinin biosynthesis encoded in Ti plasmids. In: Bopp, M., ed. *Plant Growth Substances 1985.* Springer-Verlag, Berlin, Heidelberg, New York, Tokyo. Pp. 177–184.

Shaw, G. and D. V. Wilson. 1964. Synthesis of zeatin. *Proc. Chem. Soc.* 231.

Shibaoka, H. and K. V. Thimann. 1970. Antagonisms between kinetin and amino acids. Experiments on the mode of action of cytokinins. *Plant Physiol.* **46**: 212–220.

Skene, K. G. M. 1975. Cytokinin production by roots as a factor in the control of plant growth. In: Torrey, J. G. and D. T. Clarkson, eds. *The Development and Function of Roots.* Acedemic Press. New York. Pp. 365–396.

Skoog, F. and D. J. Armstrong. 1970. Cytokinins. *Annu. Rev. Plant Physiol.* **21**: 359–384.

Skoog, F., D. J. Armstrong, J. D. Cherayil, A. E. Hampel, and R. M. Bock. 1966.

Cytokinin activity: Localization in transfer RNA preparations. *Science* **154**: 1354–1356.

Skoog, F., H. Q. Hamzi, A. M. Szweykowska, N. J. Leonard, K. L. Carraway, T. Fujii, J. P. Helgeson, and R. N. Loeppky. 1967. Cytokinins: structure/activity relationships. *Phytochemistry* **6**: 1169–1192.

Skoog, F. and C. O. Miller. 1957. Chemical regulation of growth and organ formation in plant tissues cultured *in vitro*. *Symp. Soc. Exp. Biol.* **11**: 118–131.

Skoog, F., R. Y. Schmitz, R. M. Bock, and S. M. Hecht. 1973. Cytokinin antagonists: synthesis and physiological effects of 7-substituted 3-methylpyrazolo-[4,3-*d*]pyrimidines. *Phytochemistry* **12**: 25–37.

Skoog, F., F. M. Strong, and C. O. Miller. 1965. Cytokinins. *Science* **148**: 532–533.

Sondheimer, E. and D. Tzou. 1971. The metabolism of hormones during seed germination and dormancy. II. The metabolism of 8-^{14}C-zeatin in bean axes. *Plant Physiol.* **47**: 516–520.

Sorokin, H. P. and K. V. Thimann. 1964. The histological basis for inhibition of axillary buds in *Pisum sativum* and the effects of auxins and kinetin on xylem development. *Protoplasma* **59**: 326–350.

Spiess, L. D. 1975. Comparative activity of isomers of zeatin and ribosyl-zeatin on *Funaria hygrometrica*. *Plant Physiol.* **55**: 583–585.

Steward, F. C. and A. D. Krikorian. 1971. *Plants, Chemicals and Growth*. Academic Press, New York.

Takegami, T. and K. Yoshida. 1975. Isolation and purification of cytokinin binding protein from tobacco leaves by affinity column chromatography. *Biochem. Biophys. Res. Commun.* **67**: 782–789.

Taller, B. J., N. Murai, and F. Skoog. 1987. Endogenous cytokinins in the ribosomal RNA of higher plants. *Plant Physiol.* **83**: 755–760.

Tamas, I. A. 1987. Hormonal regulation of apical dominance. In: Davies, P. J., ed. *Plant Hormones and Their Role in Plant Growth and Development*. Martinus Nijhoff Publishers, Dordrecht, The Netherlands. Pp. 393–410.

Tavares, J. and H. Kende. 1970. The effect of 6-benzylaminopurine on protein metabolism in senescing corn leaves. *Phytochemistry* **9**: 1763–1770.

Taya, Y., Y. Tanaka, and S. Nishimura. 1978. 5′-AMP is a direct precursor of cytokinin in *Dictyostelium discoideum*. *Nature (London)* **271**: 545–547.

Thimann, K. V. 1985. The senescence of detached leaves of *Tropaeolum*. *Plant Physiol.* **79**: 1107–1110.

Thimann, K. V. 1977. *Hormone Action in the Whole Life of Plants*. University of Massachusetts Press, Amherst.

Thimann K. V., T. Sachs, and K. N. Mathur. 1971. The mechanism of apical dominance in *Coleus*. *Physiol. Plantarum* **24**: 68–72.

Thomas, J. C. and F. R. Katterman. 1986. Cytokinin activity induced by thidiazuron. *Plant Physiol.* **81**: 681–683.

Torrey, J. G. 1976. Root hormones and plant growth. *Annu. Rev. Plant. Physiol.* **27**: 435–459.

Varner, J. E. and D. T. Ho. 1976. Hormones. In: Bonner, J. and J. E. Varner, eds. *Plant Biochemistry*. 3rd ed. Academic Press, New York. Pp. 713–770.

Vreman, H. J., F. Skoog, C. R. Frihart, and N. J. Leonard. 1972. Cytokinins in *Pisum* transfer ribonucleic acid. *Plant Physiol.* **49**: 848–851.

Walker, G. C., N. J. Leonard, D. J. Armstrong, N. Murai, and F. Skoog. 1974. The mode of incorporation of 6-benzylaminopurine into tobacco callus transfer ribonucleic acid. A double labeling determination. *Plant Physiol.* **54**: 737–743.

Wareing, P. F., R. Horgan, I. E. Henson, and W. Davis. 1977. Cytokinin relations in the whole plant. In: Pilet, P. E., ed. *Plant Growth Regulation.* Springer-Verlag, New York. Pp. 147–153.

Wareing, P. F. and I. D. J. Phillips. 1981. *The Control of Growth and Differentiation in Plants.* 3rd ed. Pergamon Press, New York.

Whitty, C. D. and R. H. Hall. 1974. A cytokinin oxidase in *Zea mays. Can. J. Biochem.* **52**: 789–799.

Wickson, M. and K. V. Thimann. 1958. The antagonism of auxin and kinetin in apical dominance. *Physiol. Plantarum* **11**: 62–74.

Zachau, H. G., D. Dütting, and H. Feldmann. 1966a. Nucleotide sequences of two serine-specific transfer ribonucleic acids. *Angew. Chem. Int. Edit.* **5**: 422.

Zachau, H. G., D. Dütting, and H. Feldmann, 1966b. The structures of two serine transfer ribonucleic acids. *Hoppe-Seylers Zeit. Physiol. Chem.* **347**: 212–235.

CHAPTER 5

Abscisic Acid and Related Compounds

Introduction

Regulation of the growth and development of plants is dependent on the kinds and amounts of the various hormones and on the changing sensitivity of the tissues to these substances throughout ontogeny. Yet the survival of plants as individual organisms and as populations and species over multiple generations depends also on deceleration or cessation of growth and development of resistance to stress at critical stages in the life cycles. Coping with water stress and other environmentally imposed adverse conditions by relatively rapid physiological changes is a frequent and essential manifestation of the adaptation of terrestrial plants to their often highly changeable physical environments. Longer term and more gradually developing suspensions of growth, or deceleration to barely perceptible rates, called "dormancy," also are common features of the ontogeny of seed plants. Dormancy is a phenomenon of profound biological significance, because it provides a means by which plants are enabled to survive periods of environmental conditions that would be adverse or lethal to plants in an active state of growth. In the case of annual plants, dormancy of seeds is commonplace. In biennials and perennials, dormancy of the buds of established plants and of their seeds and storage organs (e.g., bulbs, corms, and tubers) are typical events in the life cycles.

Unfortunately, considerable confusion exists concerning the terminology relating to dormancy phenomena. But, by generally acceptable connotation, "dormancy" in the broad sense refers to any temporary suspension of active growth. Dormancy that is imposed by unfavorable physical environmental conditions such as low temperature or moisture stress, which often is immediately reversible on the plant's again experiencing favorable conditions, is called "quiescence," "imposed dormancy," "temporary dormancy," or, sometimes, in the case of perennials, "summer dormancy." "True dormancy," "rest," or "innate dormancy" refers to a state of temporarily suspended growth that is due to internal conditions. In the case of true dormancy or rest, growth is suspended even under physical environmental conditions that apparently are favorable for growth. It is true dormancy or rest, hereafter called simply "dormancy," of buds and seeds, together with water stress, with which this chapter is concerned. Just as it is important to understand the environmental stimuli that evoke the onset and breaking of bud dormancy, and the stimuli involved in dormant seeds, it is of great interest to understand the changes in internal conditions that are causally related to the cycle of growth and dormancy.

Especially prominent among these changes are alterations in the balance of hormones, particularly certain growth-stimulating hormones and one or more growth-inhibiting hormones.

Actually, a quite large number of chemically diverse growth-inhibiting substances have been identified in plants (Fig. 5.1), and some of these at least probably are very significant physiologically. One such compound is jasmonic acid [3-oxo-2-(2'-pentenyl) cyclopentane acetic acid] and its methyl ester, which are constituents of many higher plants. Both of these compounds inhibit growth and promote senescence. However, presently abscisic acid (ABA) is the only inhibitor that is commonly regarded as a true hormone. Hence this chapter will be concerned chiefly with ABA and certain of its analogs. As F. T. Addicott and J. L. Lyon stated in a review in 1969, "Abscisic acid (ABA) is a plant hormone which now ranks in importance with the auxins, gibberellins, and cytokinins as a controlling factor in physiological processes."

History of Discovery

The discovery of abscisic acid (ABA) as a naturally occurring growth inhibitor occurred through independent investigations of different physiological phenomena in two different laboratories. F. T. Addicott et al. of the University of California, Davis since circa 1961 have been investigating natural substances which accelerate leaf abscission. Across the Atlantic, P. F. Wareing et al. of the University College of Wales at Aberystwyth have, for three decades or longer, been investigating natural inhibitors that appear to be causally related to bud dormancy in woody plants. By 1965 these two paths of research converged on the discovery that a single hormone was involved in both. The history of how this came about is one of the most interesting stories in the physiology and biochemistry of plant hormones.

We can logically begin the story by noting the report by Liu and Carns in 1961 that they had isolated in crystalline form from dried, mature cotton burs a substance that they first called "abscisin" (later "abscisin I") that accelerates abscission of excised debladed cotton petioles. While Liu and Carns characterized some of the properties of "abscisin," they did not determine its structure, and it is assumed that the structure was not determined later. Anyway, interest in "abscisin I" waned until 1963, when Ohkuma et al. reported on the isolation in crystalline form of another abscission-accelerating substance that they termed "abscisin II" from young cotton fruit. They described some of the chemical and physiological properties of this new substance and proposed a tentative empirical formula of $C_{15}H_{20}O_4$. In the same year, 1963, C. F. Eagles and P. F. Wareing isolated chromatographically an inhibitor from leaves of *Betula pubescens* that could be applied to buds of growing seedlings and cause them to go dormant. They proposed the term "dormin" for this unknown substance and for "other substances which appear to function as endogenous dormancy inducers."

To expedite chemical identification of "dormin" Wareing sought assistance

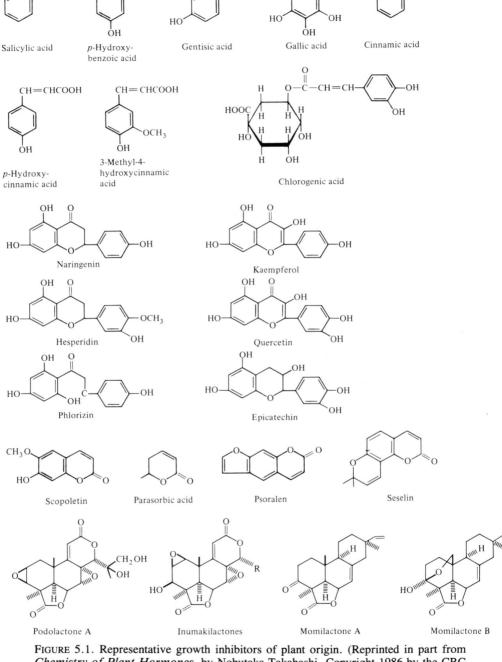

FIGURE 5.1. Representative growth inhibitors of plant origin. (Reprinted in part from *Chemistry of Plant Hormones*, by Nobutaka Takahashi. Copyright 1986 by the CRC Press, Inc., Boca Raton, Fl. Used with permission of CRC Press, Inc. and N. Takahashi.)

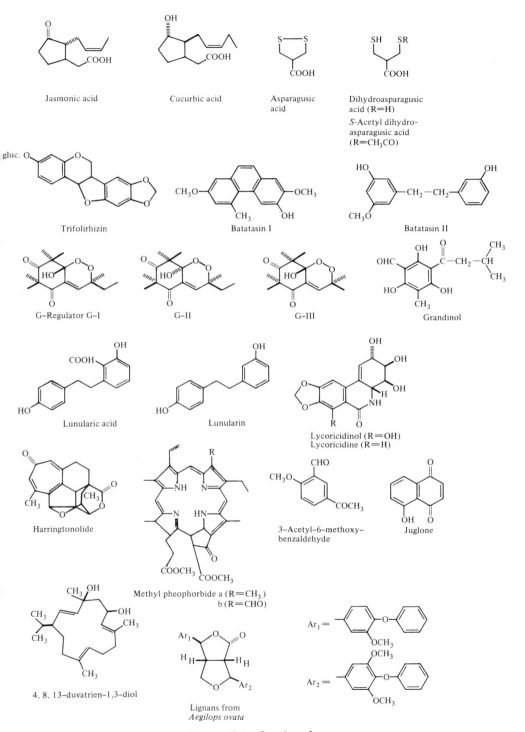

Jasmonic acid

Cucurbic acid

Asparagusic acid

Dihydroasparagusic acid (R=H)

S-Acetyl dihydro-asparagusic acid (R=CH₃CO)

Trifolirhizin

Batatasin I

Batatasin II

G-Regulator G-I

G-II

G-III

Grandinol

Lunularic acid

Lunularin

Lycoricidinol (R=OH)
Lycoricidine (R=H)

Harringtonolide

Methyl pheophorbide a (R=CH₃)
                       b (R=CHO)

3-Acetyl-6-methoxy-benzaldehyde

Juglone

4, 8, 13-duvatrien-1,3-diol

Lignans from
Aegilops ovata

FIGURE 5.1. Continued.

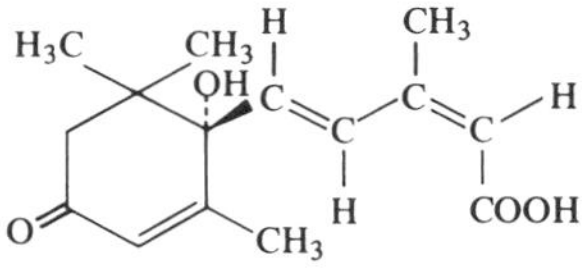

(R)-Abscisic acid

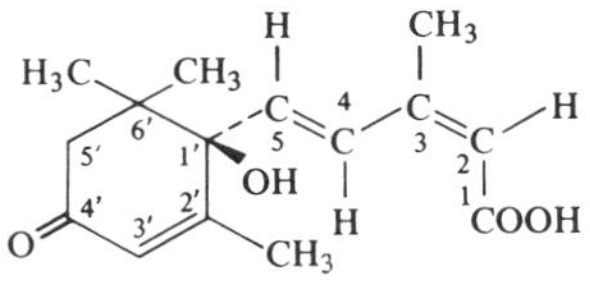

(S)-Abscisic acid

FIGURE 5.2. Structures of (R)- and (S)-abscisic acid. (S)-ABA is the naturally occurring enantiomorph.

from chemists at Shell Research Ltd. at Milstead Laboratory in England. A short time later, in 1965, J. W. Cornforth, B. V. Milborrow, G. Rayback (all of Shell), and P. F. Wareing compared the properties of dormin (from *Acer pseudoplatanus* leaves) with those of abscisin II and concluded, from a comparison of molecular weights, infrared spectra, and melting points, that the two compounds are in fact identical. A molecular structure of abscisin II was proposed by Ohkuma et al. in 1965 (Fig. 5.2), and that this structure is valid for both "abscisin II" and "dormin" was confirmed by synthesis by Cornforth et al. later that same year (1965). By 1966 B. V. Milborrow and associates in England had perfected a sensitive spectropolarimetric method for identifying (S)-abscisic acid in plant extracts and for measurement of its concentration. In 1965 Ohkuma had already synthesized some analogs of abscisic acid and by 1966, (S)-abscisic acid had been identified from several species of seed plants.

It was only natural that some confusion began to develop over the terms "dormin" and "abscisin II;" hence, the investigators in Addicott's, Wareing's and the Shell groups got together at the Sixth International Conference on Plant Growth Substances, held in Ottawa in the summer of 1967, and suggested standardized terminology, as follows: "We now propose the term abscisic acid as a reasonable and useful compromise ..." "We suggest that authors specify racemic abscisic acid ... by calling it (RS)-abscisic acid, and that the naturally occurring enantiomorph be called (S)-abscisic acid when it is necessary to draw attention to the stereochemistry" (Fig. 5.2). "As an abbreviation for abscisic acid, we propose ABA." All these proposals were adopted promptly in the literature that followed. Except where noted otherwise, all further references to ABA in this chapter will be understood to be specifically to (S)-ABA.

Chemical Characterization

ABA is a sesquiterpenoid (15-C) compound that is related, by its biogenesis, to the monoterpenes, diterpenes (including GAs), carotenoids, and triterpenes. Available evidence indicates that the naturally occurring ABA is a single enantiomorph, specifically the dextrorotatory compound, (S)-ABA [or (+)-ABA]. (R)-ABA [or (−)-ABA], which accounts for about 50% of the racemic mixtures of ABA that are made synthetically, has biological activity equal to that of (S)-ABA in most cases, an exception being stomatal closure in which (R)-ABA reportedly is inactive. Several substances related to (S)-ABA have also been found to occur in plant tissues (Fig. 5.3). Likewise, a number of synthetic analogs have been synthesized (Fig. 5.4).

The structural features of ABA and related molecules that are essential for biological activity are that (1) a free carboxyl group is essential, (2) the

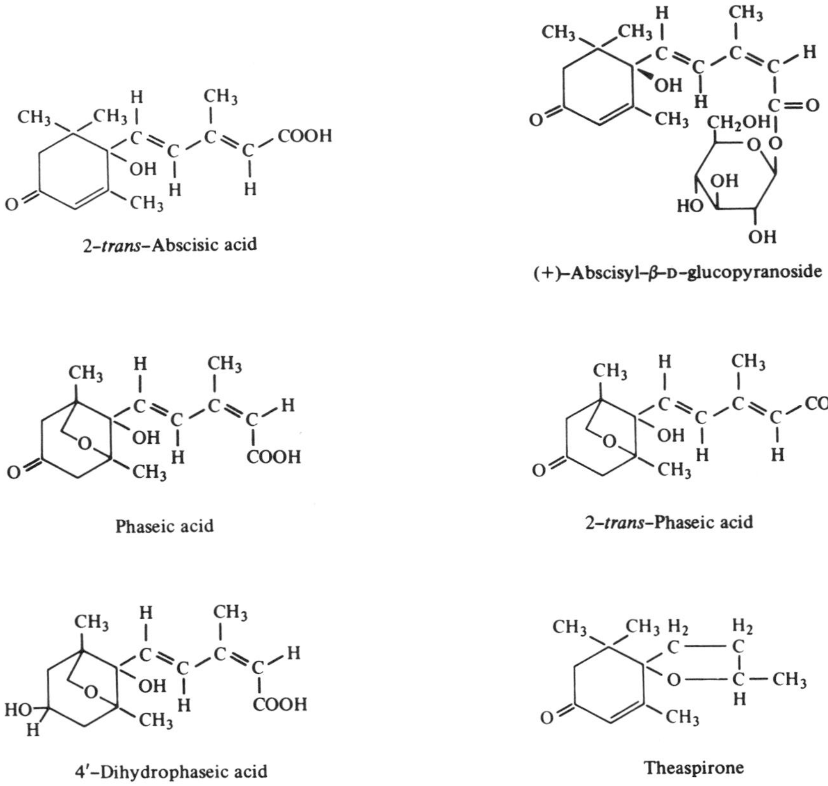

2-*trans*-Abscisic acid

(+)-Abscisyl-β-D-glucopyranoside

Phaseic acid

2-*trans*-Phaseic acid

4′-Dihydrophaseic acid

Theaspirone

FIGURE 5.3. Some naturally occurring substances related to ABA. (Reproduced, with permission, from the *Annual Review of Plant Physiology*, Volume 20. © 1969 by Annual Reviews Inc. From Addicott and Lyon, 1969.)

Structure	Physiological activity	Structure	Physiological activity
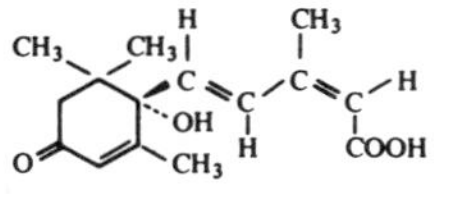 (R)–Abscisic acid	(−) Growth	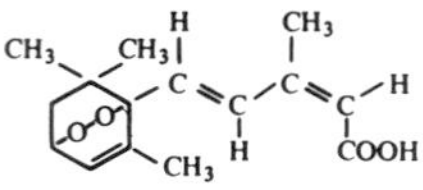	(−) Growth
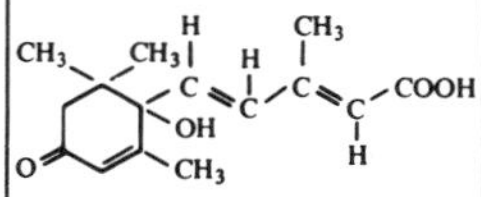 2, 4-*trans, cis*-Abscisic acid	(+) Abscission	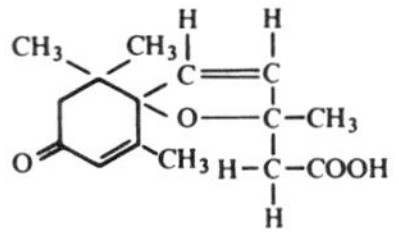	(0) Abscission
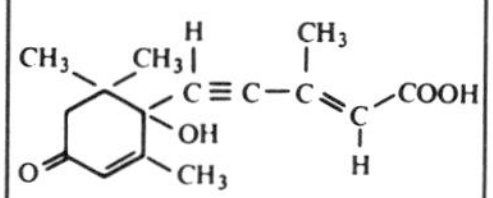	(+) Abscission	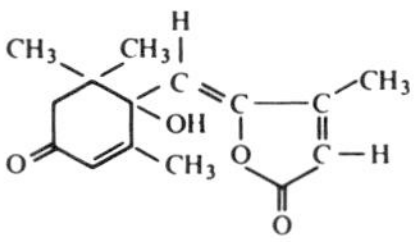	(0) Abscission
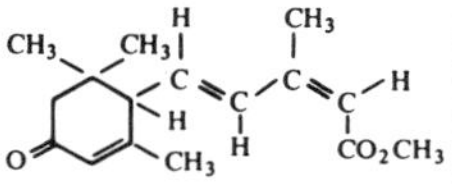	(−) Growth	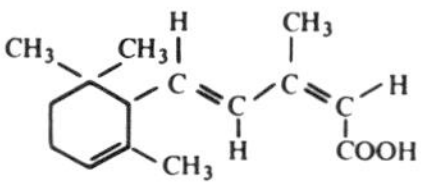	(+) Abscission (−) Growth

FIGURE 5.4. Structures and physiological activities of some synthetic substances related to (*S*)-ABA. (Reproduced, with permission, from the *Annual Review of Plant Physiology,* Volume 20. © 1969 by Annual Reviews Inc. From Addicott and Lyon, 1969.)

cyclohexane ring must contain a double bond in an α- or β-position, and (3) the configuration of the C-2 double bond must be *cis*. 2-*trans*-ABA occasionally shows activity, but this is due to its isomerization in light. Likewise, the activity of esters is due to the hydrolysis of the ester bond and formations of the free acid. The glucose ester of ABA (abscisyl-β-D-glucopyranoside) appears to be an effective storage form of ABA from which the free acid can readily be liberated by hydrolysis *in vivo*.

Biosynthesis and Other Features of Metabolism

There is very good evidence that ABA is synthesized from mevalonate via farnesyl pyrophosphate (Milborrow, 1974b; Robinson and Ryback, 1969), an intermediate in the isoprenoid biosynthetic pathway (Fig. 5.5). The derivation of GAs and ABA from farnesyl pyrophosphate is of particular interest from the

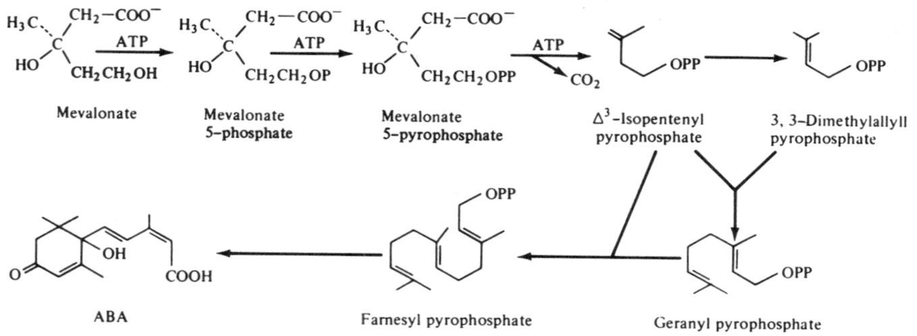

FIGURE 5.5. Proposed pathway for the biosynthesis of (S)-abscisic acid from mevalonate.

standpoint of plant hormone physiology and biochemistry. In a later section we shall see that GAs and ABA generally have opposing physiological and biochemical effects. Thus it is of particular interest to elucidate the enzymic regulation that determines the relative rates of biosynthesis of GAs and ABA from the common intermediate, farnesyl pyrophosphate.

It appears that ABA, like at least some GA, biosynthesis occurs in chloroplasts. B. V. Milborrow (1974d) developed a cell-free enzyme system from lysed chloroplasts (and etioplasts) of avocado (*Persea gratissima*) fruit that biosynthesized abscisic acid from mevalonate. It cannot be said with certainty that all the ABA synthesis in avocado fruit occurs in plastids, but it is evident that at least the majority does. Milborrow obtained similar results with cell-free enzyme extracts of bean (*Phaseolus vulgaris*) and avocado leaves. Milborrow (1974a) did, however, present evidence that two pools of ABA can exist in ripening avocado fruit, raising questions about the compartmentation of this hormone.

A possible alternate pathway of biogenesis of ABA that has been described is from a carotenoid, specifically violaxanthin, via xanthoxin as an intermediate (Fig. 5.6). Xanthoxin has been demonstrated to be formed *in vitro* by high-intensity photoconversion of violaxanthin. However, this reaction has not been reported to occur *in vivo*. In other investigations, xanthoxin was observed to be formed by the lipoxygenase-catalyzed cleavage of violaxanthin. By whatever means it is formed, xanthoxin is of great interest. Xanthoxin does occur naturally in some plants, and growth inhibitory activity has been attributed to the compound. And in one very interesting case, there was indirect evidence that xanthoxin was rapidly converted to ABA in tomato shoots.

However, there still is uncertainty as to the details of the ABA biosynthetic pathway following carotenoid cleavage. B. V. Milborrow reported recently that in tomato shoots xanthoxin was not a precursor of ABA. However, he stated that his results did not exclude the involvement of a carotenoid in ABA biosynthesis. In contrast, Sindhu and Walton (1987) reported that xanthoxin was converted to ABA in cell-free extracts from the leaves of *Phaseolus vulgaris, Pisum sativum, Zea mays,* and *Cucurbita maxima,* as well as roots of *Phaseolus*

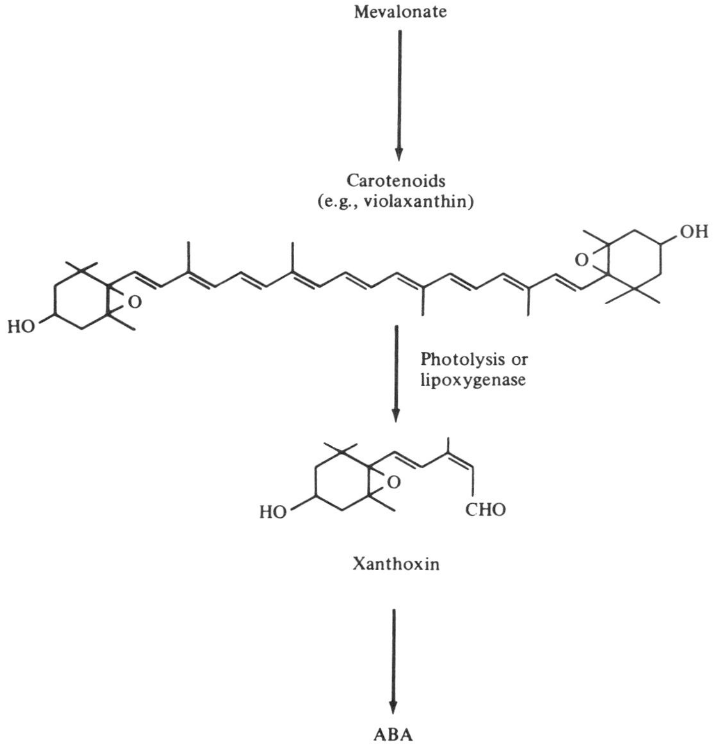

FIGURE 5.6. Possible alternate pathway of biogenesis of (S)-ABA via a carotenoid and xanthoxin as intermediates. (Redrawn, with permission, from Varner, J. E. and D. T. Ho. 1976. Hormones. In: Bonner, J. and J. E. Varner, eds. *Plant Biochemistry*, 3rd ed., pp. 713–770. Copyright by Academic Press, Inc., New York.)

vulgaris. Creelman et al. (1987) proposed that the precursors to stress-induced ABA are xanthophylls, and that a xanthophyll lacking an oxygen function at C-6 (carotenoid numbering scheme) plays a crucial role in ABA biosynthesis in *Xanthium* roots. The significance of this possible alternate pathway obviously deserves more research.

Three metabolic reactions have been demonstrated for ABA in plants. One is the easily reversible formation of a glucose ester (abscisyl-β-D-glucopyranoside) and the other two are the formations of the oxidation products phaseic acid and dihydrophaseic acid (Fig. 5.7), both of which are essentially inactive derivatives, via 6′-hydroxymethyl ABA.

One of the most dramatic features about ABA metabolism is a large and rapid increase in the endogenous levels when leaves of terrestrial higher plants are exposed to any of a variety of conditions that induce stress, including mineral starvation, flooding, injury, and drought (Fig. 5.8). ABA increases plant resistance to such stresses. The increase in ABA begins within minutes after the

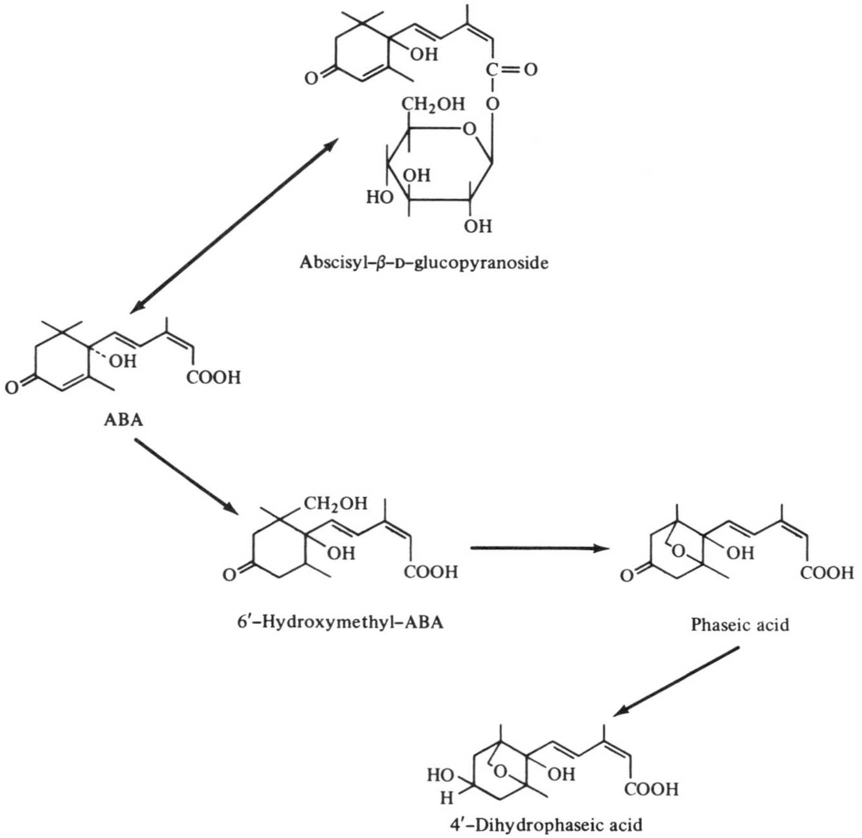

FIGURE 5.7. Metabolic reactions of (S)-ABA.

initiation of wilting, and may become 10-fold or greater than the concentration normally present in unwilted leaves. Furthermore, the high ABA levels decrease rapidly on watering of water-stressed plants (Fig. 5.9). Phaseic acid and dihydrophaseic acid also increase in the leaves of water-stressed terrestrial plants, and in some cases—e.g., bean (*Phaseolus vulgaris*) leaves—the concentrations of these two oxidation products are higher than that of ABA, although this evidently is not true in all cases, e.g., castor bean (*Ricinus communis*). Results of investigations of beans and some other species indicate that the elevation of ABA levels in wilted leaves is due to *de novo* synthesis, rather than release from a conjugate such as ABA glucose ester. The key metabolic changes that occur in water-stressed bean leaves are (1) water stress triggers an increased rate of oxidation of ABA to phaseic and dihydrophaseic acids; (2) ABA concentration increases and becomes nearly constant as the increased rates of synthesis and metabolism become approximately equal; and

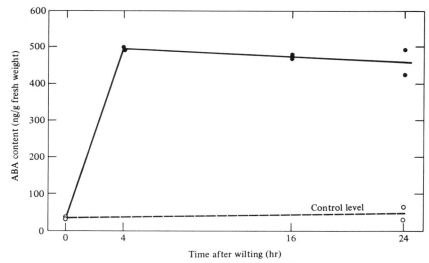

FIGURE 5.8. Increase in ABA content in wilting bean (*Phaseolus vulgaris*) leaves. Dashed line indicates level of ABA in control plants. (Redrawn, with permission, from Harrison and Walton, 1975.)

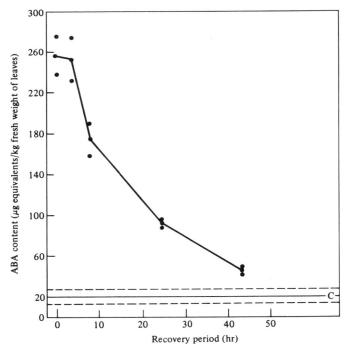

FIGURE 5.9. Decrease in ABA content of Brussels sprouts as they recover from wilt (44% leaf-water deficit). C = average ABA content of nonwilted plants. (Redrawn, with permission, from Wright and Hiron, 1972.)

(3) ABA concentration decreases as water potential increases and turgor is regained, due to decreased rates of ABA synthesis relative to metabolism (Harrison and Walton, 1975).

Milborrow (1974a) observed three patterns of regulation of ABA in different organs of avocado: (1) in the leaf there are relatively low normal levels of ABA (20 $\mu g/kg$) and on wilting there is an increase in biosynthesis until the overall concentration reaches approximately 2600 $\mu g/kg$; (2) in ripening fruit there are already high concentrations of ABA (4000–6000 $\mu g/kg$), but in spite of this there is rapid biosynthesis, as determined by incorporation of mevalonate, and the ABA content and biosynthesis of fruit slices are unaffected by wilting; (3) in roots there are low ABA concentrations (27 $\mu g/kg$), and there is very little effect of wilting. Walton et al. (1976) obtained different results with roots of bean (*Phaseolus vulgaris*). These authors observed that ABA accumulates in water-deficient roots and decreases the resistance of roots to water uptake, thus further protecting the leaves against desiccation.

Zeevaart (1977) investigated the sites of (*S*)-ABA synthesis and metabolism in castor bean (*Ricinus communis*) by analyzing the levels of ABA, phaseic acid, and dihydrophaseic acid in the shoot tips, mature leaves, and phloem sap of water-stressed and nonstressed plants. He found that water stress caused increases in the concentrations of ABA, phaseic acid, and dihydrophaseic acid in all three materials analyzed. Dihydrophaseic acid, but not ABA, accumulated in shoot tips, indicating that ABA is metabolized rapidly, that the shoot tips are effective sinks for ABA translocated in the phloem sap, and that ABA is metabolized relatively rapidly in shoot tips. When young and mature leaves were excised and allowed to wilt, the ABA increased in both, demonstrating that leaves have the capacity to synthesize ABA at an early stage of development.

Natural Occurrence of ABA

Originally ABA was found in the leaves of woody plants, *Acer pseudoplatanus* and *Betula pubescens,* and in cotton fruits. Rapidly it was found to occur also in other seed plants. Thus far, in fact, the hormone has been found in all angiosperms and gymnosperms that have been investigated, as well as in at least one representative each from among ferns, horsetails and mosses, where concentrations occur in the range of 1 nM to 1 μM. Some algae and liverworts reportedly contain a chemically related but different growth inhibitor called "lunularic acid" (Fig. 5.10), and so far it looks as if fungi and bacteria are devoid of ABA-like inhibitors.

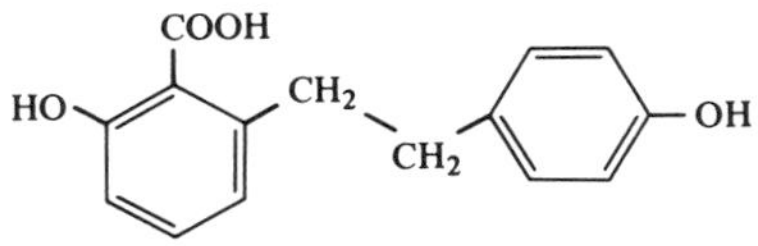

FIGURE 5.10. Structure of lunularic acid, an ABA-like substance found in some algae and liverworts.

Physiological Effects

Like all other plant hormones, ABA has multiple physiological effects, affecting growth and development literally throughout the ontogeny of seed plants. In this section, we shall be concerned with only six of the major physiological roles, namely in (1) stomatal regulation, (2) bud dormancy, (3) seed dormancy, (4) abscission, (5) regulation of synthesis of proteins in barley aleurone, and (6) seed maturation.

Stomatal Regulation

In recent years it has been discovered that ABA plays a major role in regulating closure of stomates, and thereby creating a protective mechanism toward water stress. The initial observation leading to this discovery was made as early as 1968 when it was noted that spray applications of exogenous ABA at low concentrations (e.g., 1 μM) could reduce transpiration, and by 1971 it was appreciated that this was due to closure of the stomates. Hence the possibility was raised that endogenous ABA might be involved in stomatal regulation and thus in the water relations of plants.

Onset of stomatal closure is quite rapid, occurring within approximately 3 to 9 minutes after the application of exogenous ABA to the cut bases of leaves of such plants as corn (*Zea mays*), *Rumex obtusifolia,* and sugarbeet (*Beta vulgaris*). Moreover, the concentration of endogenous ABA increases sharply in water-stressed leaves during the lag period (10 to 15 minutes) prior to stomatal closure. ABA causes the guard cells to become "leaky" with respect to K^+, the guard cells lose K^+ and turgor, and the stomates close.

Actually the total process of stomatal regulation involves the concentrations of both ABA and CO_2 in the guard cells, and the response of the guard cells to either substance depends on the concentration of the other. When, for example, water stress develops, ABA concentration in the guard cells increases, the guard cells lose K^+ and hence turgor, the stomates close—thus protecting the plant against desiccation—and CO_2 increases. Conversely, if the plant is watered and water stress subsides, ABA concentration in the guard cells decreases, K^+ and turgor are regained, CO_2 concentration decreases, and the stomates open—thus providing entry for CO_2 into the leaf and photosynthesis.

Bud Dormancy

The hypothesis that endogenous growth inhibitors may regulate bud dormancy was first put forth by T. Hemberg (1949a, b) who found that dormancy of buds of ash (*Fraxinus*) trees and potato tubers was correlated with a relatively high concentration of a growth inhibitor (or inhibitors) of unknown identity (Figs. 5.11. and 5.12). Kawase (1961) reported similar results for young birch (*Betula pubescens*) plants and confirmed further the effect of short days (10 hour) and long nights on inducing bud dormancy (Fig. 5.13).

Before discussing further the role of growth inhibitors in bud dormancy, it

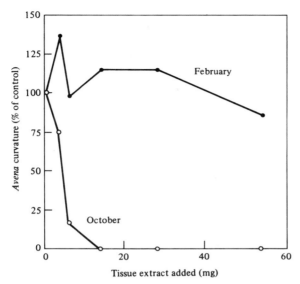

FIGURE 5.11. Growth inhibitor content of bud scales of dormant (October) and nondormant (February) buds of ash (*Fraxinus*). (Redrawn from data of Hemberg, 1949a, in: *Plant Growth and Development,* 2nd ed., by A. C. Leopold and P. E. Kriedemann. Copyright © 1975 by McGraw-Hill Book Company. Used with permission of McGraw-Hill Book Company.)

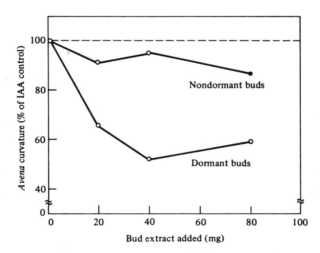

FIGURE 5.12. Growth inhibitor content of dormant and nondormant buds of potato tubers. Activity of extracts was measured as inhibition of auxin action in the *Avena* Coleoptile Curvature Test. (Redrawn from data of Hemberg, 1949b, in: *Plant Growth and Development* by A. C. Leopold. Copyright © 1964 by McGraw-Hill Book Company. Used with permission of McGraw-Hill Book Company.)

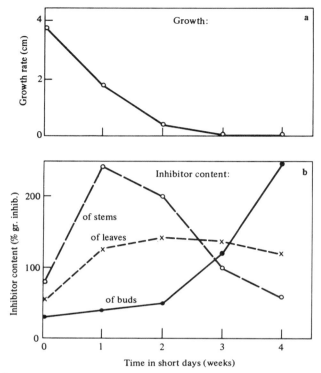

FIGURE 5.13. Induction of bud dormancy, **a**, by short days (10 hours) and long nights (14 hours) in birch (*Betula pubescens*), and correlated changes in growth inhibitor content, **b**, of leaves, stems, and buds. (Redrawn, with permission, from data of Kawase, 1961.)

will be well to consider some generalizations about the phenomenon. Most, indeed probably all, perennial plants form resting buds at some stage in the annual cycle of growth. In temperate-zone woody plants dormant buds are formed typically in late summer or autumn, and the whole tree or shrub enters a dormant phase. Some temperate-zone woody species [e.g., birch (*Betula pubescens*) and larch (*Larix decidua*)] can be maintained in continuous growth for 18 months or longer under long-day conditions in a warm greenhouse but others cannot, and eventually all temperate-zone species set dormant buds even when kept under physical environmental conditions that evidently are optimum for growth.

Resting buds also are formed by tropical species, but the dormancy behavior of these plants generally is very different from that of temperate-zone species. The latter are adapted to a physical environment in which critical environmental factors change predictably with the passing of the seasons. Both temperature and photoperiod change in repeating patterns, the latter with absolute precision at all points on Earth. In native tropical species, adapted to equatorial habitats where photoperiod changes very little (or not at all) and where adverse environmental conditions are more sporadic and typically the extremes are less severe, bud dormancy is under endogenous control (see Fig. 1.11). This behavior illustrates an important generalization about true dormancy, namely, that dormancy typically is not a response to adverse physical environmental conditions but rather an adaptive phenomenon that precedes and prepares plants for survival during exposure to adverse conditions. And, most interestingly, not all the shoots of a given tropical tree enter dormancy at the same time, so that both dormant and growing shoots occur on the tree at the same time. This behavior—endogenous regulation of dormancy—common in perennial species indigenous to tropical regions probably is the evolutionarily most primitive condition. The environmental conditioning and responses seen in species indigenous to higher and lower latitudes are believed to be evolved, more advanced types of behavior that were essential to the evolution and northward and southward migrations of species from centers of origin of their progenitors in tropical areas.

One of the most important factors affecting and controlling the induction of bud dormancy in temperate-zone woody plants is day length. In the majority of woody plants studied so far (angiosperms and gymnosperms), long days promote vegetative growth and short days bring about the cessation of extension growth and the onset of bud dormancy. This photoperiodic induction of bud dormancy shows obvious parallels with photoperiodic phenomena in herbaceous plants. Phytochrome is the photoreceptor, and red/far-red reversibility in the "night break" effect has been demonstrated for seedlings of larch. The leaves commonly are the photoreceptors, but bud scales are also in some species (e.g., *Betula pubescens*).

There are, however, serious questions concerning the role of photoperiodism in the onset of dormancy of many woody plants, for a number of common cultivated fruit trees (e.g., *Pyrus, Malus, Prunus*) and certain other species (e.g., *Fraxinus excelsior* and other species in the *Oleaceae*) show very little response to day length. They may go dormant in summer when days still are relatively long. However, a word of caution is advisable about this. Through the efforts of horticulturists, many species of fruit trees are grown in temperate-zone areas that are substantially different from the areas to which the species are indigenous. It should not surprise us therefore to observe that species such as *Malus* (apple) and *Pyrus* (pear) species of Mediterranean origin do not behave like native species when grown in temperate regions. Also, even in species in which young trees exhibit conspicuous photoperiodic responses in the development of bud dormancy, older trees may not. Young trees of poplar and larch, for example, continue primary shoot growth into September or October, and go dormant in response to increasing night length. But older trees cease growth much earlier in the season, often in June, July, or August, when photoperiods are still long. What then controls the onset of bud dormancy in such plants? Adverse environmental conditions may play a role. That is, in many cases, the trees actually pass from a state of "predormancy" or "quiescence" into true dormancy or rest. Thus, although short-day conditions are not required for the initial formation of buds, they may play an important role in the ultimate development of deep dormancy.

Emergence from (breaking of) dormancy of perennial plants of temperate regions generally is conditioned by low temperature. Most woody plants need to be exposed to a period of winter chilling to overcome dormancy (e.g., in general, 1–10°C for 260 to >1000 hours). Some species (e.g., *Acer pseudoplatanus* and *Betula pubescens*) can be induced to resume growth by transfer to long photoperiods, but this is not true for the majority of species of temperate-zone species. Although chilling is necessary to break dormancy of the buds, warm temperatures are necessary for the actual resumption of growth of the buds after the chilling period. Frequently the chilling requirements are met by January, but the buds fail to resume growth until later. They remain in a state of quiescence or postdormancy for awhile. Thus, the time of bud burst is normally determined by the return of warmer conditions.

Much of our present understanding of the hormonal regulation of bud dormancy stems from the extensive research by P. F. Wareing and his associates at the University College of Wales at Aberystwyth. They have used mainly *Betula pubescens* (birch) and *Acer pseudoplatanus* (sycamore maple). In both species dormancy can be induced by short photoperiods and long nights, and dormancy can be broken by long photoperiods and short nights (without cold treatment). They and others have shown that short-day conditions, which induce dormancy, result in increased levels of growth inhibitor in the leaves and buds of some species (Fig. 5.14). The inhibitory effect on the growth of buds exerted by leaves maintained under short-day conditions strongly suggests the translocation of inhibitor from the leaves to the buds. Furthermore, Wareing and his associates extracted inhibitor (putative ABA) from leaves of *Betula pubescens* plants grown under short-day conditions, applied the extract to birch seedlings [also sycamore maple and black currant (*Ribes nigrum*)] maintained under long days, and the result was that the seedlings ceased growth and set resting buds complete with bud scales just as if they were caused to go dormant by photoperiodic induction. Moreover, they showed, as have numerous other investigators using other species, that the application of GA caused breaking of bud dormancy (whether dormancy had been naturally photoinduced or induced by the application of exogenous ABA).

The results obtained by Wareing and others with several species, chiefly *Betula pubescens* and *Acer pseudoplatanus,* indicate that ABA is causally involved in the regulation of the cycle of growth and dormancy. Thus P. F. Wareing put forth the concept that the annual cycle of bud growth and dormancy is regulated by a *balance* between endogenous growth inhibitors and GAs. It seems probable that induction of dormancy in at least some cases is brought about by high ABA and low GA levels, whereas the converse is true for the emergence from dormancy. Investigations based on analyses of xylem sap (e.g., peach and willow) and buds [e.g., apple, beech (Fig. 5.14) and birch] have revealed correlations between comparatively high ABA levels and the dormant period in several species of woody plants. In some cases the level of ABA in the

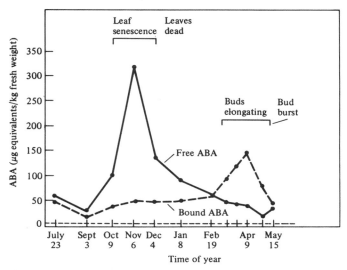

FIGURE 5.14. Seasonal changes in free and bound ABA content of beech (*Fagus sylvatica*) buds in England. (Redrawn, with permission, from Wright, 1975.)

xylem sap is 10 times higher during dormancy than during the growing period. Leaf buds do not open until well after ABA levels start to decrease. Moreover, it has been reported repeatedly that GA increases in many plant materials during chilling and long photoperiods. Thus it is very tempting to conclude that induction of bud dormancy by short days may be due mainly to high ABA levels, and that dormancy release may be determined primarily by increased GA levels.

But the conditions described above evidently do not apply in all cases; that is, even in all species of woody plants native to temperate regions. In some species—birch, sycamore maple, and red maple—it was reported that short-day treatment did not induce increases in endogenous ABA. In some other reported cases exogenous [^{14}C]ABA displayed little movement from the leaves to which it was applied, and direct applications of ABA to shoot tips stopped extension growth but did not cause normal set of terminal buds. In view of these results, some researchers have concluded that ABA may not actually play a causal role in bud dormancy. Obviously, although ABA and certainly GAs are strongly implicated, more research is needed to elucidate fully the hormonal regulation of bud dormancy in woody plants generally.

Seed Dormancy

Seeds of many species of angiosperms and gymnosperms will not germinate immediately after ripening even under conditions of moisture, temperature, and oxygen tension generally favorable for growth. Such dormancy, like dormancy of buds, is of obvious survival value to plants, since it tends to restrict germination to environmental circumstances suitable for seedling establishment and survival. Different seeds in one seed crop often vary in their degree of dormancy, causing germination of the crop to occur over an extended period of time and thus increasing the probability that at least part of them will germinate under conditions favorable for survival.

Seed dormancy results from particular conditions prevailing internally in the seed. Several causes of dormancy have been recognized, and seeds of numerous species exhibit two or more of them simultaneously. Among the many causes of seed dormancy that have been described are (1) impermeability of seed coats to water and gases, (2) immaturity of the embryo, (3) need for "after-ripening" in dry storage, (4) mechanical resistance of seed coats, (5) presence of inhibitors found either in the seed coats, dry accessory structures, or, in the case of seeds contained in fleshy fruits, in the tissues surrounding the seeds, (6) special requirement for light or its absence, and (7) requirement for chilling in the hydrated condition.

The most common cause of seed dormancy in woody plants native to temperate regions is a requirement for chilling of the hydrated seeds. Such seeds commonly germinate promptly and uniformly only after they have become hydrated and exposed to low temperature (0–10°C) for a period of a few to several weeks. When these conditions are supplied artificially, the practice is referred to as stratification.

This dormancy mechanism, by far the most common seed dormancy mechanism in seeds of tree species, has obvious important survival value to many plants, in that it protects against winterkilling of seedlings. However, only in the last two decades have we begun to understand the effect of stratification and the dormant state at the biochemical level. Attention will focus mainly on two case histories of such studies. E. Sondheimer et al. (1968) investigated the ABA levels in relation to germination in seeds of two species of ash (*Fraxinus*). They used seeds of *Fraxinus americana*, a temperate-zone species, and *Fraxinus ornus*, a species of Mediterranean origin whose seeds do not display a stratification requirement. The results (Table 5.1) were extremely

TABLE 5.1. (*S*)-Abscisic acid content of seeds and pericarps of *Fraxinus americana* and *Fraxinus ornus*.[a]

Biological material	Germination as $\Sigma 10$[b]	Total (*S*)-ABA present in sample	
		μmol/kg	μg/1000 seeds or pericarps
F. americana			
Dormant seed	0–30	1.7	11
Chilled seed	700–950	0.6	3.3
Pericarp from dormant samara	—	2.8	24
Pericarp from chilled samara	—	1.8	15
F. ornus			
Seed	750	0.4	2.6
Pericarp	—	0.8	1.6

[a] Data from Sondheimer et al. (1968), with permission.
[b] Germination data are the sums of the daily percentages of germination for a 10-day period from the time of imbibition. Thus $\Sigma 10 = 1000$ would mean 100% germination during the first day.

interesting. ABA was present in all tissues of both species, but the highest concentration was found in the seed and pericarp of dormant seeds of *Fraxinus americana*. During chilling (stratification) of seeds of *Fraxinus americana,* the ABA levels decreased 37% in the pericarp and 68% in the seed. And, of particular note, the ABA concentration in the seeds of *Fraxinus americana,* after chilling, was as low as the normally low level found in seeds of *Fraxinus ornus*. In the case of *Fraxinus americana* it seems clear that ABA in the seed has a regulatory role in germination, and that the most consequential result of stratification is to cause a decrease in endogenous ABA in the seed. Sondheimer et al., incidentally, concluded that it is unlikely that the ABA in the pericarp of *Fraxinus americana* seeds functions significantly in the regulation of dormancy.

Investigations by Sondheimer et al. showed further that other hormones besides ABA probably are involved in the regulation of the process of seed germination and seedling growth in *Fraxinus americana*. They noted, for example, that exogenous GA alone was effective in nullifying the effect of exogenous ABA on germination; however, leaf development and chlorophyll synthesis were less than normal. Exogenous kinetin, in combination with GA, promoted normal-appearing leaf development and chlorophyll synthesis. Kinetin by itself would not counteract the effect of ABA on germination. Exogenous ABA does, of course, prevent germination in nondormant seeds of both *Fraxinus americana* and *Fraxinus ornus*.

Another species with a stratification requirement that has been investigated extensively is *Corylus avellana* (known as hazelnut in Europe and the eastern United States and as filbert in the western United States). Exogenous GA is quite effective in causing nonstratified dormant seeds to germinate (Fig. 5.15), as is true generally for seeds with stratification requirements. In this case, chilling evoked an increase in endogenous GA, which occurred rapidly after stratified seeds were returned to a more normal germination temperature (Fig. 5.16). Moreover, Amo-1618 (an inhibitor of GA biosynthesis) counteracted the effect of chilling. An investigation of dormant *Corylus avellana* seeds by Bradbeer (1968) showed that while the dormant seeds contain inhibitor, ABA, if present at all, accounts for less than 1% of the total inhibitory activity. Hence, in the case of *Corylus avellana* it has been concluded that the breaking of dormancy by stratification may be attributed to the synthesis of GA.

Again, in the case of seeds having chilling requirements, as in the case of bud dormancy, the concept of balance between growth-promoting and growth-inhibiting hormones arises. In seeds evidently the balance can be altered in different ways during or immediately subsequent to stratification, either by a decrease of inhibitor (as in *Fraxinus americana*), increase in promoter (as in *Corylus avellana*), and perhaps by both such changes in other species.

Abscission

The role of ABA as a natural regulator of abscission of organs—leaves, fruits, bud scales, floral structures, branches, etc.—is not yet clear, although the hormone was discovered in part because of its capacity to cause abscission of

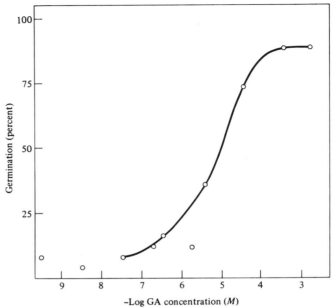

FIGURE 5.15. Effect of GA on germination of dormant filbert (*Corylus avellana*) seeds in the absence of stratification. (Redrawn, with permission, from Bradbeer and Pinfield, 1967.)

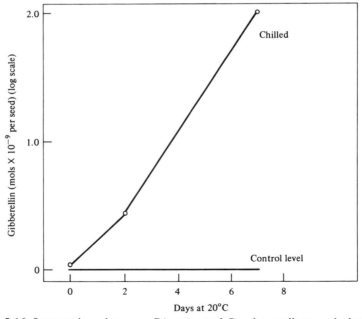

FIGURE 5.16. Increase in endogenous GA content of *Corylus avellana* seeds during the week following stratification for 28 days at 5°C. (Drawn from data of Ross and Bradbeer, 1968.)

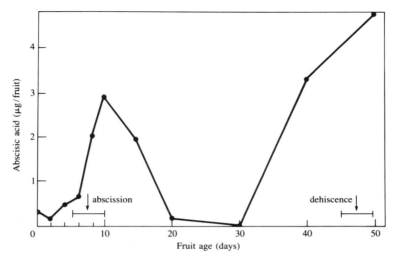

FIGURE 5.17. Abscisic acid in cotton fruit during the course of development. Note the correlation of high ABA content with young-fruit abscission and with fruit dehiscence. (Redrawn in slightly modified form, by permission, from Davis and Addicott, 1972.)

young cotton (*Gossypium hirsutum*) fruits. L. A. Davis and F. T. Addicott (1972) measured ABA in cotton fruits by means of gas–liquid chromatography. They found that high levels of ABA occurred in direct correlation with abortion and abscission of young fruit, and with senescence and dehiscence of mature fruit (Fig. 5.17). Young fruit that abscised late in the fruiting season contained twice as much ABA as young fruit that abscised early in the season. Hence, at least in cotton fruits, ABA seems to be very much involved in the hormonal control of abscission. Addicott (1982) listed numerous other cases where a hormonal role of ABA as an abscission accelerator was seen, and many other cases in which exogenous ABA caused abscission. However, there are investigations as well in which a correlation between ABA and abscission was not observed. Hence our knowledge on this subject must still be considered incomplete.

Undoubtedly, Addicott (1982) was correct when he stated that we cannot think of a single hormone as being the controlling factor in abscission, but that "we must think in terms of a balance of hormonal influences reflecting the physiological status on each side of the abscission zone." There evidently is at work in the control of abscission a complex balance among auxin, GAs, cytokinins, ethylene and ABA, with auxin being the principal natural retardant and ethylene and ABA being the principal accelerators. IAA seems to have an overriding capacity to inhibit abscission in the face of promotive factors. There is increasing evidence of the involvement of ABA, once levels of IAA have decreased sufficiently to permit it to act. Ethylene can promote abscission in many situations, but in some cases abscission can proceed in the virtual absence of ethylene (Addicott, 1982).

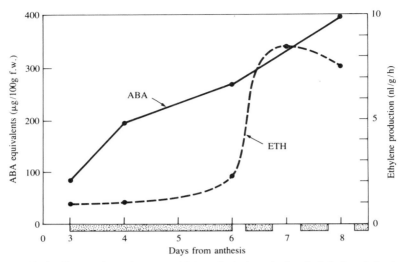

FIGURE 5.18. Changes in endogenous ABA and ethylene during dark-induced abscission of young cotton fruit. Treatment was started 3 days after anthesis. Dark treatment and subsequent light periods are indicated by the bars at the base of the figure. The average time to abscission was 5 days from the start of the dark treatment. (Based on data of Vaughan and Bate, 1977; redrawn, with permission, from Addicott, 1982.)

Ethylene is a very potent accelerator of abscission; however, physiologists are still divided about its role in natural abscission. Several good correlations have been reported between ethylene levels and abscission. One interesting study, which involved both ABA and ethylene, was conducted by Vaughan and Bate (1977). They induced abscission of young cotton fruits by placing the plants in complete darkness for 3 days. Plants were returned to the greenhouse, and abscission occurred about 2 days later. ABA concentrations in the young fruit doubled during the first day in darkness, tripled by the end of the third day, and increased to four times the initial concentration at the time of abscission on the fifth day (Fig. 5.18). Ethylene showed little change during the first 3 days, but increased sharply during the 2 days before abscission. ABA appeared to be correlated with the initiation of abscission, while ethylene was correlated with the terminal, separatory activity. There is further discussion of the role of ethylene in abscission in Chapter 6.

Regulation of Synthesis of Proteins in Barley Aleurone

As has been pointed out previously, GA_3 and ABA antagonize each other in regulating the synthesis and secretion of α-amylase in barley aleurone layers. When ABA is added to aleurone layers at the same time as GA_3, the expression of the GA_3-induced α-amylase is inhibited at the transcriptional level. However, if ABA is added 16 hours later than GA_3, when α-amylase mRNA reaches the peak level, the effect of ABA on α-amylase expression is mainly on the stability of mRNA. A metabolite of ABA, phaseic acid, can mimic the effect of ABA on the stability of α-amylase mRNA, and there is evidence that, indeed, phaseic acid is likely the active component in the ABA action. Besides the repression of α-amylase, Ho et al. have found very recently that ABA induces the synthesis of at least 16 polypeptides including an α-amylase inhibitor (21 kDa), a lectin-like protein (35 kDa), and two acid-soluble basic proteins (27 and 29 kDa). The cDNA of the 27-kDa, ABA-induced proteins has been cloned and characterized. More is expected to be learned about the potential role of this protein in the action of ABA.

Thus, like GA, ABA *suppresses* the synthesis of certain proteins in barley aleurone tissue and *induces* the synthesis of other proteins. Indeed, the addition of ABA to barley aleurone layers induces the expression of several proteins. Ho and associates have isolated ABA-inducible cDNA clones by differential screening of a λgt 10 cDNA library with mRNA isolated from tissue treated with or without ABA. Northern analysis showed that one of the cDNAs hybridizes to a 1.2-kb mRNA in ABA-treated layers. The induction of this mRNA appeared as early as 40 minutes after the addition of ABA, reached a maximum at 4 to 8 hours, and was present for greater than 48 hours. An ABA concentration as low as 10^{-9} M induced this mRNA; higher concentrations (up to 10^{-4} M) induced higher levels of the mRNA. Hybrid-select translation followed by two-dimensional electrophoresis indicated that the mRNA encoded a basic 27-kDa polypeptide. The significance of this and other ABA-induced mRNAs is under investigation by Ho et al.

Seed Maturation

A very interesting story has developed regarding the role of ABA in the maturation of at least some kinds of seeds. Embryo development ceases during the maturation of seeds of most angiosperms. During this phase, the seeds become desiccated and enter a period of arrested development that serves to prevent them from germinating before environmental conditions are favorable for seedling establishment. Precocious germination of embryos during their development in seeds would risk desiccation and death. Several lines of evidence suggest that ABA plays an important role in suppression of precocious germination. Exogenous ABA can block both visible germination and the appearance of germination-specific enzymes in cultured embryos. Moreover, analyses of ABA have revealed direct correlations between seed dormancy and high levels of endogenous ABA. Viviparous mutants exhibit a decrease in ABA levels or ABA sensitivity. Vivipary can also be induced by treatments that either inhibit ABA synthesis or deplete endogenous pools of ABA (Finklestein et al., 1985).

Studies chiefly with cereals have shown that there appear to be both embryo-specific and germination-specific gene sets the expression of which is regulated by ABA. Steady-state levels of embryo-specific messages present in mature embryos decrease rapidly soon after the onset of normal germination, while those characteristic of the germination pathway increase. Exogenously applied ABA blocks this transition, and both the synthesis of germination-specific messages and germination are inhibited (Williamson and Quatrano, 1988). Suppression of the germination-specific gene set during normal embryogenesis appears to be due, at least in part, to the presence of high levels of ABA present in the developing seed. With the onset of desiccation, when ABA levels begin to decrease in the seed, a general decrease in gene expression due to the lack of water probably also prevents expression of the germination genes.

There are then two important actions of ABA in cereal grains. One is the suppression of the germination-specific gene set during normal embryogenesis. The other is the enhancement of the synthesis of certain mRNAs coded for by both embryogenesis-specific and germination-specific gene sets.

State of Chromatin in Dormant Tissues and the Mechanism of Action of ABA

Dormant tissue is alive, it respires, and in some cases it grows very slowly. Yet clearly a dormant tissue or organ is idle; it does not grow normally, and obviously there is some mechanism which restricts growth.

Dorothy Tuan and James Bonner (1964) advanced the hypothesis "that in the dormant cell the genetic material (the DNA) is completely, or nearly completely, repressed." According to this concept, of course, the processes of DNA transcription, the production of messenger RNA, and the production of enzymes and structural proteins essential for growth and metabolism, all are repressed. Tuan and Bonner conducted investigations to test the hypothesis. Specifically they compared (1) the rates of RNA synthesis by dormant and nondormant buds of potato tubers; and (2) the capacities of the chromatin of such buds to support DNA-dependent RNA synthesis.

The results were highly revealing. Dormant buds possessed only a very limited capacity for both DNA and DNA-dependent RNA synthesis, as did the isolated chromatin from dormant buds, compared to buds in which dormancy was broken by treatment of the tubers with ethylene chlorohydrin. It was, therefore, concluded that the DNA of the buds of dormant potato tubers is largely in a repressed state, and that breaking of dormancy is accompanied by depression of the genetic material.

Similar results were reported from investigations of dormant and nondormant seeds of *Corylus avellana* by Jarvis et al. (1968). They confirmed that dormant seeds require stratification and that exogenous GA can replace chilling. Using intact embryos and isolated chromatin, they found that GA first increased the amount of DNA available for transcription and secondarily caused an increase in RNA synthesis.

The molecular mechanism of action of ABA presently is unknown. However, there are major biochemical effects of the hormone that are well documented. One is an alteration of plasma membranes, which is manifested in a change in bioelectric potential across the membranes and a leakiness and efflux of K^+, which is involved in some actions of ABA, e.g., stomatal regulation that was discussed previously. Another major effect is inhibition of RNA and protein synthesis. ABA counteracts many of the effects of GA, and certain effects also of auxins and cytokinins. For example, ABA counteracts the effect of GA in the induction of hydrolases in barley aleurone layers (Fig. 3.33), and over a wide range of concentrations ABA apparently behaves as a competitive inhibitor of GA. But recently it has been reported that, in fact, GA and ABA each can induce the synthesis of certain mRNAs and proteins and inhibit the synthesis of others.

References

Ackerson, R. C. 1980. Stomatal response of cotton to water stress and abscisic acid as affected by water stress history. *Plant Physiol.* **65**: 455–459.

Addicott, F. T. 1970. Plant hormones in the control of abscission. *Biol. Rev.* **45**: 485–524.

Addicott, F. T. 1982. *Abscission.* University of California Press, Berkeley, Los Angeles, London.

Addicott, F. T. and J. L. Lyon. 1969. Physiology of abscisic acid and related substances. *Annu. Rev. Plant Physiol.* **20**: 139–164.

Addicott, F. T., J. L. Lyon, K. Ohkuma, W. E. Thiessen, H. R. Carns, O. E. Smith, J. W. Cornforth, B. V. Milborrow, G. Ryback, and P. F. Wareing. 1968. Abscisic acid: A new name for abscisin II (dormin). *Science* **159**: 1493.

Alvim, R., E. W. Hewett, and P. F. Saunders. 1976. Seasonal variation in the hormone content of willow. I. Changes in abscisic acid content and cytokinin activity in the xylem sap. *Plant Physiol.* **57**: 474–476.

Alvim, R., S. Thomas, and P. F. Saunders. 1978. Seasonal variation in the hormone content of willow. II. Effect of photoperiod on growth and abscisic acid content of trees under field conditions. *Plant Physiol.* **62**: 779–780.

Audus, L. J. 1975. Geotropism in roots. In: Torrey, J. G. and D. T. Clarkson, eds. *The Development and Function of Roots.* Academic Press, New York. Pp. 327–363.

Boyer, G. L. and J. A. D. Zeevaart. 1982. Isolation and quantitation of β-D-glucopyranosyl abscisate from leaves of *Xanthium* and spinach. *Plant Physiol.* **70**: 227–231.

Bradbeer, J. W. 1968. Studies in seed dormancy. IV. The role of endogenous inhibitors and gibberellin in the dormancy and germination of *Corylus avellana* L. seeds. *Planta* **78**: 266–276.

Bradbeer, J. W. and N. J. Pinfield. 1967. Studies in seed dormancy. III. The effects of gibberellin on dormant seeds of *Corylus avellana* L. *New Phytol.* **66**: 515–523.

Bray, E. A. and J. A. D. Zeevaart. 1985. The compartmentation of abscisic acid and β-D-glucopyranosyl abscisate in mesophyll cells. *Plant Physiol.* **79**: 719–722.

Chen, T. H. H. and L. V. Gusta. 1983. Abscisic acid-induced freezing resistance in cultured plant cells. *Plant Physiol.* **73**: 71–75.

Chrispeels, M. J. and J. E. Varner. 1966. Inhibition of gibberellic acid induced formation of α-amylase by abscisin II. *Nature (London)* **212**: 1066–1067.

Chrispeels, M. J. and J. E. Varner. 1967. Hormonal control of enzyme synthesis: on the mode of action of gibberellic acid and abscisin in aleurone layers of barley. *Plant Physiol.* **42**: 1008–1016.

Colquhoun, A. J. and J. R. Hillman. 1975. Endogenous abscisic acid and the senescence of leaves of *Phaseolus vulgaris* L. *Z. Pflanzenphysiol.* **76**: 326–332.

Cornforth, J. W., B. V. Milborrow, and G. Ryback. 1965. Synthesis of (±)-abscisin II. *Nature (London)* **206**: 715.

Cornforth, J. W., B. V. Milborrow, and G. Ryback. 1966. Identification and estimation of (+)-abscisin II ('dormin') in plant extracts by spectropolarimetry. *Nature (London)* **210**: 627–628.

Cornforth, J. W., B. V. Milborrow, G. Ryback, and P. F. Wareing. 1965. Identity of sycamore 'dormin' with abscisin II. *Nature (London)* **205**: 1269–1270.

Cornish, K. and J. A. D. Zeevaart. 1984. Abscisic acid metabolism in relation to water stress and leaf age in *Xanthium strumarium*. *Plant Physiol.* **76**: 1029–1035.

Creelman, R. A., D. A. Gage, J. T. Stults, and J. A. D. Zeevaart. 1987. Abscisic acid biosynthesis in leaves and roots of *Xanthium strumarium*. *Plant Physiol.* **85**: 726–732.

Creelman, R. A. and J. A. D. Zeevaart. 1984. Incorporation of oxygen into abscisic acid and phaseic acid from molecular oxygen. *Plant Physiol.* **75**: 166–169.

Daie, J. and W. F. Campbell. 1981. Response of tomato plants to stressful temperatures. Increase in abscisic acid concentrations. *Plant Physiol.* **67**: 26–29.

Dathe, W., H. Rönsch, A. Preiss, W. Schade, G. Sembdner, and K. Schreiber. 1981. Endogenous plant hormones of the broad bean, *Vicia faba* L. (−)-Jasmonic acid, a plant growth inhibitor in pericarp. *Planta* **153**: 530–535.

Davis, L. A. and F. T. Addicott. 1972. Abscisic acid: correlations with abscission and with development in the cotton fruit. *Plant Physiol.* **49**: 644–648.

Dennis, F. G., Jr. and L. J. Edgerton. 1961. The relationship between an inhibitor and rest in peach flower buds. *Proc. Am. Soc. Hort. Sci.* **77**: 107–116.

Downs, R. J. and H. A. Borthwick. 1956. Effects of photoperiod on growth of trees. *Bot. Gaz.* **117**: 310–326.

Eagles, C. F. and P. F. Wareing. 1964. The role of growth substances in the regulation of bud dormancy. *Physiol. Plantarum* **17**: 697–709.

Eagles, C. F. and P. F. Wareing. 1964. The role of growth substances in the regulation of bud dormancy. *Physiol, Plantarum* **17**: 697–709.

El-Antably, H. M. M. 1975. Redistribution of endogenous indoleacetic acid, abscisic acid and gibberellins in geotropically stimulated *Ribes nigrum* shoots. *Physiol. Plantarum* **34**: 167–170.

El-Antably, H. M. M., P. F. Wareing, and J. Hillman. 1967. Some physiological responses to d,l abscisin (dormin). *Planta* **73**: 74–90.

Evenari, M. 1949. Germination inhibitors. *Bot. Rev.* **15**: 153–194.

Evenari, M. 1957. The physiological action and biological importance of germination inhibitors. *Symp. Soc. Exp. Biol.* **11**: 21–43.

Finkelstein, R. R., K. M. Tenbarge, J. E. Shumway, and M. L. Crouch. 1985. Role of ABA in maturation of rapeseed embryos. *Plant Physiol.* **78**: 630–636.

Firn, R. D., R. S. Burden, and H. F. Taylor. 1972. The detection and estimation of the growth inhibitor xanthoxin in plants. *Planta* **102**: 115–126.

Firn, R. D. and J. Friend. 1972. Enzymatic production of the plant growth inhibitor, xanthoxin. *Planta* **103**: 263–266.

Frankland, B. and P. F. Wareing. 1962. Changes in endogenous gibberellins in relation to chilling of dormant seeds. *Nature (London)* **194**: 313–314.

Gepstein, S. and K. V. Thimann. 1980. Changes in the abscisic acid content of oat leaves during senescence. *Proc. Natl. Acad. Sci. U.S.A.* **77**: 2050–2053.

Gianfagna, T. A. and S. Rachmiel. 1986. Changes in gibberellin-like substances of peach seed during stratification. *Physiol. Plantarum* **66**: 154–158.

Grantz, D. A., T.-H. D. Ho, S. J. Uknes, J. M. Cheeseman, and J. S. Boyer. 1985. Metabolism of abscisic acid in guard cells of *Vicia faba* L. and *Commelina communis* L. *Plant Physiol.* **78**: 51–56.

Greathouse, D. C., W. M. Laetsch, and B. O. Phinney. 1971. The shoot-growth rhythm of a tropical tree, *Theobroma cacao. Am. J. Bot.* **58**: 281–286.

Guinn, G. and D. L. Brummett. 1987. Concentrations of abscisic acid and indoleacetic acid in cotton fruits and their abscission zones in relation to fruit retention. *Plant Physiol.* **83**: 199–202.

Harrison, M. A. and P. F. Saunders. 1975. The abscisic acid content of dormant birch buds. *Planta* **123**: 291–298.

Harrison, M. A. and D. C. Walton. 1975. Abscisic acid metabolism in water-stressed bean leaves. *Plant Physiol.* **56**: 250–254.

Hartung, W. and H. Gimmler. 1982. The compartmentation of abscisic acid (ABA), of ABA-biosynthesis, ABA-metabolism and ABA-conjugation. In: Wareing, P. F., ed. *Plant Growth Substances 1982.* Academic Press, London. Pp. 324–333.

Hartung, W., B. Heilmann, and H. Gimmler. 1981. Do chloroplasts play a role in abscisic acid synthesis? *Plant Sci. Lett.* **2**: 235–242.

Heide, O. M. 1974. Growth and dormancy in Norway spruce ecotypes (*Picea abies*). I. Interaction of photoperiod and temperature. *Physiol. Plantarum* **30**: 1–12.

Hein, M. B., M. L. Brenner, and W. A. Brun. 1984. Concentrations of abscisic acid and indole-3-acetic acid in soybean seeds during development. *Plant Physiol.* **76**: 951–954.

Hemberg, T. 1949a. Growth inhibiting substances in buds of Fraxinus. *Physiol. Plantarum* **2**: 37–44.

Hemberg, T. 1949b. Significance of growth inhibiting substances and auxins for the rest-period of the potato tuber. *Physiol. Plantarum* **2**: 24–36.

Hemberg, T. 1958. The significance of the inhibitor β complex in the rest period of the potato tuber. *Physiol. Plantarum* **11**: 615–626.

Hemberg, T. 1961. Biogenous inhibitors. *Handbuch Pflanzenphysiol.* **14**: 1162–1183.

Hendershott, C. H. and L. F. Bailey. 1955. Growth inhibiting substances in extracts of dormant flower buds of peach. *Proc. Am. Soc. Hort. Sci.* **65**: 85–92.

Hendershott, C. H. and D. R. Walker. 1959. Identification of a growth inhibitor from extracts of dormant peach flower buds. *Science* **130**: 798–799.

Ho, T.-H. D. 1979. On the mode of action of abscisic acid in barley aleurone cells. *Plant Physiol.* (Suppl.) **63**: 79.

Hocking, T. J. and J. R. Hillman. 1975. Studies on the role of abscisic acid in the initiation of bud dormancy in *Alnus glutinosa* and *Betula pubescens. Planta* **125**: 235–242.

Ingersoll, R. B. and O. E. Smith. 1970. Movement of (RS)-abscisic acid in the cotton explant. *Plant Physiol.* **45**: 576–578.

Jacobsen, J. V., T. J. V. Higgins, and J. A. Zwar. 1979. Hormonal control of endosperm function during germination. In: Rubenstein, J., ed. *The Plant Seed—Development, Preservation and Germination.* Academic Press, New York. Pp. 241–262.

Jarvis, B. C., B. Frankland, and J. H. Cherry. 1968. Increased DNA template and RNA polymerase associated with the breaking of seed dormancy. *Plant Physiol.* **43**: 1734–1736.

Juniper, B. E. 1976. Geotropism. *Annu. Rev. Plant Physiol.* **27**: 385–406.

Kahn, A. A. 1968. Inhibition of gibberellic acid-induced germination by abscisic acid and reversal by cytokinins. *Plant Physiol.* **43**: 1463–1465.

Kawase, M. 1961. Growth substances related to dormancy in *Betula. Proc. Am. Soc. Hort. Sci.* **78**: 532–544.

Ketring, D. L. 1973. Germination inhibitors. *Seed Sci. Tech.* **1**: 305–324.

Kriedemann, P. E., B. R. Loveys, G. L. Fuller, and A. C. Leopold. 1972. Abscisic acid and stomatal regulation. *Plant Physiol.* **49**: 842–847.

Lenton, J. R., V. M. Perry, and P. F. Saunders. 1972. Endogenous abscisic acid in relation to photoperiodically induced bud dormancy. *Planta* **106**: 13–22.

Leopold, A. C. 1964. *Plant Growth and Development.* McGraw-Hill Book Company. New York.

Leopold, A. C. and P. E. Kriedemann. 1975. *Plant Growth and Development.* 2nd ed. McGraw-Hill Book Company, New York.

Lipe, W. N. and J. C. Crane. 1966. Dormancy regulation in peach seeds. *Science* **153**: 541–542.

Mayer, A. M. and Y. Shain. 1974. Control of seed germination. *Annu. Rev. Plant Physiol.* **25**: 167–193.

Milborrow, B. V. 1967. The identification of (+)-abscisin II [(+) dormin] in plants and measurement of its concentrations. *Planta* **76**: 93–113.

Milborrow, B. V. 1974a. Biosynthesis of abscisic acid and its regulation. In: *Plant Growth Substances 1973.* Hirokawa Publishing Company, Tokyo. Pp. 384–395.

Milborrow, B. V. 1974b. Chemistry and biochemistry of abscisic acid. In: Runeckles, V. C., E. Sondheimer, and D. C. Walton, eds. *The Chemistry and Biochemistry of Plant Hormones. Vol. 7. Recent Advances in Phytochemistry.* Academic Press, York. Pp. 57–91.

Milborrow, B. V. 1974c. The chemistry and physiology of abscisic acid. *Annu. Rev. Plant Physiol.* **25**: 259–307.

Milborrow, B. V. 1974d. Biosynthesis of abscisic acid by a cell-free system. *Phytochemistry* **13**: 131–136.

Norman, S. M., R. D. Bennett, S. M. Poling, V. P. Maier, and M. D. Nelson. 1986. Paclobutrazol inhibits abscisic acid biosynthesis in *Cercospora rosicola. Plant Physiol.* **80**: 122–125.

Norman, S. M., S. M. Poling, V. P. Maier, and E. D. Orme. 1983. Inhibition of abscisic acid biosynthesis in *Cercospora rosicola* by inhibitors of gibberellin biosynthesis and plant growth retardants. *Plant Physiol.* **71**: 15–18.

Ohkuma, K. 1965. Synthesis of some analogs of abscisin II. *Agr. Biol. Chem.* **29**: 962–964.

Ohkuma, K., F. T. Addicott, O. E. Smith, and W. E. Thiessen. 1965. The structure of abscisin II. *Terahedron Lett.* **29**: 2529–2535.

Ohkuma, K., J. L. Lyon, F. T. Addicott, and O. E. Smith. 1963. Abscisin II, an abscission-accelerating substance from young cotton fruit. *Science* **142**: 1592–1593.

Pearson, J. A. and P. F. Wareing. 1969. Effect of abscisic acid on activity of chromatin. *Nature (London)* **221**: 672–673.

Perry, T. O. and H. Hellmers. 1973. Effects of abscisic acid on growth and dormancy of two races of red maple. *Bot. Gaz.* **134**: 283–289.

Phillips, I. D. J. and P. F. Wareing. 1958a. Effect of photoperiodic conditions on the level of growth inhibitors in *Acer pseudoplatanus. Naturwissenschaften* **45**: 317.

Phillips, I. D. J. and P.F. Wareing. 1958b. Studies in dormancy of sycamore. I. Seasonal changes in the growth-substance content of the shoot. *J. Exp. Bot.* **9**: 350–364.

Phillips, I. D. J. and P. F. Wareing. 1959. Studies in dormancy of sycamore. II. The effect of daylength on the natural growth-inhibitor content of the shoot. *J. Exp. Bot.* **10**: 504–514.

Pilet, P. E. 1977. Growth inhibitors in growing and geostimulated maize roots. In: Pilet, P. E., ed. *Plant Growth Regulation.* Springer-Verlag, New York. Pp. 115–128.

Powell, L. E. 1987. The hormonal control of bud and seed dormancy in woody plants. In: Davies, P. J., ed. *Plant Hormones and Their Role in Plant Growth and Development.* Martinus Nijhoff Publishers, Dordrecht, The Netherlands. Pp. 539–552.

Pryce, R. J. 1972. The occurrence of lunularic and abscisic acids in plants. *Phytochemistry* **11**: 1759–1761.

Quatrano, R. S. 1987. The role of hormones during seed development. In: Davies, P. J., ed. *Plant Hormones and Their Role in Plant Growth and Development.* Martinus Nijhoff Publishers, Dordrecht, The Netherlands. Pp. 494–514.

Robinson, D. R. and G. Ryback. 1969. Incorporation of tritium from [(4R)-4-^{3}H]mevalonate into abscisic acid. *Biochem. J.* **113**: 895–897.

Robinson. P. M. and P. F. Wareing. 1964. Chemical nature and biological properties of the inhibitor varying with photoperiod in sycamore (*Acer pseudoplatanus*). *Physiol. Plantarum* **17**: 314–323.

Robinson, P. M., P. F. Wareing, and T. H. Thomas. 1963. Isolation of the inhibitor varying with photoperiod in *Acer pseudoplatanus*. *Nature (London)* **199**: 875–876.

Ross, J. D. and J. W. Bradbeer. 1968. Concentrations of gibberellin in chilled hazel seeds. *Nature (London)* **220**: 85–86.

Samet, J. S. and T. R. Sinclair. 1980. Leaf senescence and abscisic acid in leaves of field-grown soybean. *Plant Physiol.* **66**: 1164–1168.

Samish, R. M. 1954. Dormancy in woody plants. *Annu. Rev. Plant Physiol.* **5**: 183–204.

Schrempf, M., R. L. Satter, and A. W. Galston. 1976. Potassium-linked chloride fluxes during rhythmic leaf movement of *Albizzia julibrissin*. *Plant Physiol.* **58**: 190–192.

Sibaoka, T. 1969. Physiology of rapid movements in higher plants. *Annu. Rev. Plant Physiol.* **20**: 165–184.

Sindhu, R. K. and D. C. Walton. 1987. Conversion of xanthoxin to abscisic acid by cell-free preparations from bean leaves. *Plant Physiol.* **85**: 916–921.

Sondheimer, E. and E. C. Galson. 1966. Effects of abscisin II and other plant growth substances on germination of seeds with stratification requirements. *Plant Physiol.* **41**: 1397–1398.

Sondheimer, E., D. S. Tzou, and E. C. Galson. 1968. Abscisic acid levels and seed dormancy. *Plant Physiol.* **43**: 1443–1447.

Sondheimer, E. and D. C. Walton. 1970. Structure-activity correlations with compounds related to abscisic acid. *Plant Physiol.* **45**: 244–248.

Steadman, J. R. and L. Sequeira. 1970. Abscisic acid in tobacco plants. Tentative identification and its relation to stunting induced by *Pseudomonas solanacearum*. *Plant Physiol.* **45**: 691–697.

Takahashi, N., ed. 1986. *Chemistry of Plant Hormones.* CRC Press, Boca Raton, Florida.

Taylor, H. F. and T. A. Smith. 1967. Production of plant growth inhibitors from xanthophylls: a possible source of dormin. *Nature (London)* **215**: 1513–1514.

Thomas, T. H., P. F. Wareing, and P. M. Robinson. 1965. Action of sycamore dormin as a gibberellin antagonist. *Nature (London)* **205**: 1270–1272.

Tuan, D. Y. H. and J. Bonner. 1964. Dormancy associated with repression of genetic activity. *Plant Physiol.* **39**: 768–772.

Ueda, J. and J. Kato. 1980. Isolation and identification of a senescence-promoting substance from wormwood (*Artemisia absinthium* L.). *Plant Physiol.* **66**: 246–249.

Ueda, J. and J. Kato. 1981. Promotive effect of methyl jasmonate on oat leaf senescence in the light. *Z. Pflanzenphysiol.* **103**: 357–359.

Varner, J. E. and D. T. Ho. 1976. Hormones. In: Bonner, J. and J. E. Varner, eds. *Plant Biochemistry.* 3rd ed. Academic Press, New York. Pp. 713–770.

Vaughan, A. K. F. and G. C. Bate. 1977. Changes in the levels of ethylene, abscisic-acid-like substances and total non-structural carbohydrate in young cotton bolls in relation to abscission induced by a dark period. *Rhod. J. Agric. Res.* **15**: 51–63.

Vegis, A. 1964. Dormancy in higher plants. *Annu. Rev. Plant Physiol.* **15**: 185–224.

Vick, B. A. and D. C. Zimmerman. 1984. Biosynthesis of jasmonic acid by several plant species. *Plant Physiol.* **75**: 458–461.

Walton, D. C. 1980. Biochemistry and physiology of abscisic acid. *Annu. Rev. Plant Physiol.* **31**: 453–489.

Walton, D. C. 1987. Abscisic acid biosynthesis and metabolism. In: Davies, P. J., ed. *Plant Hormones and Their Role in Plant Growth and Development.* Martinus Nijhoff Publishers, Dordrecht, The Netherlands. Pp. 113–131.

Walton, D. C., M. A. Harrison, and P. Coté. 1976. The effects of water stress on abscisic-acid levels and metabolism in roots of *Phaseolus vulgaris* L. and other plants. *Planta* **131**: 141–144.

Walton, D. C. and E. Sondheimer. 1972. Metabolism of 2-^{14}C($\pm$)-abscisic acid in excised bean axes. *Plant Physiol.* **49**: 285–289.

Walton, D. C., G. S. Soofi, and E. Sondheimer. 1970. The effects of abscisic acid on growth and nucleic acid synthesis in excised embryonic bean axes. *Plant Physiol.* **45**: 37–40.

Wareing, P. F. 1954. Growth studies in woody species. VI: The locus of photoperiodic perception in relation to dormancy. *Physiol. Plantarum* **7**: 261–277.

Wareing, P. F. 1956. Photoperiodism in woody plants. *Annu. Rev. Plant Physiol.* **7**: 191–214.

Wareing, P. F. 1969. The control of bud dormancy in seed plants. *Symp. Soc. Exp. Biol.* **23**: 241–262.

Wareing, P. F. and D. L. Roberts. 1956. Photoperiodic control of cambial activity in *Robinia pseudacacia* L. *New Phytol.* **55**: 356–366.

Wareing, P. F. and P. F. Saunders. 1971. Hormones and dormancy. *Annu. Rev. Plant Physiol.* **22**: 261–288.

Wilkins, H. and R. L. Wain. 1975. Abscisic acid and the response of the roots of *Zea mays* L. seedlings to gravity. *Planta* **126**: 19–23.

Williamson, J. D. and R. S. Quatrano. 1988. ABA-regulation of two classes of embryo-specific sequences in mature wheat embryos. *Plant Physiol.* **86**: 208–215.

Wright, S. T. C. 1975. Seasonal changes in the levels of free and bound abscisic acid in blackcurrent (*Ribes nigrum*) buds and beech (*Fagus sylvatica*) buds. *J. Exp. Bot.* **26**: 161–174.

Wright, S. T. C. and R. W. P. Hiron. 1972. The accumulation of abscisic acid in plants during wilting and under other stress conditions. In: Carr, D. J., ed. *Plant Growth Substances 1970.* Springer-Verlag, New York. Pp. 291–298.

Yamane, H., J. Sugawara, Y. Suzuki, E. Shimamura, and N. Takahashi. 1980. Synthesis of jasmonic acid related compounds and their structure-activity relationships on the growth of rice seedlings. *Agric. Biol. Chem.* **44**: 2857–2864.

Zeevaart, J. A. D. 1977. Sites of abscisic acid synthesis and metabolism in *Ricinus communis* L. *Plant Physiol.* **59**: 788–791.

Zeevaart, J. A. D. 1980. Changes in the levels of abscisic acid and its metabolites in excised leaf blades of *Xanthium strumarium* during and after water stress. *Plant Physiol.* **66**: 672–678.

Zeevaart, J. A. D. 1983. Metabolism of abscisic acid and its regulation in *Xanthium* leaves during and after water stress. *Plant Physiol.* **71**: 477–481.

Zeevaart, J. A. D. and G. L. Boyer. 1984. Accumulation and transport of abscisic acid and its metabolites in *Ricinus* and *Xanthium*. *Plant Physiol.* **74**: 934–939.

CHAPTER 6

Ethylene

Historical Background

General recognition of ethylene, a simple hydrocarbon gas (C_2H_4), as a plant hormone has come about only relatively recently, although it has been known for more than three-fourths of a century that the gas has numerous interesting effects on growth and development. D. N. Neljubow, a Russian physiologist, evidently was the first scientist to write about ethylene action on plants. In 1901 he recorded that ethylene, which he identified as a component of illuminating gas, causes the classical "triple response" of etiolated pea seedlings. This response includes inhibition of stem elongation, increase in radial expansion (swelling) of stems, and a horizontal orientation of stems to gravity (see Fig. 6.13). Some years later (1910) H. H. Cousins advised the Jamaican government that oranges and bananas should not be stored together on ships because some unidentified volatile agent would cause the bananas to ripen prematurely.

Several workers reported in the years between about 1917 and 1937 that ethylene stimulates fruits to ripen. During the years from about 1933 to 1937, several investigators showed further that ethylene is produced by many plant materials, especially fleshy fruits. Indeed, the observations of several responses of plants to exogenous ethylene and the observation that a variety of vegetative tissues as well as fleshy fruits produce ethylene led W. Crocker, A. E. Hitchcock, and P. W. Zimmerman of the Boyce Thompson Institute for Plant Research in New York to suggest in 1935 that ethylene might be an endogenous growth regulator, and that it be regarded as a ripening hormone. Unfortunately, however, interest in ethylene among plant physiologists did not become intense or even widespread until nearly 24 years later. In good part the reason for comparatively little attention to ethylene was the lack of availability of techniques to analyze the relatively small amounts of the gas in plant tissues generally. About 1959 interest in the biochemistry and physiology intensified when Stanley P. Burg, Harlan K. Pratt, and others began applying the newly developed technique of flame ionization gas chromatography to ethylene analyses. This advanced the limit of analysis nearly a millionfold over the older manometric techniques that had been used previously. By flame ionization gas chromatography, it is possible to measure a fraction of a part per billion, or as little as 10^{-3} pmol, in a small gas sample.

Still the general acceptance of ethylene as a hormone was rather slow to develop. For, as Burg (1962) observed, "it (ethylene) would be unique in that it is a vapor at physiological temperatures so that its production, accumulation, transport, and function would involve special problems not encountered in other areas of hormone physiology." Unlike some other hormones, ethylene apparently does not undergo directed transport. Hence it somehow accomplishes a hormonal, integrative function by diffusing rapidly through tissues. But by 1969, Pratt and Goeschl could say that, "With very sensitive instruments and very careful technique, it has become possible to show that ethylene is an endogenous growth regulator in plants."

Resurgence and great expansion of interest in ethylene biochemistry and physiology occurred in the late 1950s, and this has been a very active and productive field of research ever since. Ethylene now is understood to be a natural product of plant metabolism, which is produced by healthy as well as senescent and diseased tissues, and which exerts regulatory control or influence over plant growth and development generally throughout ontogeny. Its interactions with other kinds of hormones are numerous and complex. Among the most diverse physiological effects that have been reported are

1. Stimulation of ripening of fleshy fruits.
2. Stimulation of leaf abscission.
3. "Triple response" of etiolated legume seedlings—reduced stem elongation, radial swelling of stems, and agravitropism or diagravitropism of stems.
4. Inhibition of leaf and terminal bud expansion in etiolated seedlings.
5. Tightening of the epicotyl or hypocotyl hook of etiolated dicot seedlings.
6. Inhibition of root growth.
7. Increase in membrane permeability.
8. Stimulation of adventitious root formation.
9. Stimulation of flowering in pineapple.
10. Inhibition of lateral bud development.
11. Causes various types of flowers to fade.
12. Interference with polar auxin transport.
13. Causes epinasty of leaves.
14. Participation in normal root gravitropism.

Ethylene and Fruit Ripening

For many years ethylene physiology remained primarily the province of those interested in fruit physiology. Several practical applications developed from this research, among them being the process of "gas storage" or "controlled atmosphere" (CA) storage for the preservation of fruits, which was developed by Kidd and West in 1933. It had been determined that oxygen is necessary for ethylene action and that carbon dioxide inhibits action of the gas. Basically the process of "gas storage" consists of storing fruits in an atmosphere rich in carbon dioxide (5 to 10%), low in oxygen (1 to 3%), and with as little ethylene as possible. Gas storage is achieved commercially by storing fruits in an airtight room at low temperature. The oxygen is naturally depleted and the carbon dioxide increased by respiration. Ethylene produced by the fruits is absorbed on brominated charcoal filters. The effects of the controlled atmosphere are

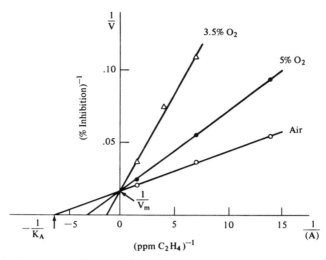

FIGURE 6.1. Lineweaver–Burk double reciprocal plot (percentage inhibition^{-1} versus concentration of ethylene^{-1}) at various levels of O_2, illustrating positive effect of O_2 on ethylene action. Data are based on inhibition of growth of pea stem sections by ethylene. (Redrawn, with permission, from Burg and Burg, 1967a.)

explainable on the grounds that oxygen is essential for ethylene action (Fig. 6.1) and carbon dioxide is a competitive inhibitor of ethylene (Fig. 6.2).

Numerous changes are associated with the ripening of fleshy fruits, including (1) softening, (2) hydrolysis of storage materials, (3) changes in pigmentation and flavor, and (4) changes in respiration rate. The fleshy fruit is more complex in its senescence than other plant organs, and all of the changes that occur do not represent deterioration, but in a real sense fruit ripening may be regarded as a special case of organ senescence. In many kinds of fleshy fruits, there are marked changes in respiration rate after the fruits mature (reach final size). In some species there is a decrease in respiration rate in the mature fruit, followed by a large increase during the time of ripening (the respiratory climacteric), and a final decrease in respiration rate as the fruit enters a senescent decline (Fig. 6.3). Examples of fruits that show a climacteric are avocado, banana, pear, apple, and mango. Not all fleshy fruits exhibit a respiratory climacteric, however. There are in fact two types of nonclimacteric fruits: (1) those that show a steady respiration rate during ripening (e.g., orange, lemon, fig); and (2) those that show a decline in respiration rate during the ripening period (e.g., pepper).

The induction of ripening and senescent changes induced in many kinds of fruits by ethylene has been repeatedly demonstrated over the past 50 years. However, the role of ethylene in the metabolism of fruits, and particularly in the respiratory climacteric, has been vigorously debated. As recently as 1960, some authors argued that ethylene is a by-product rather than the cause of ripening. Their argument was based on their findings, largely by relatively insensitive

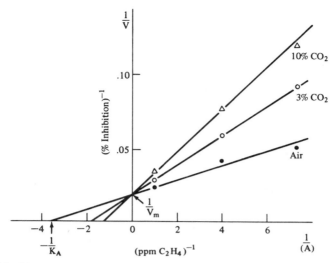

FIGURE 6.2. Lineweaver–Burk double reciprocal plot (percentage inhibition^{-1} versus concentration of ethylene^{-1}) at various CO_2 concentrations, showing competitive inhibition of ethylene action by CO_2. Data are based on inhibition of growth of pea root sections by ethylene. (Redrawn, with permission, from Chadwick and Burg, 1967.)

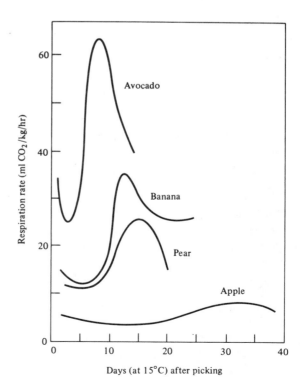

FIGURE 6.3. The course of the climacteric rise in fruit respiration during the post-harvest period. (Reproduced, with permission, from the *Annual Review of Plant Physiology,* Volume 1. © 1950 by Annual Reviews Inc. From Biale, 1950.)

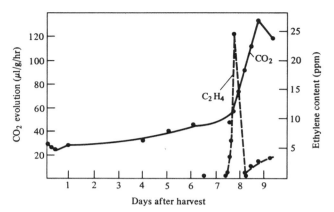

FIGURE 6.4. Relationship between endogenous ethylene content and respiration rate in banana. (Redrawn, with permission, from Burg, S. P. and E. A. Burg. 1965b. Ethylene action and the ripening of fruits. *Science* **148**: 1190–1196, 28 May 1965. Copyright 1965 by the American Association for the Advancement of Science.)

manometric techniques, that ethylene is not always present in large enough quantity prior to the climacteric to stimulate ripening. On the other hand, Stanley P. Burg of the University of Miami, Miami, Florida, in the years after circa 1959, presented what many regard as convincing evidence that ethylene is, in fact, the hormone that initiates fruit ripening. Burg utilized consistently the technique of flame ionization gas chromatography in his ethylene analyses. He showed that even fruits that do not normally exhibit a respiratory climacteric can be induced to do so by applying ethylene (e.g., lemon and orange). Moreover, Burg and others have shown that for the climacteric fruits studied, there is a rise in the intercellular content of ethylene to a level that stimulates these fruits to ripen—well in advance of the respiratory climacteric (Fig. 6.4). But actually there is considerable variation among species, and H. K. Pratt and J. D. Goeschl (1969) stated that fleshy fruits fall into several classes as regards their ripening response to ethylene. In most fleshy fruits, the onset of ripening is preceded by an increase in ethylene production. In a few species (e.g., avocado and mango), significant concentrations of ethylene may be present some time before ripening, but the response to ethylene is inhibited until after the fruit is harvested. In apple, it appears that a step in ethylene synthesis may be inhibited by a product from the parent tree. When detached from the tree the inhibitor presumably disappears and the rate of ethylene production rises. In another group, including certain kinds of banana, a presumably effective ethylene concentration may be present in the unripe fruit, but the fruit is insensitive to that concentration at that stage of development. As the fruit matures, it becomes more sensitive, and ripening ensues when sensitivity matches the endogenous ethylene content. In several annual crops (e.g., tomato and melons) ethylene production, ripening, and senescence occur at about the same age from anthesis, regardless of age at harvest. It appears that, in these fruits, the basic ethylene-forming mechanism exists and operates at a low level, but is not fully activated until the fruit reaches a critical physiological age.

Evidently ethylene causes two major types of biochemical effects in the process of fruit ripening. The oldest evidence is for changes in membrane permeability, leading to altered compartmentation and release of enzymes associated with such ripening processes as respiration and breakdown of acids and wall constituents. Permeability changes reportedly do occur in cellular membranes preceding or during ripening, as indicated by a leakage of solutes from the cells and also an increase in free space and filling of intercellular spaces with juice. Yet whether, in fact, ethylene causes permeability changes remains an unresolved question. Some investigators have argued that if there is a direct effect of ethylene on one or more membranes, it is more likely to be a specific effect involving only certain solutes rather than a general increase in permeability. Legge et al. (1986) investigated changes in molecular organization of lipid bilayers in the membranes in ripening tomato fruit by fluorescence depolarization after labeling with fluorescent lipid-soluble probes. The fluorescent labels were partitioned into isolated protoplasts and purified plastids from fruit at various stages of senescence. They found evidence that plasma membranes but not plastid membranes did undergo deterioration during ripening. Meanwhile, as regards fruit ripening, even though increased membrane permeability may not be the direct cause, such permeability changes may make possible the metabolic processes that collectively comprise ripening. And, in any event, of equal or perhaps greater importance are the biochemical changes correlated with the observed increases in protein (enzyme) synthesis that occur during the climacteric.

Clearly there is an increase in protein content during the climacteric (Fig. 6.5). It is suggested by available evidence that ripening processes are correlated with the formation of enzymes that catalyze the biochemical changes occurring during ripening. If mature fruit tissues are treated with inhibitors of protein synthesis, ripening is prevented parallel with the reduction in protein synthesis (Fig. 6.6). Ethylene synthesis itself apparently is dependent on protein synthesis at the early climacteric stage, but less so once the climacteric has progressed.

There is no doubt that fruit tissues respond (directly or indirectly) to ethylene by exhibiting increases in the activities of enzymes that catalyze ripening reactions—so do other target tissues respond likewise to ethylene—and in many cases the increases in enzyme activity probably are the result of de novo enzyme synthesis, rather than activation of preexisting enzyme. But a yet unanswered question is whether ethylene acts directly to evoke new enzyme production. Interpretation of results with inhibitors of RNA and protein synthesis leaves some uncertainty, because it could be merely that RNA and protein synthesis are essential to maintain the cells in a state competent to respond to ethylene. Moreover, there are some responses to ethylene, besides fruit ripening, that

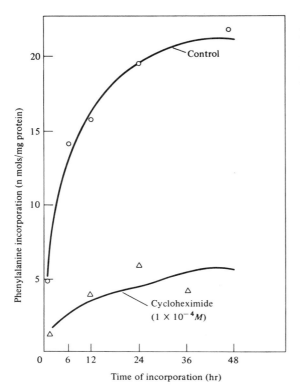

FIGURE 6.5. Effect of cyclo-heximide on incorporation of [^{14}C]phenylalanine into protein in midclimacteric pears. (Redrawn with modifications, by permission, from Frenkel et al., 1968.)

occur apparently in too brief a time to allow for the direct involvement of RNA and protein synthesis. Obviously, additional research is required to elucidate the primary action of ethylene in responsive tissues.

As reviewed by Biggs et al. (1986), stimulation of protein synthesis appears to be an essential component of the ripening process as judged by incorporation of radioactive amino acids and inhibition of ripening following treatment with protein synthesis inhibitors. The activities of several enzymes change during ripening of climacteric fruits, including tomato polygalacturonase, which is synthesized *de novo* and accumulates as ripening proceeds. In addition, changes in the levels of specific mRNAs during tomato and avocado fruit ripening have been reported. Patterns of polypeptides translated *in vitro* on mRNAs from early stages of tomato fruit development differ from those translated on mRNAs from later stages of fruit development. Several tomato fruit mRNAs have been cloned recently.

Biggs et al. (1986) investigated changes in the populations of proteins and mRNAs during normal tomato fruit development and the extent to which those changes differ in fruit from three abnormally ripening-impaired mutants, namely *rin, nor,* and *nr.* They found that the relative predominance of many proteins changed during tomato fruit development, and that several proteins that

FIGURE 6.6. Effect of cycloheximide on ripening in pears. (Redrawn with modifications, by permission, from Frenkel et al., 1968.)

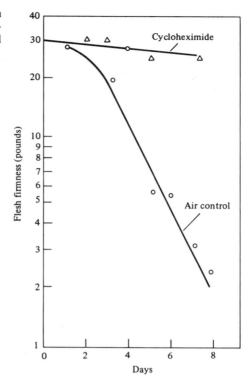

accumulate during normal ripening either failed to increase or increased at reduced levels in the three ripening mutants. Analyses of mRNA populations from normal and mutant phenotypes indicated even more extensive changes in gene expression during ripening.

Interaction between Auxin and Ethylene

Zimmerman and Wilcoxon first reported in 1935 that exogenous auxin stimulates ethylene production in several plants. Since about 1964, several workers have confirmed and extended that observation. Physiological concentrations of IAA and other auxins stimulate ethylene formation in the roots, stems, flowers, fruits, and leaves of all plants that have been examined. By circa 1966 an hypothesis had developed that rapidly gained wide support, which attributes many of the effects of auxin to its influence on ethylene production.

The response of plant tissues to auxin varies with the concentration of auxin applied. In most cases, an optimum auxin concentration is found, with lower concentrations being promotive and higher concentrations inhibitory, or at least supraoptimal (Fig. 6.7). The reason for the growth inhibition (or submaximal

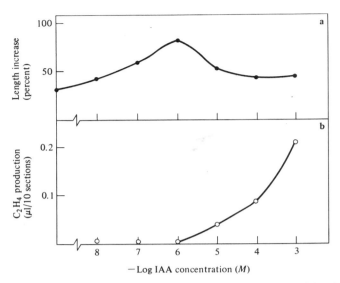

FIGURE 6.7. Growth response, **a**, and ethylene production, **b**, in etiolated pea stem sections treated with different concentrations of auxin during 18 hours in sealed flasks. (Redrawn with modification, by permission, from Burg and Burg, 1966b.)

response) was for a long time unknown. But since circa 1966 Stanley P. Burg and others have shown that in some systems auxin itself only promotes growth, and that auxin per se is never inhibitory. What happens is that at certain critical concentrations of auxin, which are different for each tissue, the production of ethylene is induced, and it is the ethylene that directly inhibits growth (Fig. 6.7). Thus, in some cases the inhibition of growth, e.g., elongation of etiolated pea stem sections (Fig. 6.7), etiolated sunflower stem sections, and roots (Fig. 6.8) by high concentrations of IAA evidently can be completely accounted for by auxin-stimulated ethylene production. However, the situation with some materials, e.g., sections of etiolated corn and oat coleoptiles, is different. Auxin stimulates these tissues to evolve ethylene, but the production does not continually increase as the IAA concentration is raised. Instead, ethylene formation and growth both reach their highest rates at about 10^{-5} M IAA, and they both decline as the auxin concentration is increased further. Hence it seems that auxin-induced growth inhibitions may occur for basically different reasons in coleoptiles and seedlings such as sunflower and pea. Sections of light-grown pea stems also behave quite differently than etiolated stem sections. Growth rate does not saturate at any concentration of IAA; that is, there is no sharp optimum (Fig. 6.9). Green stem sections are less responsive to ethylene than etiolated seedlings and produce ethylene at a slower rate in the presence of 10^{-5} or 10^{-4} M IAA. Consequently, even though green tissue swells when it responds to its own ethylene, the inhibition of elongation due to auxin-induced ethylene formation is never sufficient to offset the auxin-induced increase in length.

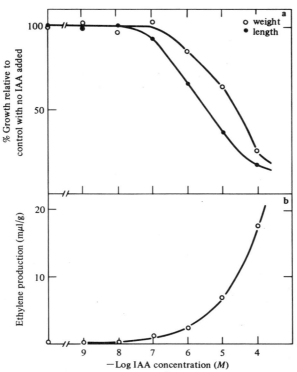

FIGURE 6.8. Growth inhibition, **a**, and ethylene production, **b**, in apical segments of pea roots treated with different concentrations of auxin. (Redrawn, with permission, from Chadwick and Burg, 1967.)

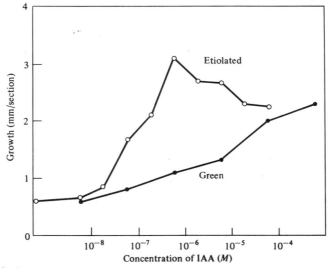

FIGURE 6.9. Growth responses of etiolated and light-grown pea stem sections to different concentrations of IAA. (Redrawn from data of Galston and Kaur, 1961.)

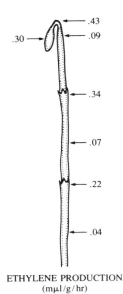

.43
.30
.09

.34

.07

.22

.04

ETHYLENE PRODUCTION
(mμl/g/hr)

FIGURE 6.10. Distribution of ethylene production along the etiolated pea stem. Value for the apical leaves is calculated from the difference in rates between intact apical pieces and similar tissue from which the leaves were excised. (Redrawn, in slightly modified form, with permission, from Burg and Burg, 1968.)

Ethylene and Apical Dominance

More than 50 years ago IAA was shown to be the cause of the dominance of the apical bud over lateral buds (see also Chapter 4, pp. 186–188). However, Stanley P. Burg and Ellen A. Burg of the University of Miami School of Medicine showed in 1968 that the inhibition of growth of lateral buds in decapitated pea seedlings was not due to auxin itself but to the ethylene that auxin induces to be synthesized. There were several lines of evidence, including (1) the nodal regions of etiolated pea seedlings are sites of localized ethylene synthesis (Fig. 6.10); (2) the production of ethylene at the upper node is reduced 24% one day after decapitation, although this reduction could not be detected in nodal regions possessing buds but lacking leaf scales; and (3) at all concentrations of IAA tested there was a close correlation between the intensity of ethylene production and the bud inhibition that results. Interestingly also, but not surprisingly, kinetin counteracted the inhibitory actions of ethylene and IAA on bud growth.

Twenty years later W. Russell and Kenneth V. Thimann of the University of California at Santa Cruz extended the observations of Burg and Burg. They performed experiments on decapitated *Vicia faba* plants and found that (1) continued exposure to ethylene (15 ppm) duplicated completely the bud inhibition exerted by IAA applied to the cut surface; (2) maintaining decapitated plants, treated with IAA, under reduced pressure (approximately 65% atmospheric pressure) decreased the auxin inhibition, while controls in air showed complete inhibition; and (3) injecting aminoethoxyvinyl glycine (AVG), an inhibitor of ethylene biosynthesis (0.5 mM), into the nodal tissue also largely annulled the inhibition due to applied IAA. Hence they concluded that "ethylene indeed is the 'second messenger' for the primary stimulus of IAA in apical dominance."

As Burg and Burg noted in 1968, "The finding that ethylene mediates the inhibition of bud growth caused by IAA does not clarify the role of auxin in apical dominance; it simply adds an intervening step between IAA (the stimulus) and inhibition (the response) which it causes."

In contrast, Harrison and Kaufman (1982) investigated the release of lateral buds (tillers) from apical dominance in oats (*Avena sativa*). Their studies on endogenous ethylene production in intact *Avena* shoots and the effects of ethylene and IAA on tiller bud growth indicated to them that IAA-induced ethylene production is most likely not involved in maintenance of apical dominance in these shoots. Rather, their evidence was interpreted to indicate that ethylene may act by promoting the swelling phase of tiller growth during release of tiller buds from quiescence and by stimulating kinetin-promoted elongation in the tiller shoots following the swelling phase of growth.

Inhibition of Root Growth and Role in Root Gravitropism

Previously it was pointed out that roots are extremely sensitive to auxin, inhibition of root sections occurring at concentrations of auxin that promote growth of stem sections, and that auxin inhibition of root growth was directly correlated with auxin-induced production of ethylene. Evidence reported by Chadwick and Burg (1970) strongly indicates that, indeed, auxin-induced inhibition of root growth is mediated entirely by auxin-stimulated formation of ethylene *in vivo*, as well as in root sections incubated with auxin *in vitro*. Moreover, ethylene—rather than auxin—may be the agent which directly mediates gravitropism of roots—but not of stems.

According to the Cholodny–Went theory of gravitropism, gravitropism results from asymmetric distribution of endogenous auxin under the stimulus of gravity, such that the ratio of concentrations of endogenous auxin in the gravistimulated organs becomes approximately 30:70 in the upper and lower halves, respectively. Roots being more sensitive to auxin than stems, the additional increment of auxin on the lower side of the root is sufficient to cause inhibition of growth on the lower side relative to the upper. Thus gravistimulated roots curve downward when a growth inhibitory concentration of auxin accumulates in their lower side. Conversely, in gravistimulated stems only differential growth promotion results from the asymmetric distribution of endogenous auxin.

The involvement of ethylene in root gravitropism helps to overcome several objections which have been raised against application of the Cholodny–Went theory to roots. More than two decades ago Audus and Brown noted that during gravitropism of roots growth of the upper side is, in fact, first promoted then inhibited. They proposed that auxin per se is not the active inhibitor, but rather an inhibitor which moves from the lower side to the upper side of a gravistimulated root. Ethylene fits this concept well, since the ethylene produced in the lower side, as a result of unequal distribution of auxin, would be expected eventually to diffuse to the upper side to some extent, and thus account for the slight inhibition of growth that subsequently occurs there. Moreover, results of investigations of ethylene transport by Burg and associates help to eliminate a second objection to the Cholodny–Went theory, namely, that the observed 30:70 distribution of auxin in horizontally oriented roots is insufficient to account for the amount of growth inhibition on the lower side of the root, since there are indications that low concentrations of applied IAA cause very extensive initial inhibition, followed by some recovery, and that the distribution between the upper and lower sides can approach 10:90 rather than the 30:70 previously reported for gravistimulated stems.

But it must not be too hastily concluded that auxin-induced ethylene is the only growth-inhibiting substance involved in root gravitropism for the Cholodny–Went theory has been severely challenged on two major grounds: (1) several investigators have not found consistent differences in the IAA concentrations in the upper and lower halves of gravistimulated roots; and (2) there is mounting evidence for the involvement of one or more growth inhibitors, one of which is ABA. ABA is found in the root cap, it can be shown to be redistributed in the root cap, it moves basipetally in the root extension zone, and it does in fact inhibit growth in the extension zone. Available evidence shows that the inhibitor is produced only in the root cap and is laterally transported only in the root cap and not in the elongating zone. In view of these collective findings, the role of IAA, hence also IAA-induced ethylene, presently is uncertain.

Role in Emergence of Dicot Seedlings

Protection of the tender shoot tip from injury during seed germination and emergence is an important feature of all seed plants. In species that exhibit hypogean germination (e.g., *Pisum sativum*, garden pea) (Fig. 6.11) the cotyledons remain below ground (hypocotyl does not elongate) and the emergent plumule is arched or recurved near the apex, so that the shoot tip is pushed to the soil surface in a protected position. The common garden bean (*Phaseolus vulgaris*), on the other hand, displays epigean germination, according to which the hypocotyl elongates, forming a hook or arch, and the shoot tip is protected as it is pulled upward by the elongating hypocotyl. In both cases, establishment of symmetrical growth in the elongating shoot causes the epicotyl (or plumular) or hypocotyl hook to be straightened out as the seedlings emerge above ground.

It turns out that this is yet another of the many aspects of growth and development in which ethylene has an important hormonal role. In the case of the etiolated pea seedling, for example, there is a localized production of ethylene in the plumule and plumular hook portion of the epicotyl at a rate of approximately 6 μl kg^{-1} hour^{-1}. This rate is sufficient to yield physiologically

FIGURE 6.11. Light-grown, **a**, and etiolated, **b**, seedlings of pea (*Pisum sativum*). (Adapted from *The Living Plant* by Peter Martin Ray. Copyright 1963 by Holt, Rinehart & Winston, Inc. Redrawn by permission of Holt, Rinehart & Winston.)

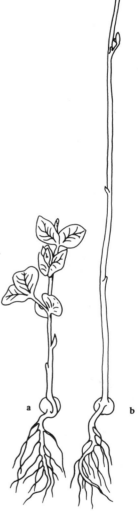

active concentrations of ethylene within the tissue. The localized production of ethylene, in fact, is responsible for formation and maintenance of the hook in the etiolated seedling. As the etiolated seedling emerges naturally from soil, or is experimentally irradiated with white light, there is a transient decrease in ethylene production in the hook region and establishment of symmetrical growth about the shoot (Fig. 6.12). Moreover, as the etiolated tissue greens it becomes less sensitive to the ethylene that is produced. In determining the action spectrum for the plumular expansion, red light (in the region of 660 nm) was most

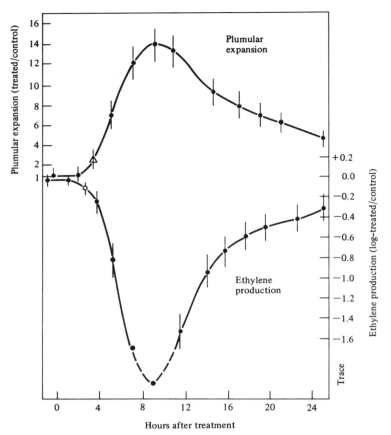

FIGURE 6.12. Effect of a 1-minute exposure to red light (40,000 ergs cm^{-1} total dose) on plumular expansion and hook opening and ethylene production in the upper part of the epicotyls of etiolated pea seedlings. (Redrawn, with permission, from Goeschl et al., 1967.)

effective. Moreover, far-red irradiation (approximately 730 nm) immediately following the red decreased the red effect to the level achieved by the far-red alone. These results suggested that the ethylene production mechanism is mediated by phytochrome (Chapter 8) and that ethylene acts as a hormone in the phytochrome control of plumular expansion. This explanation applies equally well to seedlings that develop hypocotyl hooks, such as bean.

Role in Abscission

Mention was made previously of the possible role of ethylene in abscission (see Chapter 5). There is considerable evidence that ethylene has some natural role in the regulation of the rate of abscission, but it is not clear whether ethylene is the primary inducer. The majority of available data on hormonal control of abscission are consistent with the following hypothetical scheme, which was developed recently by Michael S. Reid of the University of California at Davis (Reid, 1985):

- A gradient of auxin from the subtended organ to the plant axis maintains the abscission zone in a nonsensitive state. This gradient is itself maintained by factors which inhibit senescence of the organ. Thus, auxins, cytokinins, light, and good nutrition all tend to reduce or delay abscission.
- Reduction or reversal of the auxin gradient causes the abscission zone to become sensitive to ethylene. Application of auxin proximal to the abscission zone, removal of the leaf blade, or treatments which accelerate its senescence (shading, poor nutrition, ethylene) therefore hasten abscission. Ethylene, or stresses which enhance its production, also may hasten abscission by reducing auxin synthesis and/or by interfering with its transport from the leaf. Where ABA stimulates abscission it may do so by stimulating ethylene production, or by interfering with the production, transport, or action of auxin.
- Once sensitized, the cells of the abscission zone respond to low concentrations of ethylene, whether exogenous or endogenous, by the rapid production and secretion of hydrolytic enzymes and subsequent shedding of the subtended organ. [Citations to original papers omitted.]

The preceding hypothetical scheme is consistent with much of the evidence regarding regulation of the processes resulting in eventual abscission of plant organs. However, some investigators have expressed reservations about the view that ethylene plays a major, hormonal role in natural abscission. For example, F. T. Addicott (1982) concluded that ethylene probably is not essential to abscission. The three major lines of evidence that ethylene *does* play a hormonal role in natural abscission have been summarized recently by Reid (1985), as follows:

First, ethylene production increases prior to abscission in many abscising plant organs, although abscission can occur in some tissues without increased ethylene production. Perhaps what is involved in the latter cases is a changing sensitivity to ethylene during ontogeny, such that sensitivity increases with age, as applies in ripening fruits. In the case of many plant responses to hormones, there is increasing evidence for a changing sensitivity to hormones being at least as important as changes in endogenous concentrations of hormones. Moreover, as Addicott emphasized, regulation of abscission surely involves a balance among several hormones.

Second, treatment of a wide array of plant species with ethylene and with ethylene-releasing compounds stimulates abscission dramatically (Fig. 6.15). Abeles (1973) noted that in some plants application of ethylene did not induce leaf abscission. He suggested that in those cases the ultimate control of abscission might depend on factors other than ethylene. Perhaps the cells in the abscission zone were not sensitized to ethylene under the conditions used. In such cases also, one can draw parallels with fruit physiology (Reid, 1985). Avocado, a classical climacteric fruit, is, as we have seen, insensitive to ethylene while still attached to the parent tree.

Third, inhibitors of ethylene binding (2,5-norbornadiene), biosynthesis (analogs of rhizobitoxin) and action (silver ion) all have been reported to inhibit abscission (Reid, 1985).

Decoteau and Craker (1987) investigated the role of ethylene in the light control of leaf abscission in mung bean (*Vigna radiata* L.) cuttings. They stated that the importance of ethylene in abscission and the fact that light treatments may regulate ethylene production suggest that control of ethylene production or action could be a component in the mechanisms by which light regulates abscission. Their results were interesting. Red light inhibited and far-red light promoted abscission as compared with dark controls. However, changes in the rate of abscission could not be correlated with changes in rate of ethylene production. Moreover, reducing ethylene synthesis *in vivo* with aminoethoxy-vinylglycine did not alter the effects of red or far-red light on abscission. They concluded that far-red light appeared to increase and red light appeared to decrease tissue sensitivity to ethylene.

Effects of Ethylene on Planes of Cell Expansion

Several of the effects of ethylene involve alteration of the normal planes of cell growth. Radial swelling or abnormal radial expansion of stems—such as is seen in the "triple response" of etiolated pea seedlings (Fig. 6.13) and probably also in leaf epinasty and in the herbicidal action of some auxin-type herbicides—and roots result from inhibited elongation and increased radial expansion of individual cells (Fig. 6.14). Sometimes "root hair-like" outgrowths and even adventitious roots are formed from the cells. Radial expansion of cells that normally expand mainly by elongation is correlated with an alteration by ethylene of the orientation of the cellulose microfibrils in the cell wall. The newly deposited microfibrils at the inner surface of the wall are oriented in a more longitudinal plane than normal, restricting cell elongation but permitting radial expansion. An effect on auxin transport may also be involved in some of the abnormal growth caused by ethylene.

Other Effects of Ethylene

There are several other effects of ethylene that will be considered only briefly. One of these is of long historical interest and considerable economic importance. This is stimulation of flowering in mangos and bromeliads such as pineapple.

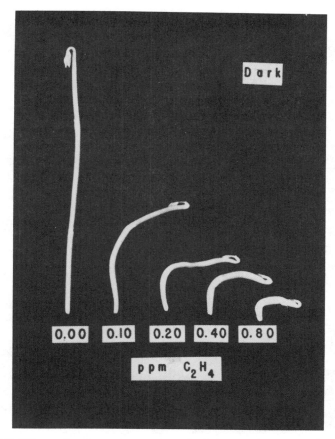

FIGURE 6.13. Effects of various concentrations of ethylene on the growth of etiolated pea seedlings, showing the portions of the seedlings that developed during the 48-hour treatment. (From Goeschl and Pratt, 1968.)

For many years, the synthetic auxin naphthaleneacetic acid (NAA) was sprayed on pineapple fields to stimulate flowering. It turns out that it is NAA-stimulated ethylene that is the actual active hormone involved in the response.

Leaf abscission is another process which is stimulated by ethylene (Fig. 6.15), and many chemicals that are utilized as defoliants, including some auxin-type herbicides such as 2,4-D, act by inducing ethylene formation.

Practical applications of ethylene on a commercial scale really got underway only as recently as about 1963, at which time a new water-soluble, slow ethylene-releasing growth regulator was introduced. Patented as "Ethrel" and given the trivial name "ethephon," the compound is 2-chloroethylphosphonic acid (Fig. 6.16). Ethephon decomposes spontaneously in aqueous solution and in

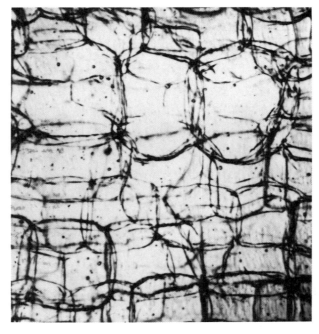

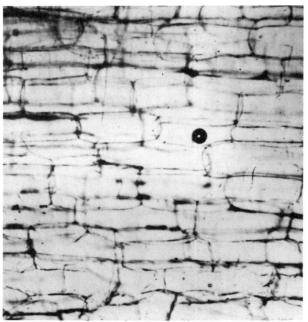

FIGURE 6.14. Longitudinal sections of the cortical region of etiolated pea seedling stems. Left, control; right, ethylene-treated. (From Stewart et al., 1974, by permission; photos courtesy of M. Lieberman, 1978.)

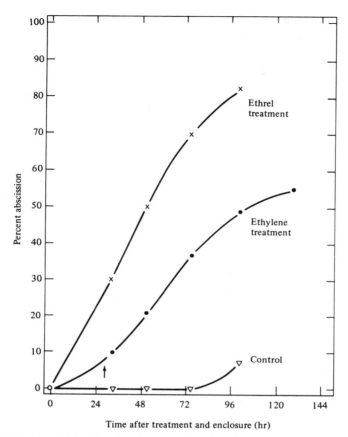

FIGURE 6.15. Abscission of leaves of cotton plants following treatment with ethylene and Ethrel (2-chloroethylphosphonic acid). (Redrawn, with permission, from Morgan, 1969.)

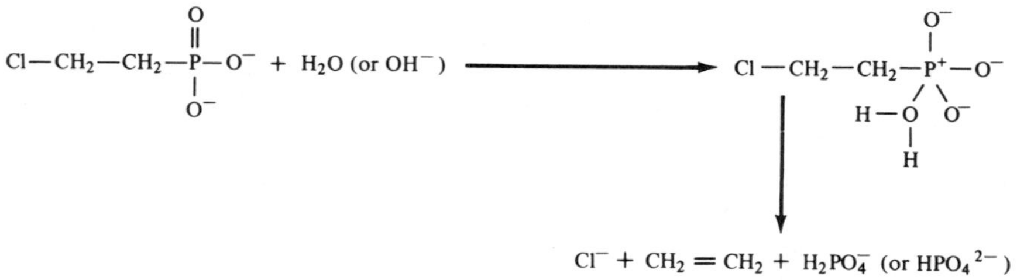

$$Cl-CH_2-CH_2-\overset{\overset{\displaystyle O}{\|}}{\underset{\underset{\displaystyle O^-}{|}}{P}}-O^- + H_2O \ (or \ OH^-) \longrightarrow Cl-CH_2-CH_2-\overset{\overset{\displaystyle O^-}{|}}{P^+}-O^-$$

$$Cl^- + CH_2=CH_2 + H_2PO_4^- \ (or \ HPO_4^{2-})$$

FIGURE 6.16. 2-Chloroethylphosphonic acid, a synthetic growth regulator that decomposes spontaneously in aqueous solution and in plant tissues to yield ethylene and phosphoric acid. (Redrawn, with permission, from Varner, J. E. and D. T. Ho. 1976. Hormones. In: Bonner, J. and J. E. Varner, eds. *Plant Biochemistry*. 3rd ed. pp. 713–770. Copyright by Academic Press, Inc., New York.)

tissues to yield ethylene. Within a few years after its introduction, this synthetic plant growth regulator was being utilized extensively on certain fruit and vegetable crops, including pineapple and cucumbers.

Ethylene Biosynthesis and Mechanism of Action

Ethylene evidently is much more restricted in its occurrence than the other kinds of plant hormones. It probably is ubiquitous among seed plants, but its distribution among other phyletic groups is limited. For certain, some bacteria and fungi produce ethylene, and ethylene production by certain fungal pathogens is directly involved in the host–pathogen interaction and the disease symptoms that develop in the host.

Ethylene is synthesized from methionine via S-adenosylmethionine as an intermediate, according to the pathway proposed from investigations with apple tissue (Fig. 6.17). Methionine is activated by ATP to give S-adenosylmethionine, which then is cleaved to give 5′-S-methyl-5′-thioadenosine and a cyclic amino acid, 1-aminocyclopropane-1-carboxylic acid. The latter intermediate next is cleaved to give ethylene, formate, and ammonium ion. To complete the cycle, the nucleoside can donate its methylthio group to a 4-carbon acceptor (homoserine) to reform methionine.

As discussed in a previous section, the mechanism of action of ethylene is at present virtually unknown. Some of the effects of ethylene—such as promoting leaf abscission—involve *de novo* synthesis of particular enzymes (e.g., cellulase). However, it remains unknown as to how—whether directly or indirectly—ethylene stimulates enzyme synthesis. Moreover, the effects on membranes are incompletely resolved. Considerable evidence indicates that the hormone binds to a metal (perhaps copper) in whatever action it undergoes, but the details and metabolic consequences of such binding have yet to be elucidated.

There have been several reports that suggest that white light inhibits ethylene production (e.g., Gepstein and Thimann, 1980). However, in all of these reported cases, excised leaf segments were incubated in closed flasks, and the ethylene samples were withdrawn after varying periods of time. Bassi and Spencer (1983) monitored the effect of light on ethylene biosynthesis in three plant species using two different systems—leaf segments incubated in closed flasks versus intact plants in a flow-through open system. Experiments were conducted both in the presence and absence of 1-aminocyclopropane-1-carboxylic acid (ACC). They found that light did strongly inhibit ethylene production when cut leaf segments were incubated in the presence of ACC in closed flasks. When measurements were made with intact plants in a continuous flow system, the rate of ethylene production was almost identical in light and darkness, in the three plants studied. They concluded that the effect of light on cut leaf segments incubated in the presence of ACC in closed flasks can be attributed to the techniques used for these measurements, and that light has little effect on ethylene production by intact plants in an open system.

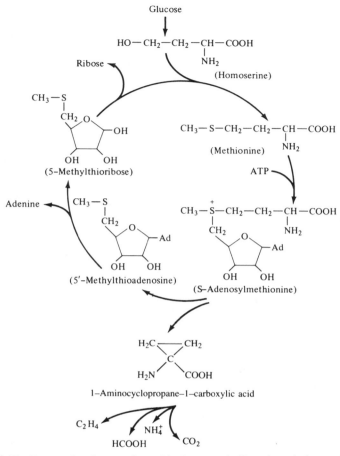

FIGURE 6.17. Proposed scheme of methionine metabolism in relation to ethylene biosynthesis in apple. (Redrawn, with permission, from Adams and Yang, 1977.)

Ethylene Receptor

A group of investigators at Aberystwyth has purified and characterized an ethylene binding protein (EBP) from developing cotyledons of *Phaseolus vulgaris*. It was found to be a glycoprotein having a molecular weight of 42,000. It was purified from Triton X-100 and β-octylglucoside solubilized membrane preparations using two anion-exchange steps. It appeared as a single band on SDS–PAGE and two-dimensional gel electrophoresis. The characteristics of binding including affinity, specificity, rate constants of association and dissociation, location, and the effect of silver suggested that the observed binding activity represents an ethylene receptor. Antibodies have been raised against purified ethylene binding protein, and these antibodies are being used to probe putative receptors in abscission zones and pea epicotyls.

References

Abeles, F. B. 1966a. Auxin stimulation of ethylene evolution. *Plant Physiol.* **41**: 585–588.

Abeles, F. B. 1966b. Effect of ethylene on auxin transport. *Plant Physiol.* **41**: 946–948.

Abeles, F. B. 1967a. Inhibition of flowering in *Xanthium pennsylvanicum* Walln. by ethylene. *Plant Physiol.* **42**: 608–609.

Abeles, F. B. 1967b. Mechanism of action of abscission accelerators. *Physiol. Plantarum* **20**: 442–454.

Abeles, F. B. 1972. Biosynthesis and mechanism of action of ethylene. *Annu. Rev. Plant Physiol.* **23**: 259–292.

Abeles. F. B. 1973. *Ethylene in Plant Biology*. Academic Press, New York.

Abeles, F. B. 1986. Role of ethylene in *Lactuca sativa* cv 'Grand Rapids' seed germination. *Plant Physiol.* **81**: 780–787.

Abeles, F. B. and H. E. Gahagan III. 1968. Abscission: The role of ethylene, ethylene analogues, carbon dioxide, and oxygen. *Plant Physiol.* **43**: 1255–1258.

Abeles, F. B. and R. E. Holm. 1966. Enhancement of RNA synthesis, protein synthesis, and abscission by ethylene. *Plant Physiol.* **41**: 1337–1342.

Abeles, F. B. and J. M. Ruth. 1972. Mechanisms of hormone action. Use of deuterated ethylene to measure isotopic exchange with plant material and the biological effects of deuterated ethylene. *Plant Physiol.* **49**: 669–671.

Adams, D. O. and S. F. Yang. 1977. Methionine metabolism in apple tissue. Implication of *S*-adenosylmethionine as an intermediate in the conversion of methionine to ethylene. *Plant Physiol.* **60**: 892–896.

Adams, D. O. and S. F. Yang. 1979. Ethylene biosynthesis: Identification of 1-aminocyclopropane-1-carboxylic acid as an intermediate in the conversion of methionine to ethylene. *Proc. Natl. Acad. Sci. U.S.A.* **76**: 170–174.

Addicott, F. T. 1970. Plant hormones in the control of abscission. *Biol. Rev.* **45**: 485–524.

Andreae, W. A., M. A. Venis, F. Jursic, and T. Dumas. 1968. Does ethylene mediate root growth inhibition by indole-3-acetic acid? *Plant Physiol.* **43**: 1375–1379.

Apelbaum, A. and S. P. Burg. 1971. Altered cell microfibrillar orientation in ethylene-treated *Pisum sativum* stems. *Plant Physiol.* **48**: 648–652.

Apelbaum, A. and S. F. Yang. 1981. Biosynthesis of stress ethylene induced by water deficit. *Plant Physiol.* **68**: 594–596.

Bassi, P. K. and M. S. Spencer. 1983. Does light inhibit ethylene production in leaves? *Plant Physiol.* **73**: 758–760.

Beaudry, R. M., N. Paz, C. C. Black, and S. J. Kays. 1987. Banana ripening: implications of changes in internal ethylene and CO_2 concentrations, pulp fructose 2,6-bisphosphate concentration, and activity of some glycolytic enzymes. *Plant Physiol.* **85**: 277–282.

Bennett, A. B., G. M. Smith, and B. G. Nichols. 1987. Regulation of climacteric respiration in ripening avocado fruit. *Plant Physiol.* **83**: 973–976.

Beyer, E. M., Jr. 1979. [^{14}C] Ethylene metabolism during leaf abscission in cotton. *Plant Physiol.* **64**: 971–974.

Beyer, E. M., Jr. and P. W. Morgan. 1970. Effect of ethylene on the uptake, distribution, and metabolism of indoleacetic acid-1-^{14}C and -2-^{14}C and naphthaleneacetic acid-1-^{14}C. *Plant Physiol.* **46**: 157–162.

Biale, J. B. 1950. Postharvest physiology and biochemistry of fruits. *Annu. Rev. Plant Physiol.* **1**: 183–206.

Biggs, M. S., R. W. Harriman, and A. K. Handa. 1986. Changes in gene expression during tomato fruit ripening. *Plant Physiol.* **81**: 395–403.

Blake, T. J., D. M. Reid, and S. B. Rood. 1983. Ethylene, indoleacetic acid and apical dominance in peas: A reappraisal. *Physiol. Plantarum* **59**: 481–487.

Burg, S. P. 1962. The physiology of ethylene formation. *Annu. Rev. Plant Physiol.* **13**: 265–302.

Burg, S. P. 1968. Ethylene, plant senescence, and abscission. *Plant Physiol.* **43**: 1503–1511.

Burg, S. P. 1973. Ethylene in plant growth. *Proc. Natl. Acad. Sci. U.S.A.* **70**: 591–597.

Burg, S. P. and E. A. Burg. 1962. Role of ethylene in fruit ripening. *Plant Physiol.* **37**: 179–189.

Burg, S. P. and E. A. Burg. 1964. Biosynthesis of ethylene. *Nature (London)* **203**: 869–870.

Burg, S. P. and E. A. Burg. 1965a. Gas exchange in fruits. *Physiol. Plantarum* **18**: 870–884.

Burg, S. P. and E. A. Burg. 1965b. Ethylene action and the ripening of fruits. *Science* **148**: 1190–1196.

Burg, S. P. and E. A. Burg. 1965c. Relationship between ethylene production and ripening in bananas. *Bot. Gaz.* **126**: 200–204.

Burg, S. P. and E. A. Burg. 1966a. Auxin-induced ethylene formation: Its relation to flowering in the pineapple. *Science* **152**: 1269.

Burg, S. P. and E. A. Burg. 1966b. The interaction between auxin and ethylene and its role in plant growth. *Proc. Natl. Acad. Sci. U.S.A.* **55**: 262–269.

Burg, S. P. and E. A. Burg. 1967a. Molecular requirements for the biological activity of ethylene. *Plant Physiol.* **42**: 144–152.

Burg, S. P. and E. A. Burg. 1967b. Inhibition of polar auxin transport by ethylene. *Plant Physiol.* **42**: 1224–1228.

Burg, S. P. and E. A. Burg. 1968. Ethylene formation in pea seedlings: Its relation to the inhibition of bud growth caused by indole-3-acetic acid. *Plant Physiol.* **43**: 1069–1074.

Burg, S. P. and C. O. Clagett. 1967. Conversion of methionine to ethylene in vegetative tissue and fruits. *Biochem. Biophys. Res. Commun.* **27**: 125–130.

Chadwick, A. V. and S. P. Burg. 1967. An explanation of the inhibition of root growth caused by indole-3-acetic acid. *Plant Physiol.* **42**: 415–420.

Chadwick, A. V. and S. P. Burg. 1970. Regulation of root growth by auxin-ethylene interaction. *Plant Physiol.* **45**: 192–200.

Crocker, W., A. E. Hitchcock, and P. W. Zimmerman. 1935. Similarities in the effects of ethylene and the plant auxins. *Contrib. Boyce Thompson Inst.* **7**: 231–248.

Crocker, W., P. W. Zimmerman, and A. E. Hitchcock. 1932. Ethylene-induced epinasty of leaves and the relation of gravity to it. *Contrib. Boyce Thompson Inst.* **4**: 177–218.

Decoteau, D. R. and L. E. Craker. 1987. Abscission: Ethylene and light control. *Plant Physiol.* **83**: 970–972.

Dilley, D. R. 1969. Hormonal control of fruit ripening. *HortScience* **4**: 111–114.

Drouet, A. and C. Hartmann. 1979. Polyribosomes from pear fruit. Changes during ripening and senescence. *Plant Physiol.* **64**: 1104–1108.

Eisinger, W. R. and S. P. Burg. 1972. Ethylene-induced pea internode swelling. Its relation to ribonucleic acid metabolism, wall protein synthesis, and cell wall structure. *Plant Physiol.* **50**: 510–517.

Frenkel, C., I. Klein, and D. R. Dilley. 1968. Protein synthesis in relation to ripening of pome fruits. *Plant Physiol.* **43**: 1146–1153.

Galston, A. W. and P. Kaur. 1961. Comparative studies on the growth and light sensitivity of green and etiolated pea stem sections. In: McElroy, W. D. and B. Glass, eds. *Light and Life.* Johns Hopkins Press, Baltimore. Pp. 687–705.

Gepstein, S. and K. V. Thimann. 1980. The effect of light on the production of ethylene from 1-aminocyclopropane-1-carboxylic acid by leaves. *Planta* **149**: 196–199.

Gepstein, S. and K. V. Thimann. 1981. The role of ethylene in the senescence of oat leaves. *Plant Physiol.* **68**: 349–354.

Goeschl, J. D. and H. K. Pratt. 1968. Regulatory roles of ethylene in the etiolated growth habit of *Pisum sativum.* In: Wightman, F. and G. Setterfield, eds. *Biochemistry and Physiology of Plant Growth Substances.* Runge Press, Ottawa. Pp. 1229–1242.

Goeschl, J. D., H. K. Pratt., and B. A. Bonner. 1967. An effect of light on the production of ethylene and the growth of the plumular portion of etiolated pea seedlings. *Plant Physiol.* **42**: 1077–1080.

Hansen, E. 1966. Postharvest physiology of fruits. *Annu. Rev. Plant Physiol.* **17**: 459–477.

Hansen, E. and E. Hartman. 1937. Effect of ethylene and metabolic gases upon respiration and ripening of pears before and after cold storage. *Plant Physiol.* **12**: 441–454.

Harrison, M. A. and P. B. Kaufman. 1982. Does ethylene play a role in the release of lateral buds (tillers) from apical dominance in oats? *Plant Physiol.* **70**: 811–814.

Ho, T.-H. D., J. Abroms, and J. E. Varner. 1982. Effect of ethylene on the release of α-amylase through cell walls of barley aleurone layers. *Plant Physiol.* **69**: 1128–1131.

Jackson, M. B. and D. J. Campbell. 1976. Waterlogging and petiole epinasty in tomato: the role of ethylene and low oxygen. *New Phytol.* **76**: 21–29.

Jeffery, D., C. Smith, P. Goodenough, I. Prosser, and D. Grierson. 1984. Ethylene-independent and ethylene-dependent biochemical changes in ripening tomatoes. *Plant Physiol.* **74**: 32–38.

Jiao, X.-Z., W.-K. Yip, and S. F. Yang. 1987. The effect of light and phytochrome on 1-aminocyclopropane-1-carboxylic acid metabolism in etiolated wheat seedling leaves. *Plant Physiol.* **85**: 643–647.

Kang, B. G., W. Newcomb, and S. P. Burg. 1971. Mechanism of auxin-induced ethylene production. *Plant Physiol.* **47**: 504–509.

Kang, B. G., C. S. Yokum, S. P. Burg, and P. M. Ray. 1967. Ethylene and carbon dioxide: Mediation of hypocotyl hook-opening response. *Science* **156**: 958–959.

Kawase, M. 1976. Ethylene accumulation in flooded plants. *Physiol. Plantarum* **36**: 236–241.

Kende, H. and A. D. Hanson. 1976. Relationship between ethylene evolution and senescence in morning-glory flower tissue. *Plant Physiol.* **57**: 523–527.

Kidd, F. and C. West. 1933. The influence of the composition of the atmosphere upon the incidence of the climacteric in apples. *Rep. Food Invest. Board.* 51–57.

Legge, R. L., K.-H. Cheng, J. R. Lepock, and J. E. Thompson. 1986. Differential effects of senescence on the molecular organization of membranes in ripening tomato fruits. *Plant Physiol.* **81**: 954–959.

Leopold, A. C. and P. E. Kriedemann. 1975. *Plant Growth and Development.* 2nd ed. McGraw-Hill Book Company, New York.

Lieberman, M. 1975. Biosynthesis and regulatory control of ethylene in fruit ripening. A review. *Physiol. Vég.* **13**: 489–499.

Lieberman, M., A. Kunishi, L. W. Mapson, and D. A. Wardale. 1966. Stimulation of ethylene production in apple tissue slices by methionine. *Plant Physiol.* **41**: 376–382.

Ludford, P. M. 1987. Postharvest hormone changes in vegetables and fruit. In: Davies, P. J., ed. *Plant Hormones and Their Role in Plant Growth and Development.* Martinus Nijhoff Publishers, Dordrecht, The Netherlands. Pp. 574–592.

Mapson, L. W. 1970. Biosynthesis of ethylene and the ripening of fruit. *Endeavour* **29**: 29–33.

McGlasson, W. B. 1970. The ethylene factor. In: Hulme, A. C., ed. *The Biochemistry of Fruits and Their Products.* Vol. 1. Academic Press, New York. Pp. 475–519.

McKeon, T. A. and S. F. Yang. 1987. Biosynthesis and metabolism of ethylene. In: Davies, P. J., ed. *Plant Hormones and Their Role in Plant Growth and Development.* Martinus Nijhoff Publishers, Dordrecht, The Netherlands. Pp. 94–112.

Morgan, P. W. 1969. Stimulation of ethylene evolution and abscission in cotton by 2-chloroethanephosphonic acid. *Plant Physiol.* **44**: 337–341.

Morgan, P. W. and J. I. Durham. 1980. Ethylene production and leaflet abscission in *Mèlia azédarach* L. *Plant Physiol.* **66**: 88–92.

Morgan, P. W. and H. W. Gausman. 1966. Effects of ethylene on auxin transport. *Plant Physiol.* **41**: 45–52.

Ness, P. J. and R. J. Romani. 1980. Effects of aminoethoxyvinylglycine and countereffects of ethylene on ripening of Bartlett pear fruits. *Plant Physiol.* **65**: 372–376.

Nichols, S. E. and G. G. Laties. 1985. Differential control of ethylene-induced gene expression and respiration in carrot roots. *Plant Physiol.* **77**: 753–757.

Phillips, I. D. J. 1971. *Introduction to the Biochemistry and Physiology of Plant Growth Hormones.* McGraw-Hill Book Company, New York.

Pratt, H. K. and J. D. Goeschl. 1969. Physiological roles of ethylene in plants. *Annu. Rev. Plant Physiol.* **20**: 541–584.

Ray, P. M. 1963. *The Living Plant.* Holt, Rinehart & Winston, New York.

Reid, M. S. 1985. Ethylene and abscission. *HortScience* **20**: 45–50.

Reid, M. S. 1987. Ethylene in plant growth, development, and senescence. In: Davies, P. J., ed. *Plant Hormones and Their Role in Plant Growth and Development.* Martinus Nijhoff Publishers, Dordrecht, The Netherlands. Pp. 257–279.

Reid, M. S., Y. Mor, and A. M. Kofranek. 1981. Epinasty of poinsettias—the role of auxin and ethylene. *Plant Physiol.* **67**: 950–952.

Riov, J. and S. F. Yang. 1982. Effects of exogenous ethylene on ethylene production in citrus leaf tissue. *Plant Physiol.* **70**: 136–141.

Roberts, J. A. and G. A. Tucker. 1985. *Ethylene and Plant Development.* Butterworths, London.

Rubinstein, B. and F. B. Abeles. 1965. Relationship between ethylene evolution and leaf abscission. *Bot. Gaz.* **126**: 255–259.

Satler, S. O. and H. Kende. 1985. Ethylene and the growth of rice seedlings. *Plant Physiol.* **79**: 194–198.

Sisler, E. C. 1980. Partial purification of an ethylene-binding component from plant tissue. *Plant Physiol.* **66**: 404–406.

Sitrit, Y., J. Riov, and A. Blumenfeld. 1986. Regulation of ethylene biosynthesis in avocado fruit during ripening. *Plant Physiol.* **81**: 130–135.

Smith, A. R. and M. A. Hall. 1984. Mechanism of ethylene action. *Pl. Growth Regulation* **2**: 151–165.

Speirs, J., C. J. Brady, D. Grierson, and E. Lee. 1984. Changes in ribosome organization and messenger RNA abundance in ripening tomato fruits. *Aust. J. Plant Physiol.* **11**: 225–233.

Stewart, R. N., M. Lieberman, and A. T. Kunishi. 1974. Effects of ethylene and gibberellic acid on cellular growth and development in apical and subapical regions of etiolated pea seedlings. *Plant Physiol.* **54**: 1–5.

Suttle, J. C. 1986. Cytokinin-induced ethylene biosynthesis in nonsenescing cotton leaves. *Plant Physiol.* **82**: 930–935.

Suttle, J. C. and H. Kende. 1980. Ethylene action and loss of membrane integrity during petal senescence in *Tradescantia*. *Plant Physiol.* **65**: 1067–1072.

Varner, J. E. and D. T. Ho. 1976. Hormones. In: Bonner, J. and J. E. Varner, eds. *Plant Biochemistry*. 3rd ed. Academic Press, New York. Pp. 713–770.

Venis, M. A. 1984. Cell-free ethylene-forming systems lack stereochemical fidelity. *Planta* **162**: 1141–1145.

Von Abrams, G. J. and H. K. Pratt. 1967. Effect of ethylene on the permeability of excised cantaloupe tissue. *Plant Physiol.* **42**: 299–301.

Wang, C. Y. and D. O. Adams. 1982. Chilling-induced ethylene production in cucumbers (*Cucumis sativus* L.). *Plant Physiol.* **69**: 424–427.

Wang, C. Y. and W. M. Mellenthin. 1972. Internal ethylene levels during ripening and climacteric in Anjou pears. *Plant Physiol.* **50**: 311–312.

Wheeler, R. M. and F. B. Salisbury. 1981. Gravitropism in higher plant shoots. I. A role for ethylene. *Plant Physiol.* **67**: 686–690.

Wheeler, R. M., R. G. White, and F. B. Salisbury. 1986. Gravitropism in higher plant shoots. IV. Further studies on participation of ethylene. *Plant Physiol.* **82**: 534–542.

Yang, S. F. 1985. Biosynthesis and action of ethylene. *HortScience* **20**: 41–45.

Yang, S. F. and N. E. Hoffman. 1984. Ethylene biosynthesis and its regulation in higher plants. *Annu. Rev. Plant Physiol.* **35**: 155–189.

Yu, Y.-B. and S. F. Yang. 1979. Auxin-induced ethylene production and its inhibition by aminoethoxyvinylglycine and cobalt ion. *Plant Physiol.* **64**: 1074–1077.

Yu, Y.-B. and S. F. Yang. 1980. Biosynthesis of wound ethylene. *Plant Physiol.* **66**: 281–285.

Zimmerman, P. W. and F. Wilcoxon. 1935. Several chemical growth substances which cause initiation of roots and other responses in plants. *Contrib. Boyce Thompson Inst.* **7**: 209–229.

CHAPTER 7

Brassinosteroids

Introduction

It has long been known that pollen is rich in hormones, and since the 1930s the United States Department of Agriculture's Plant Hormone Laboratory at Beltsville, Maryland has investigated growth-promoting substances in pollen extracts. In 1979, for the first time, a novel plant growth regulating steroidal substance called "brassinolide" was isolated from rape (*Brassica napus*) pollen by Grove et al. Since then a large number (approximately 24, including two conjugates) of steroids related to brassinolide have been isolated and identified from various plant sources including angiosperms, gymnosperms, and an alga. These natural brassinosteroids are abbreviated BRs, and are numbered BR_1, BR_2, ..., BR_n by some authors. Mandava (1988) recommended that the numbering system be applied only to natural BRs and not to others derived by synthesis. Although some plant physiologists disagree, the BRs will be considered here as a separate group of hormones or growth substances.

Brief History of Discovery

John W. Mitchell and his associates at the USDA Plant Hormone Laboratory at Beltsville, Maryland reported in 1970 that extracts of pollen from rape (*Brassica napus*) and alder (*Alnus glutinosa*) produced an unusual growth response that combined elongation (a GA-like response) with swelling and curvature when tested in a bean first-internode bioassay. They proposed that the pollen of rape contains a new group of lipoidal hormones that they termed "brassins." In 1972, Mitchell and Gregory demonstrated that brassins could enhance crop yield, crop efficiency, and seed vigor. Then in 1975 the USDA group mounted a major effort to identify the active constituent in brassins. A large amount (500 lb) of bee-collected rape pollen was extracted with 2-propanol in 50-lb batches, and the extract was partitioned among carbon tetrachloride, methanol, and water. The methanol fraction containing the biological activity was chromatographed on a series of large silica gel columns, a process that reduced the amount of material to about 100 g. Further purification by column chromatography and high-performance liquid chromatography yielded 10 mg of crystalline material, termed "brassinolide" (Grove et al., 1979). Biological activity was monitored by the bean second-internode bioassay throughout the purification steps. Thus brassinolide (Fig. 7.1) is one specific BR, the first to be discovered. Unfortunately, two different systems for naming and numbering the BRs have

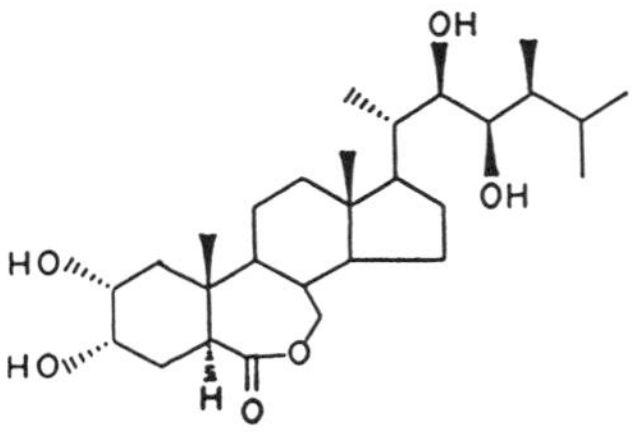

FIGURE 7.1. Structure of brassinolide, BR_1, the first brassinosteroid to be discovered.

been proposed; these may be compared in Figs. 7.2 and 7.3. Hopefully in the near future a single system will come to prevail.

Chemistry of BRs

BRs characterized from plants thus far (Figs. 7.2 and 7.3) are all 5α-cholestane derivatives and can be classified into C_{27}, C_{28}, and C_{29}, steroids, as is the case with typical phytosterols. All BRs contain a steroid nucleus (some with the oxygen function in the B ring) with a side chain at C-17 similar to the side chain in plant sterols. Other common features for all BRs, in addition to β-oriented angular C-18 and C-19 methyl groups, are (1) α-orientation of H at C-5 (A/B ring junction), (2) α-oriented hydroxyl groups at C-22 and C-23 (side chain), and (3) α-oriented hydroxyl groups (*cis*-geometry) at C-2 and C-3 in ring A of the steroid nucleus (exceptions: typhasterol contains only one hydroxyl at C-3 in the α position, and theasterol contains a β-hydroxyl at C-3 only; both compounds lack an hydroxyl at C-2).

Variations in BR structure come from different substitutions at C-24; that is, no substitution in C_{27} steroids, a methyl or an exo-methylene in C_{28} steroids, and an ethyl or an (*E*)-ethylidene in C_{29} steroids. The alkyl (methyl or ethyl) substituents at C-24 are *S*-oriented in BRs of higher plants, but a BR with a 24*R*-methyl (24-epicastasterone) has been characterized from a green alga (*Hydrodictyon reticulatum*). All BRs have a vicinal glycol in the side chain at C-22*R* and C-23*R*. With respect to B-ring oxidation, BRs can be grouped into either a 7-oxalactone, a 6-ketone, or a nonoxidized type. Structure/activity relationships have been described in detail by Yokota and Takahashi (1986) and Mandava (1988). Based on five bioassays, the following order of activity was seen: (1) lactones (highest), (2) ketones (intermediate), and (3) 6-deoxo compounds (least). Brassinolide is the most active of all known natural and synthetic BRs in any test system.

Natural Occurrence of BRs

BRS have been isolated from numerous species of angiosperms and gymnosperms and one green alga. BRs may be present in all parts of the plant, although roots evidently have not been examined. Pollen still is the richest source; pollen

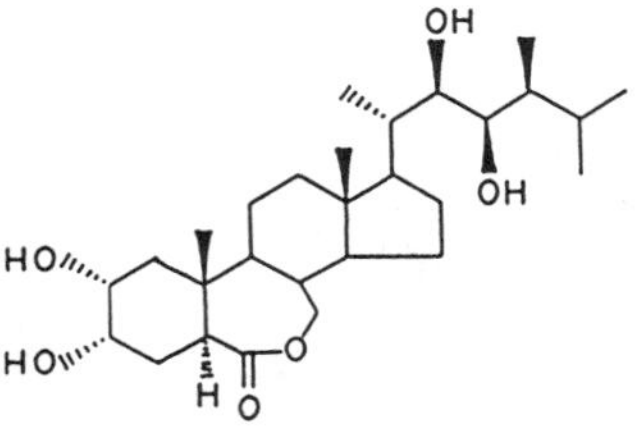

Brassinolide

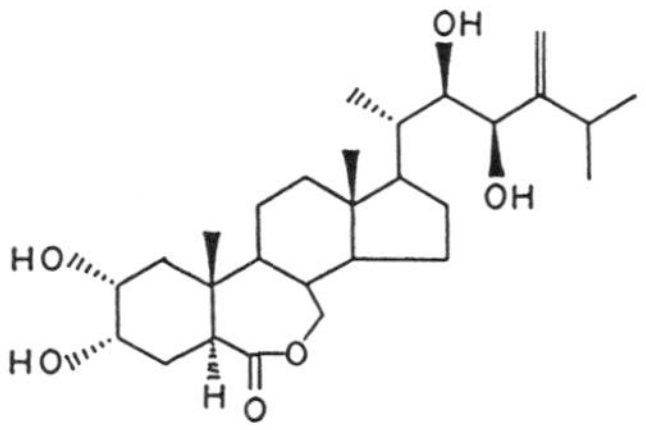

Dolicholide

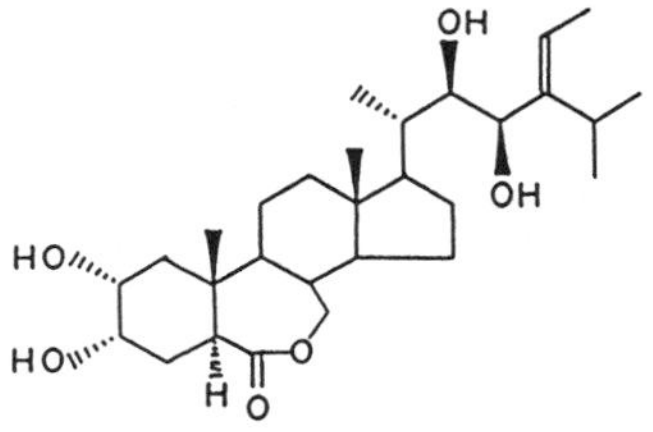

Homodolicholide

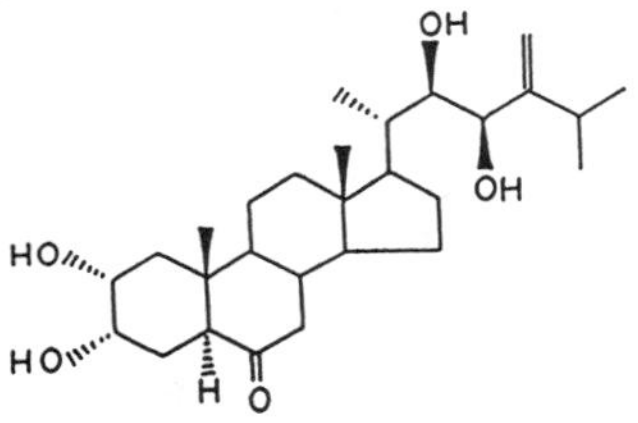

Dolichosterone

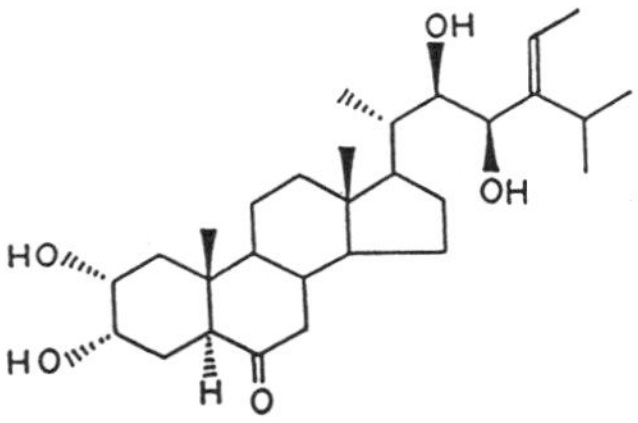

Homodolichosterone

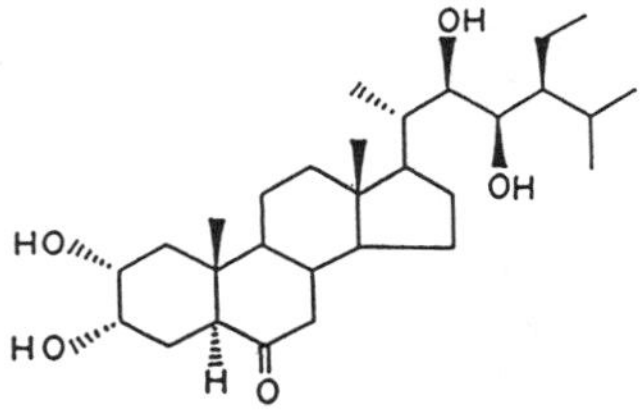

(24S)-24-Ethylbrassinone

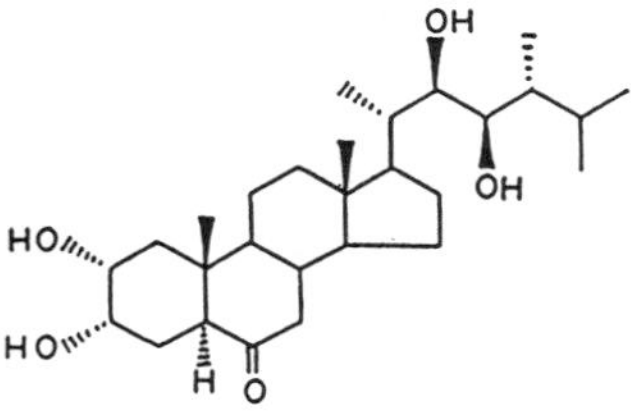

24-Epicastasterone

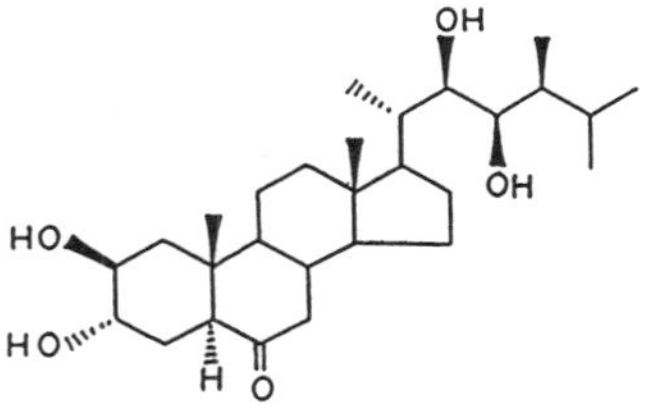

2-Epicastasterone

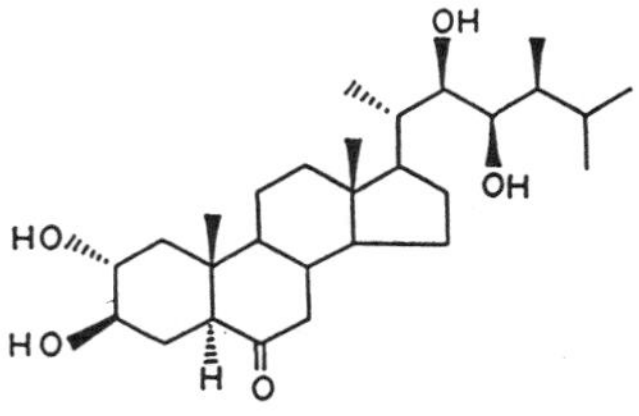

3-Epicastasterone

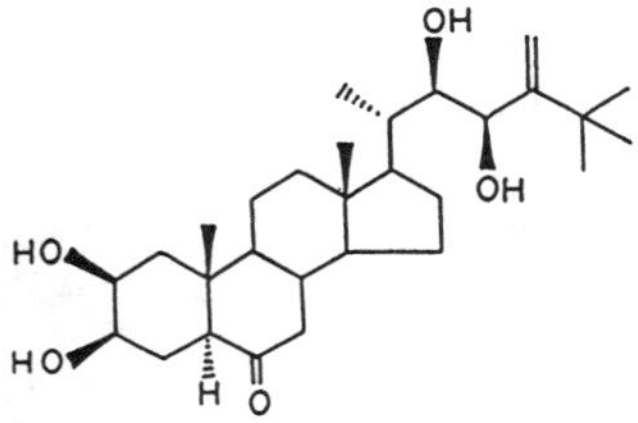

2,3-Diepi-25-methyl-dolichosterone

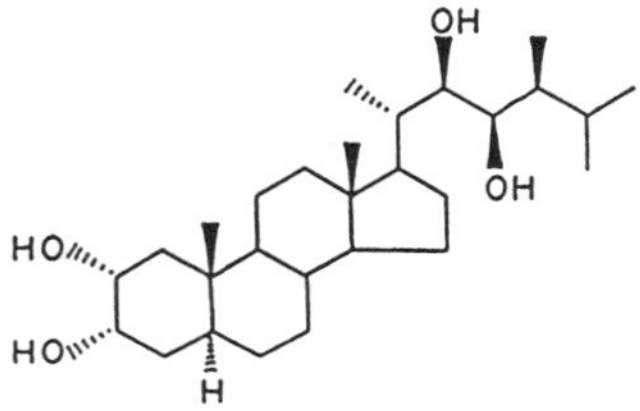

6-Deoxodihydrocastasterone

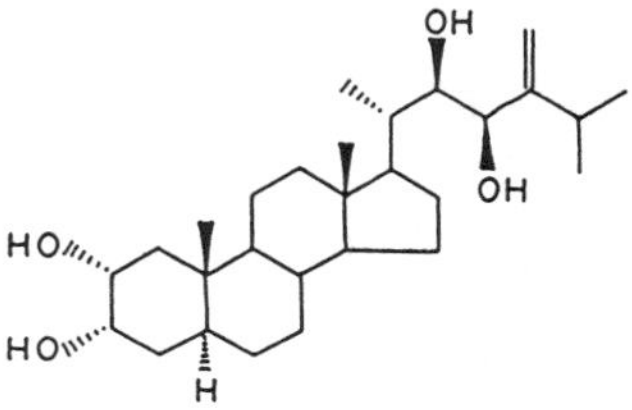

6-Deoxodihydro-dolichosterone

FIGURE 7.2. Structures of the naturally occurring brassinosteroids, including two conjugates. (Courtesy of Takao Yokota, 1988.)

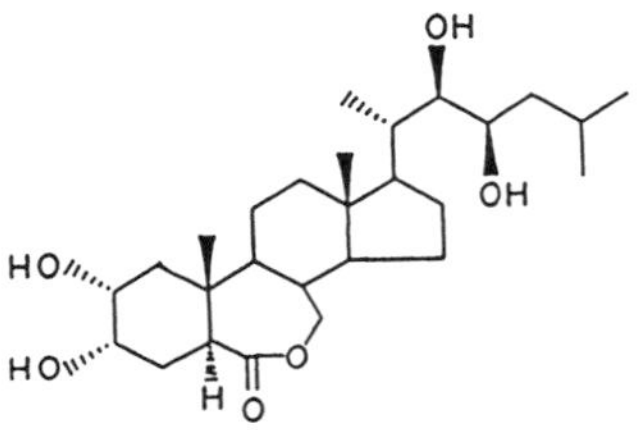

28-Norbrassinolide

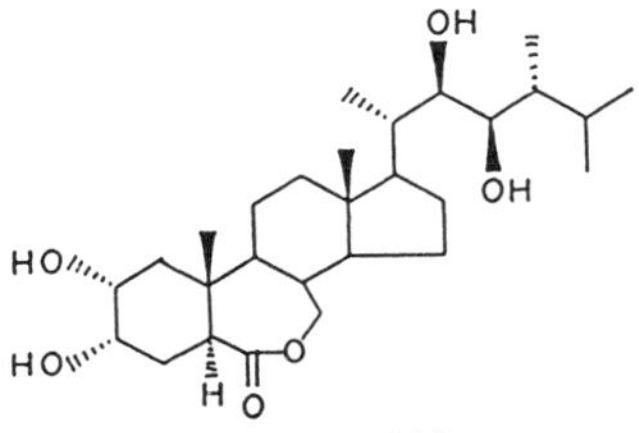

24-Epibrassinolide

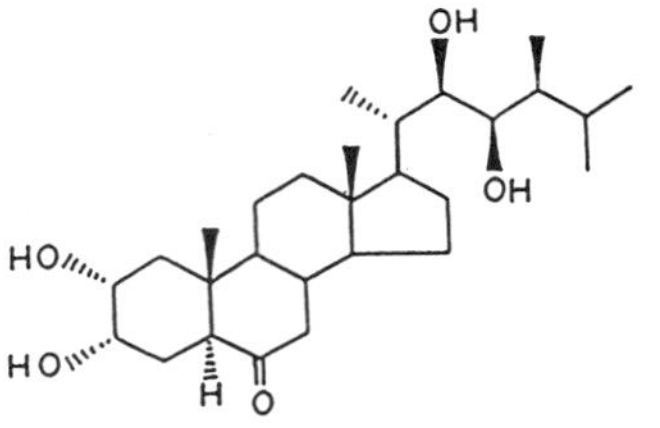

Castasterone

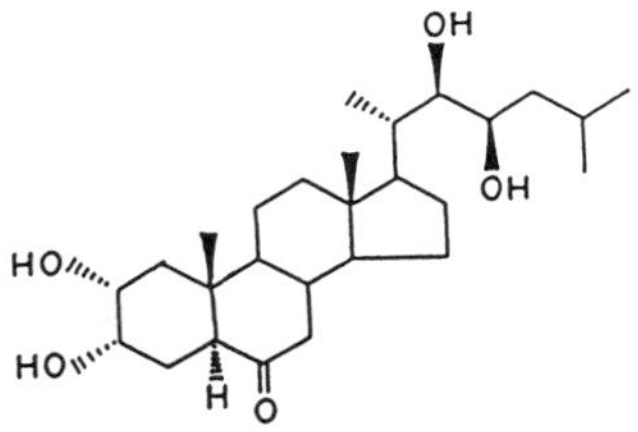

Brassinone

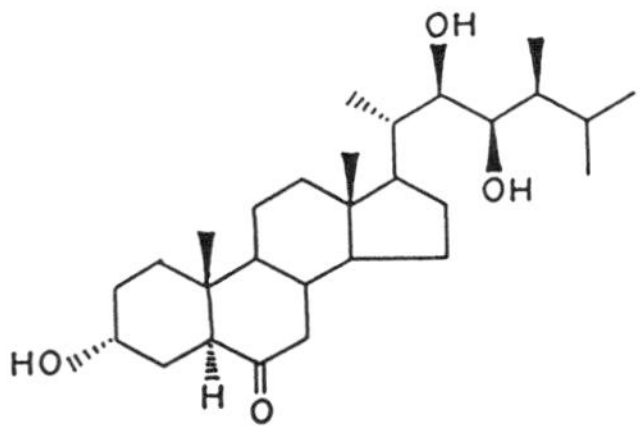

Typhasterol

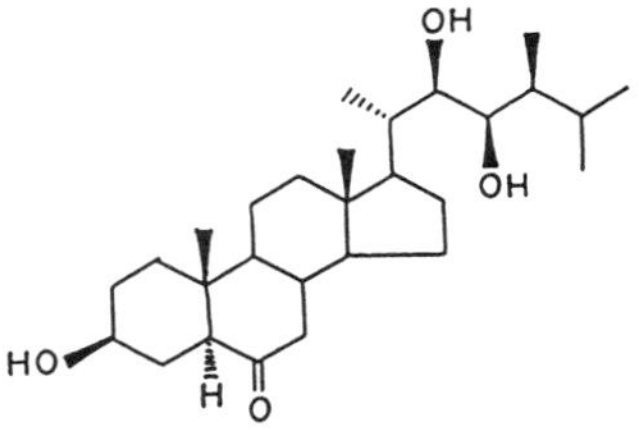

Teasterone

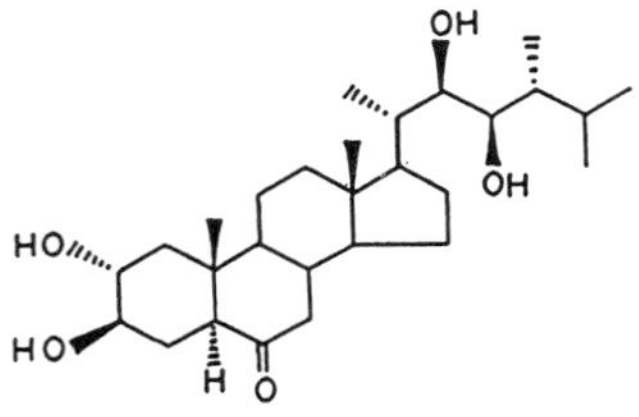

3,24-Diepicastasterone

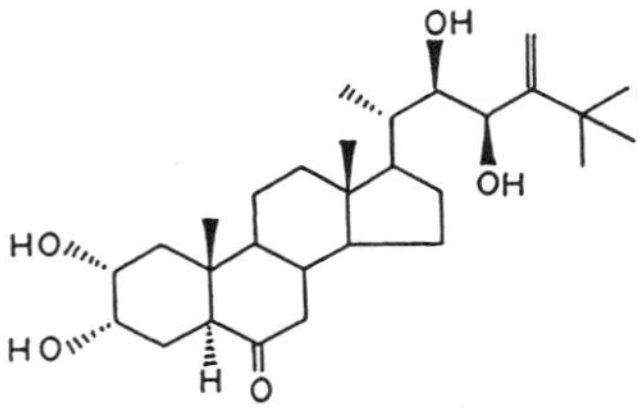

25-Methyldolichosterone

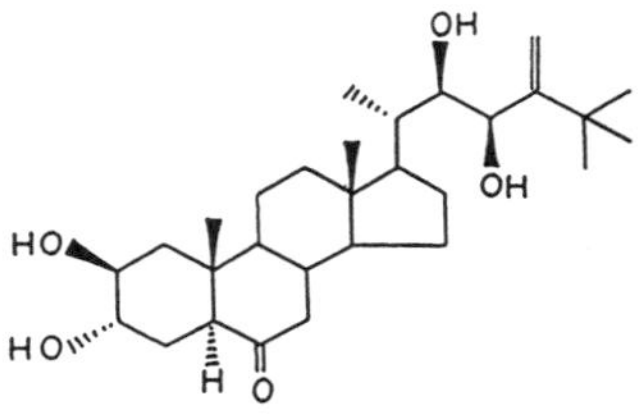

2-Epi-25-methyl-
dolichosterone

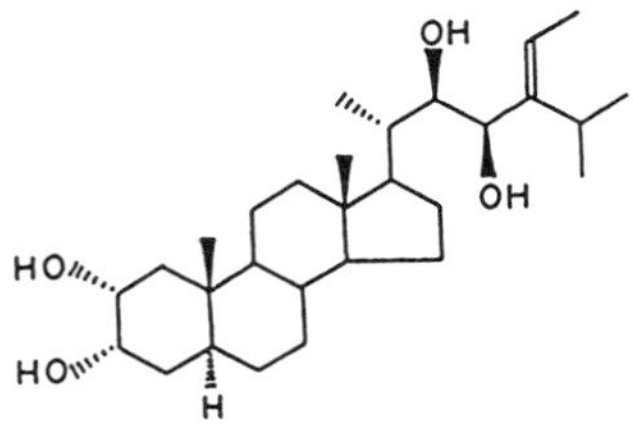

6-Deoxodihydro-
homodolichosterone

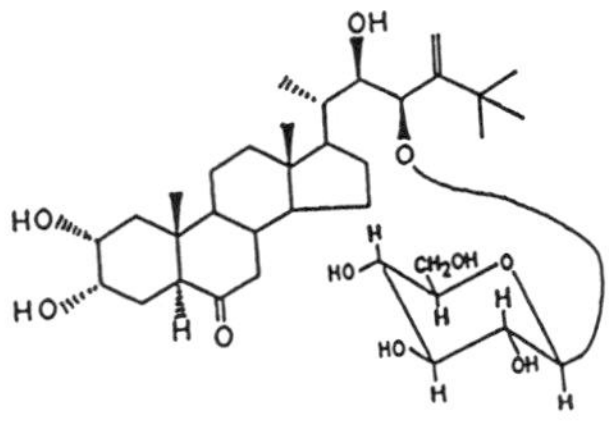

23-O-ß-D-Glucopyranoside
of 17

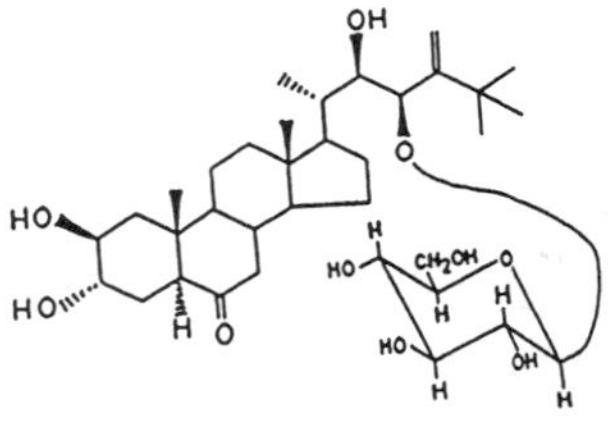

23-O-ß-D-Glucopyranoside
of 18

FIGURE 7.2. *Continued.*

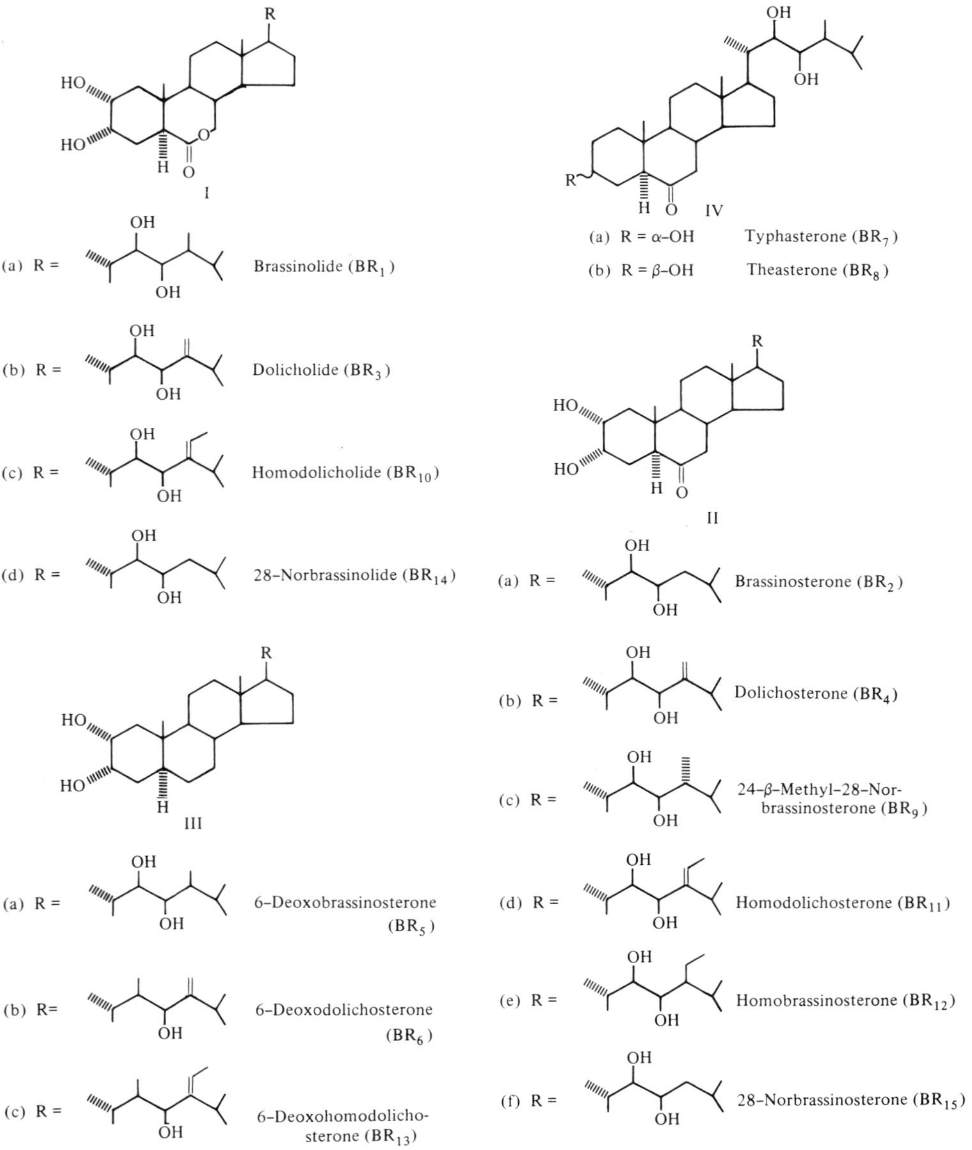

FIGURE 7.3. Naming and numbering system for 15 BRs according to Mandava (1988). (Reproduced, with permission, from the Annual Review of Plant Physiology and Plant Molecular Biology, Vol. 39 © 1988 by Annual Reviews, Inc.)

and seeds contain 1 to 1000 ng of BR per kg of tissue. Shoots contain 1 to 100 ng/kg, and fruit and leaves contain 1 to 10 ng/kg. It seems likely, according to Yokota and Takahashi (1986), that the number of BRs will increase in the future, as they have already found that immature seeds of *Phaseolus vulgaris* contain several steroidal growth substances, together with some conjugates.

Biological Effects

Brassinosteroids interact with other types of hormones (e.g., auxins and GAs) in many test systems. Yet, upon close examination, the BRs do appear to have biological activity that distinguishes them from the other known groups of plant hormones. BRs are active in various bioassays formerly thought to be specific for different types of other hormones. In some tests the effects of BRs and GAs are independent and additive (Fig.7.4), and there appears to be no interactive relationship between BRs and GAs. In many test systems BRs interact strongly with auxins, perhaps synergistically, that is, more than additively (Fig. 7.5). In systems designed to test for cytokinins, BR effects vary. ABA interacts strongly with BRs and prevents the effects BR induces. BRs alone and in combination with auxin induce the synthesis of ethylene. Generally, BRs are active at concentrations much lower (nM to pM range) than those of other types of hormones.

Now let us turn to some of the specific effects of BRs. The bioassay first used to detect and quantify brassinolide was the bean second-internode bioassay. Although both BRs and GAs cause elongation of the treated and upper internodes, BRs also characteristically evoke swelling, curvature, and splitting

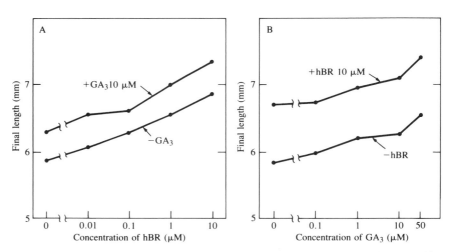

FIGURE 7.4. Interaction of 22,23(S,S)-homobrassinolide (hBR) and GA$_3$ on the elongation of cucumber hypocotyl sections. Measurements were made 24 hours after incubation. **A**, hBR concentration varied; **B**, GA$_3$ concentration varied. Effects are simply additive in both **A** and **B**. (Redrawn, with permission, from Katsumi, 1985.)

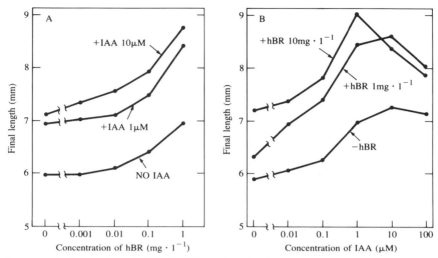

FIGURE 7.5. Interaction of 22,23(S,S)-homobrassinolide (hBR), a synthetic brassinoste-roid, and IAA on the elongation of cucumber hypocotyl sections. Measurements were made 24 hours after incubation. **A**, hBR concentration varied; **B**, IAA concentration varied. Effects of IAA and hBR are additive in **A** but are synergistic in **B**. (Redrawn, with permission, from Katsumi, 1985.)

of the second internode. Auxins and cytokinins are not detected by this bioassay. There is a rice bioassay in which BRs, but not GAs or auxins, evoke lamina inclination. Auxins evoke a response in this test, but only at much higher concentrations than those of BRs. GAs evoke a straight growth response without the bending that is characteristic for BRs. Hence the bean second-internode and rice test systems generally are regarded as bioassays specific for BRs.

BRs elicit a pronounced stem elongation response in dwarf pea epicotyls, dwarf bean apical segments, and mung bean epicotyls (Fig. 7.6) that are sensitive also to GAs but not to auxins. The effects of BRs and GAs, as noted previously, appear to be merely additive (Fig. 7.4). Ancymidol, an inhibitor of GA biosynthesis, inhibits GA-mediated growth but has no influence on the effect of BRs. BR is almost ineffective in a dwarf-rice growth test that is quite sensitive to GA.

There is also similarity between the effects of auxins and BRs in selected auxin bioassays, such as elongation of dwarf pea hook segment, maize coleoptile, azuki bean epicotyl, and retardation of dwarf bean hypocotyl hook opening. It has been reported that a strong synergism exists between auxins and BRs (see, e.g., Fig. 7.5). However, it was observed that the synergism only occurs when tissues are treated first with BR and then with IAA. The effects of BR and IAA were only additive when the order of treatment was reversed. Also, some BR-elicited growth responses, such as hypocotyl elongation and rice lamina bending, are nullified by antiauxins such as PCIB and TIBA. Mandava has reported that these observations suggest that BR enhances the sensitivity of

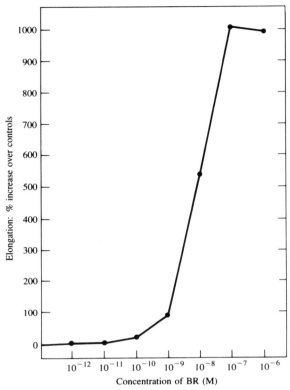

FIGURE 7.6. Effect of increasing concentrations of BR on the elongation of mung bean epicotyls after 48 hours. (Redrawn, with permission, from Gregory and Mandava, 1982.)

plant tissues to auxin, resulting in synergistic interaction with endogenous auxin, but does not itself have auxin-like functions. However, there are some observations that do not agree, such as (1) in maize root segments BR induces a significant stimulation of growth, but auxin inhibits such action; (2) BR was not active in the *Avena* curvature assay when BR was tested alone or in combination with IAA; and (3) although BR does not inhibit the initiation of adventitious root meristems in the mung bean epicotyl bioassay, it inhibits their outgrowth to form adventitious roots, apparently interfering with the action of endogenous auxin. Nevertheless, Mandava concluded that, in most tests, BR action is mediated through endogenous auxin, resulting in a synergism.

BR shows cytokinin-like activity in the cucumber cotyledon expansion assay, but it is reported to be ineffective in other cytokinin bioassays. BRs are quite active in the wheat leaf unrolling test, in which cytokinins are moderately active but auxins are inhibitory. It has been reported also that combinations of auxin and BR are more effective than those of auxin and benzyladenine in promoting callus growth of several species.

Further evidence for BR–auxin synergism comes from data on auxin-induced ethylene production in etiolated mung bean hypocotyl segments. BR stimulates this response in a synergistic manner. BR alone and in combination with auxin induces the synthesis of ethylene, perhaps between SAM and ACC.

The importance of radiant energy and spectral quality in the BR-elicited growth response has been emphasized by several authors. In soybean and mung bean seedlings the growth-promoting effect of brassinolide occurs only in the light, but not in the dark, and the most important region of light is red light. In fact, in test systems that require darkness, BR generally does not cause any growth.

Effects on Nucleic Acid and Protein Metabolism

Mandava and co-workers attempted to determine whether BR affects nucleic acid and protein metabolism (Mandava et al., 1987). First, putative inhibitors of RNA and protein synthesis were tested for their possible effects on the BR-induced response in mung bean epicotyl cuttings. The inhibitors all interfered with epicotyl growth, actinomycin D and cycloheximide to the greatest extent. Similar results were obtained with azuki bean epicotyls. However, these studies indicate only that the growth effects induced by BR, like those induced by auxins and GAs, depend upon the synthesis of nucleic acids and proteins.

In another study, Kalinich et al. (1985) found that BR treatment significantly increased RNA and DNA polymerase activities, and the synthesis of RNA, DNA, and protein in *Phaseolus vulgaris* and *Phaseolus aureus*. This finding suggests the involvement of BRs in transcription and replication during growth. However, obviously much more research will be required to elucidate the molecular mechanism of action of BRs.

Practical Applications in Agriculture

In 1972, Mitchell and his associates reported for the first time on the biological effects of brassins (crude brassinolide extract of rape pollen). They found that brassins increased growth in beans and soybeans and in woody plants such as Siberian elm. After the isolation of brassinolide and subsequent synthesis of BRs and some analogues, the USDA workers conducted limited tests in greenhouses and small field plots on a few vegetable and root crops. Significantly increased was the harvest of various vegetables and root crops (e.g., radish, lettuce, bean, pepper and potato), wheat, mustard, and barley. The growth responses occur chiefly in the slow-growing plants, and, therefore, reduce the phenotypic variability of the plants.

In Japan, homobrassinolide and brassinolide were tested for agricultural application, and some promising results were obtained. There was a significant acceleration of growth in rice resulting from treatment of seeds with homobrassinolide, and this effect was prominent under low temperature conditions. Yield increases of potato and sweet potato occurred when they were treated with brassinolide at the planting stage. Yield increases were also observed in fruits and grains when flowers of wheat, maize, rice, and tomato were treated with brassinolide and when flowers of rice, tomato, cucumber and egg plant were treated with homobrassinolide. Environmental stresses were also seen to be alleviated by treatment with brassinolide. In most all of these cases, the effective concentration range of brassinolide lay between 10^{-3} and 10^{-6} ppm, while that of homobrassinolide was much higher.

As Mandava pointed out, in summary, although these findings are encouraging, they result mainly from preliminary findings in greenhouses and limited field plots. Several parameters (e.g., formulation, timing, method of application) must be investigated further before the full potential of BRs for increasing biomass and yield of crops, and for alleviating diseases and other forms of environmental stress, can be realized.

Some Problems and Prospects

As Mandava noted in 1988, the brassinosteroid area of plant hormone research has not yet caught the attention of many plant scientists. One limiting factor at this time is the availability of brassinolide and/or its analogues from commercial sources. Yet progress may soon accelerate, because several BRs have been isolated and identified from various plants, and chemists have already devised methods for the synthesis of many of the known BRs. Moreover, structure/activity relationships have been well established.

The biosynthesis and metabolism of BRs have yet to be thoroughly investigated. The mechanism of action is unknown. Practical applications have to be more conclusively assessed. In all these areas, progress may be expected to be forthcoming in the near future.

Of utmost importance to workers in this field is the adoption of a single naming and numbering system for the BRs.

References

Arima, M., T. Yokota, and N. Takahashi. 1984. Identification and quantification of brassinolide-related steroids in the insect gall and healthy tissues of the chestnut. *Phytochemistry* **23**: 1587–1591.

Arteca, R. N., D. S. Tsai, C. Schlagnhaufer, and N. B. Mandava. 1983. The effect of brassinolide on auxin-induced ethylene production by etiolated mung bean segments. *Physiol. Plantarum* **59**: 539–544.

Braun, P. and W. Wild. 1984. The influence of brassinosteroid on growth and parameters of photosynthesis of wheat and mustard plants. *J. Plant Physiol.* **116**: 189–196.

Cohen, J. D. and W. J. Meudt. 1983. Investigations on the mechanism of the brassinosteroid response. I. Indole-3-acetic acid metabolism and transport. *Plant Physiol.* **72**: 691–694.

Donauber, J. R., A. W. Greaves, and T. C. McMorris. 1984. A novel synthesis of brassinolide. *J. Org. Chem.* **49**: 2833–2834.

Geuns, J. M. C. 1983. Plant-steroid hormones. *Biochem. Soc. Trans.* **11**: 543–548.

Gregory, L. E. 1981. Acceleration of plant growth through seed treatment with brassins. *Am. J. Bot.* **68**: 586–588.

Gregory, L. E. and N. B. Mandava. 1982. The activity and interaction of brassinolide and gibberellic acid in mung bean epicotyls. *Physiol. Plantarum* **54**: 239–243.

Grove, M. D., F. G. Spencer, W. K. Rohwedder, N. B. Mandava, J. F. Worley, J. D. Warthen, Jr., G. L. Steffens, J. L. Flippen, and J. C. Cook, Jr. 1979. A unique plant growth promoting steroid from *Brassica napus* pollen. *Nature (London)* **281**: 216–217.

Horgan, P. A., C. K. Nakagawa, and R. T. Irvin. 1984. Production of monoclonal antibodies to a steroid plant growth regulator. *Can. J. Cell Biol.* **62**: 715–721.

Ikekawa, N. and S. Takasuto. 1984. Microanalysis of brassinosteroids in plants by gas chromatography/mass spectrometry. *Mass Spectrum* **32**: 55–70.

Ikekawa, N., S. Takasuto, T. Kitsuwa, H. Saito, T. Morishita, and H. Abe. 1984. Analysis of natural brassinosteroids by gas chromatography-mass spectrometry. *J. Chromatogr.* **290**: 289–302.

Kalinich, J. F., N. B. Mandava, and J. A. Todhunter. 1985. Relationship of nucleic acid metabolism to brassinolide-induced responses in beans, *J. Plant Physiol.* **120**: 207–214.

Katsumi, M. 1985. Interaction of a brassinosteroid with IAA and GA_3 in the elongation of cucumber hypocotyl sections. *Plant Cell Physiol.* **26**: 615–625.

Krizek, D. T. and N. B. Mandava. 1982. Influence of spectral quality on the growth response of intact bean plants to brassinosteroids, a growth promoting steroidal lactone. I. Stem elongation and morphogenesis. *Physiol. Plantarum* **57**: 317–323.

Krizek, D. T. and J. F. Worley. 1973. The influence of light intensity on the internodal response of intact bean plants to brassins. *Bot. Gaz.* **134**: 147–150.

Mandava, N. B. 1988. Plant growth-promoting brassinosteroids. *Annu. Rev. Plant Physiol. Plant Mol. Biol.* **39**: 23–52.

Mandava, N. B., J. M. Sasse and J. H. Yopp. 1981. Brassinolide, a growth promoting steriodal lactone. II. Activity in selected gibberellin and cytokinin bioassays. *Physiol. Plantarum* **53**: 453–461.

Mandava, N. B., M. J. Thompson, and J. H. Yopp. 1987. Effects of selected putative inhibitors of RNA and protein synthesis on brassinosteroid-induced growth in mung bean epicotyls. *J. Plant Physiol.* **128**: 53–65.

Meudt, W. J. 1987. Investigations on the mechanism of the brassinosteroid response. VI. Effect of brassinolide on gravitropism of bean hypocotyls. *Plant Physiol.* **83**: 195–198.

Meudt, W. J. and M. J. Thompson. 1983a. Investigations on the mechanism of the brassinosteroid response. II. A modulation of auxin action. *10th Proc. Plant Growth Regul. Soc. Am.* Pp. 306–311.

Meudt, W. J. and M. J. Thompson. 1983b. Investigations on the mechanism of the brassinosteroid response. III. Techniques for potential enhancement of crop production. *10th Proc. Plant Growth Regul. Soc. Am.* Pp. 312–318.

Mitchell, J. W. and L. E. Gregory. 1972. Enhancement of overall growth, a new response to brassins. *Nature (London)* **239**: 254.

Mitchell, J. W. and G. A. Livingston. 1968. Methods of studying plant hormones and growth regulating substances. *U. S. Dept. Agric. Handb. No. 336.* GPO, Washington, D.C.

Mitchell, J. W., N. B. Mandava, J. F. Worley, J. R. Plummer, and M. V. Smith. 1970. Brassins: A new family of plant hormones from rape pollen. *Nature (London)* **225**: 1065–1066.

Mori, K. 1980. Synthesis of a brassinolide analog with high plant growth promoting activity. *Agric. Biol. Chem.* **44**: 1211–1212.

Sakakibara, M. and K. Mori. 1983. Improved synthesis of brassinolide. *Agric. Biol. Chem.* **47**: 663–664.

Sasse, J. M. 1985. The place of brassinolide in the sequential response to plant growth regulators in elongating tissue. *Physiol. Plantarum* **63**: 303–308.

Schlagnhaufer, C. D. and R. N. Arteca. 1985. Brassinosteroid-induced epinasty in tomato plants. *Plant Physiol.* **78**: 300–303.

Schlagnhaufer, C., R. N. Arteca, and J. H. Yopp. 1984. A brassinosteroid-cytokinin interaction on ethylene production by etiolated mung bean segments. *Physiol. Plantarum* **60**: 347–350.

Takeno, K. and R. P. Pharis. 1982. Brassinolide-induced bending of the lamina of dwarf rice seedlings: An auxin-mediated phenomenon. *Plant Cell Physiol.* **23**: 1275–1281.

Thompson, M. J., N. B. Mandava, J. L. Flippen-Anderson, J. F. Worley, S. R. Dutky, and W. E. Robbins. 1979. Synthesis of brassinosteroids. New plant growth promoting steriods. *J. Org. Chem.* **44**: 5002–5004.

Thompson, M. J., W. J. Meudt, N. B. Mandava, S. R. Dutky, W. R. Lusby, and D. W. Spaulding. 1982. Synthesis of brassinosteroids and relationship of structure to plant growth-promoting effects. *Steroids* **39**: 89–105.

Wada, K., H. Kondo, and S. Marumo. 1985. A simple bioassay for brassinosteroids: a wheat leaf unrolling test. *Agric. Biol. Chem.* **49**: 2249–2251.

Yokota, T., M. Arima, and N. Takahashi. 1982. Castasterone, a new phytosterol with plant-hormone activity from chestnut insect gall. *Tetrahedron Lett.* **23**: 1275–1278.

Yokota, T., M. Arima, N. Takahashi, and A. Crozier. 1985. Steroidal plant growth regulators, castasterone and typhasterol (2-deoxycastasterone) from shoots of sitka spruce (*Picea sitchensis*). *Phytochemistry* **24**: 1333–1335.

Yokota, T., M. Arima, N. Takahashi, S. Takatsumo, N. Ikekawa, and T. Takematsu. 1983. 2-Deoxycastasterone, a new brassinolide-related bioactive steroid from *Pinus* pollen. *Agric. Biol. Chem.* **47**: 2419–2420.

Yokota, T., J. Baba, S. Koba, and N. Takahashi. 1984. Purification and separation of eight steroidal plant growth regulators from *Dolichos lablab* seed. *Agric. Biol. Chem.* **48**: 2529–2543.

Yokota, T., S. Koba, S. K. Kim, S. Takasuto, N. Ikekawa, M. Sukakibara, K. Okada, K. Mori, and N. Takahashi. 1987. Diverse structural variations of the brassinosteroids in *Phaseolus vulgaris* seeds. *Agric. Biol. Chem.* **51**: 1625–1631.

Yokota, T. and N. Takahashi. 1986. Chemistry, physiology and agricultural application of brassinolide and related steroids. In: Bopp, M., ed. *Plant Growth Substances 1985.* Springer-Verlag, Berlin, Heidelberg, New York, Tokyo. Pp. 129–138.

Yopp, J. H., G. C. Colclasure, and N. B. Mandava. 1979. Effect of brassin-complex on auxin and gibberellin mediated effects in the morphogenesis of the etiolated bean epicotyl. *Physiol. Plantarum* **46**: 247–254.

Yopp, J. H., N. B. Mandava, and J. Sasse. 1981. Brassinolide, a growth promoting steroidal lactone. I. Activity in selected auxin bioassays. *Physiol. Plantarum* **53**: 445–452.

CHAPTER 8

Phytochrome

Introduction

Phytochrome very definitely is not a hormone. It is a pigment. However, no book on plant hormones could be regarded as complete without detailed consideration of phytochrome. For phytochrome is involved in the regulation of growth and development literally throughout the ontogeny of seed plants, from the germination of some kinds of seeds to floral initiation in many species of angiosperms. There are multiple important interactions of the phytochrome system with hormone metabolism and hormone action and complex interactions as well between the phytochrome system and endogenous circadian rhythms.

History of Discovery and Modern Description

Phytochrome is a water-soluble chromoprotein that exists in two definitive, photointerconvertible forms—P_r, which absorbs light maximally in the red region (λ_{max} = 666 nm[1]), and P_{fr}, which absorbs maximally in the near far-red (λ_{max} = 730 nm)—and which serves as the photoreceptor in all the "red, far-red reversible photoreactions" of plants.

The definitive discovery of phytochrome in 1959 by H. A. Borthwick, W. L. Butler, S. B. Hendricks, K. H. Norris, and H. W. Siegelman at U.S.D.A. laboratories at Beltsville, Maryland constitutes one of the most significant achievements in plant physiology. And how fitting it is that phytochrome ultimately was discovered at the same laboratories where W. W. Garner and H. A. Allard discovered photoperiodism some 39 years earlier. The story of how phytochrome was discovered should be appreciated by all students of plant biology.

A very fundamental advance in plant photobiology, which dates from circa 1935, was the discovery that various plant responses to light are elicited maximally by red light, and that certain of these actions of red light are reversible by far-red light. For example, it has been known since the early reports by Flint and McAlister (1935, 1937) that the germination of some kinds of seeds that require light (e.g., lettuce, *Latuca sativa* var. Grand Rapids) is promoted by red light and suppressed by far-red light. Detailed action spectra for the germination response of light-sensitive lettuce seeds showed maximum promotion near 660 nm and maximum suppression near 730 nm. Additionally, it

---

[1] Etiolated *Avena* phytochrome.

TABLE 8.1. Sample class data showing the effects of red and far-red irradiation on the germination of Grand Rapids lettuce seeds[a].

Treatment	Percentage germination
Dark control	5
Red	93
Red–far-red	11
Red–far-red–red	89

[a] Seeds were hydrated for 11 hours in darkness at room temperature. Working under a blue safelight in a darkroom, 50 seeds each were placed on moist filter paper in four Petri dishes. One group of seeds then was placed in darkness, and the others were irradiated using incandescent light passed through red and far-red glass interference filters. Exposures to red light were 4 minutes, and to far-red light, 8 minutes. All groups were returned to darkness after irradiation, and percentages germination were determined 48 hours after irradiation.

was found that the effects of red light were experimentally reversible by immediately following a red light treatment with far-red irradiation. By repeatedly alternating brief red and far-red treatments, it was seen that the light applied last determined whether the seeds germinated or not (Table 8.1). Within approximately a decade, it was discovered that interruption of a long dark period inhibited or prevented flowering of short-day plants and promoted flowering of long-day plants, and that the action spectrum for this "night-break" or "night interruption" phenomenon was identical to that for promotion of light-sensitive seed germination. Moreover, red, far-red reversibility also obtained in the "night-break" phenomenon.

The simplest interpretation of these photoresponses—one photoperiodic, one nonphotoperiodic—was that a single photoreceptor pigment was involved that could exist in two photointerconvertible forms:

$$P_r \xrightleftharpoons[\text{far-red light}]{\text{red light}} P_{fr}$$

Since the energy requirements for the photoreactions were relatively quite low (as compared to photosynthesis, for example), it seemed certain that the effective wavelengths brought about changes only in molecular configuration of the pigment per se and did not involve transfer of energy to another system. Furthermore, since red light potentiates responses and far-red nullifies the action of red light, it was hypothesized (correctly, as it turns out) that only one form (the far-red light absorbing form, P_{fr}) of the pigment should be active, the other inactive. Thus, it came to be understood that the response of a plant to red light (that is, in the red, far-red reversible responses) results from the action of P_{fr} thereby generated. Response to far-red results from the absence of action of P_{fr}.

Attempts to demonstrate directly the pigment in plants and to extract it from plant tissues began in earnest in the mid-1950s. The problem was understandably difficult for various reasons: (1) the pigment occurs in very low concentrations in plant tissues (e.g., 10^{-8} to $10^{-7} M$); (2) passing a beam of red or far-red light through a sample of the pigment (*in vivo* or *in vitro*) to detect or measure one form of the pigment converts it to the other form; (3) detection and measurement *in vivo* are complicated by the fact that plant tissues are optically very dense, highly light-scattering materials; (4) the presence of chlorophyll in green tissues and the conversion of protochlorophyll to chlorophyll on irradiation of etiolated tissues pose obvious problems also; and (5) the probability that the pigment was a chromoprotein made efforts to reintroduce extracts into living tissues seem unpromising, so bioassay was excluded. Thus, a special spectrophotometer capable of measuring very slight differences in absorption in very dense materials had to be built so advantage could be taken of the reversible absorbance by the pigment of red and far-red light. The instrument that was developed was a dual-wavelength difference spectrophotometer, that permitted a direct measurement of the ΔOD between 660 and 730 nm.

Finally, in 1959, the definitive discovery of the pigment, then named "phytochrome," was made by W. L. Butler, K. H. Norris, H. W. Siegelman, and S. B. Hendricks at Beltsville. They successfully detected phytochrome *in vivo* in the cotyledons of etiolated turnip seedlings and in shoots of etiolated corn seedlings. They also extracted and partially purified the pigment from etiolated corn seedlings. The first rigorous efforts to purify phytochrome extensively were by Siegelman and Firer (1964), who achieved about 60-fold purification by conventional methods of protein purification. With methods used subsequently, purifications to homogeneity are routinely achieved. The pigment can easily be seen on columns during purification and changes color on irradiation with red and far-red light. P_r is blue and P_{fr} is blue-green.

Absorption spectra for partially purified P_r and P_{fr}—measured after forcing irradiation with far-red and red light, respectively—show absorption peaks for P_r at about 666 nm in the red, and for P_{fr} at about 730 nm (Fig. 8.1, bottom). Notably also, both P_r and P_{fr} show significant absorption in the blue region, the absorption maxima being at about 370 and 400 nm, respectively. The action spectra for the phototransformations of P_r and P_{fr} are as predicted (Fig. 8.2). With narrow-band far-red light, P_{fr} can be almost completely transformed to P_r. However, because of the overlapping of the absorption spectra in the region of 600 to 700 nm, the most effective attainable band of red light transforms only about 80% of P_r to P_{fr}.

Basically four major transformation reactions have been described for phytochrome (Fig 8.3). Reactions I and II, described previously, both are first-order, photochemical reactions. Reaction IV, a zero-order destruction or decay reaction, seems to be ubiquitous in all tissues containing phytochrome. It results in a decrease in total photoreversible phytochrome, probably in fact degradation of the protein moiety. Dark reversion of P_{fr} to P_r, reaction III, has been demonstrated in various dicots but as yet has not been demonstrated conclusively in monocots.

The quantum efficiencies for the two photoreactions are such that reaction I is energetically over twice as efficient as reaction II (Fig. 8.2). The action of blue light is very much less effective than red and far-red light in reactions I and II.

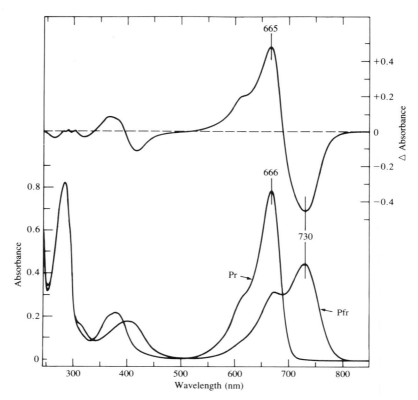

FIGURE 8.1. Absorbance and difference spectra of 124-kDa phytochrome ($A_{666}/A_{280} =$ 0.97) from etiolated *Avena* purified by the Affi-Gel Blue affinity procedure in 100 mM potassium phosphate, 5 mM Na$_4$EDTA, and 14 mM 2-mercaptoethanol, final pH 7.8 (4°C). Absorbance spectra were measured at 3°C after saturating red (P$_{fr}$) and far-red (P$_r$) irradiation. The difference spectrum was determined by subtracting the spectrum of P$_{fr}$ from that of P$_r$. Phytochrome was >95% homogeneous. (Reprinted, with permission, from R. D. Vierstra and P. H. Quail, 1983a. Biochemistry 22: 2498–2505. Copyright (1983) American Chemical Society.)

Red light has been reported to be 100 times more effective than blue *in vivo* in converting P$_r$, and far-red light is 25 times more efficient than blue in converting P$_{fr}$. *In vitro*, blue light is relatively more effective than *in vivo* but still markedly less effective than red and far-red light.

As noted previously, P$_{fr}$ is the active form of phytochrome, and P$_r$ is metabolically inert. Thus, in phytochrome-mediated photoresponses, a response to irradiation with red light results from the action of P$_{fr}$ that is thereby generated or maintained. Conversely, an effect of far-red light results from the absence of action of P$_{fr}$—not from any action of P$_r$. Notably, white light or sunlight has a net red-light effect, even though such radiation of course contains wavelengths

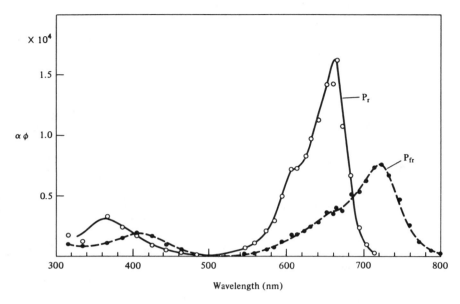

FIGURE 8.2. Action spectra of photochemical transformations of P_r and P_{fr}. The extinction coefficient, α, is in liter mol^{-1} cm^{-1}, and the quantum yield, ϕ, is in moles Einstein^{-1}. (Reproduced, with permission, from the *Annual Review of Plant Physiology, Volume 16.* © 1965 by Annual Reviews Inc. From Siegelman and Butler, 1965.)

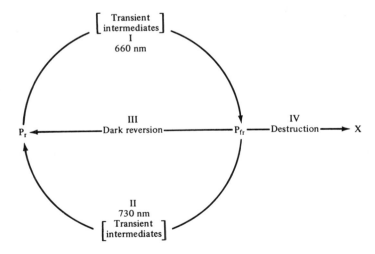

FIGURE 8.3. Phytochrome transformation reactions.

absorbed by both forms of phytochrome. The main reason for this is the greater quantum efficiency of reaction I than reaction II.

Phytochrome is a dimeric chromoprotein, with each of the two subunits consisting of a single, linear tetrapyrrole chromophore covalently linked to a cysteine residue (cysteine 321 in *Avena* and cysteine 322 in *Cucurbita*) in the N-terminal domain of a polypeptide of 120 to 127 kDa, depending on plant species. A comparison among plant species indicates that the relative molecular mass (M_r) of native phytochrome is variable: *Zea* = 127,000; *Secale* and *Avena* = 124,000; *Pisum* = 121,000; and *Cucurbita* = 120,000. The *in vitro* phototransformation difference spectrum for native phytochrome from each species is similar to that observed in each case and is indistinguishable from that described for native *Avena* phytochrome. Moreover, the difference minima between the red- and far-red absorbing forms of the pigment are all at 730 nm, and the spectral change ratios are near unity. All behave similarly during proteolysis by endogenous proteases.

Phytochrome from etiolated *Avena* is a homogeneous species with a monomeric molecular mass of 124 kDa, and it has been characterized extensively. Etiolated *Avena* phytochrome exists as a nonglobular dimer in solution. Biochemical, immunochemical, and nucleotide-sequence analyses have yielded valuable information on a number of structural features of the molecule potentially related to its mechanism of action. These features include regions of the polypeptide involved in photoconversion-induced conformational changes, in interactions with the chromophore, and in dimerization. The chromophore, phytochromobilin, is a bilitriene or open-chain tetrapyrrole that is closely similar to but distinct from the chromophore of c-phycocyanin and allophycocyanin (Fig. 8.4). Available evidence suggests that there probably is one chromophore per monomer of phytochrome. Upon phototransformation of P_r to P_{fr}, a proton migration from ring I to its ethylidene group occurs, causing a shift in a double bond, and the loss of another hydrogen atom. Subtle changes evidently occur also in the protein, which probably accounts for P_{fr} being active and P_r being inactive.

Elich and Lagarias (1987) reported on the biosynthesis of phytochromobilin from 5-aminolevulinic acid and biliverdin in oats (*Avena sativa* L.) and proposed the scheme depicted in Fig. 8.5. It has been found that the synthesis of phytochrome chromophore is not tightly coupled to the synthesis of the apoprotein.

Some recent data show that there is a kind of phytochrome that predominates in green oat tissue that is different from the well-characterized phytochrome in etiolated *Avena* tissue (Tokuhisa and Quail, 1987 and references cited therein). Green-oat phytochrome, extracted in nondenaturing buffers, when compared with etiolated-oat phytochrome, exhibited an altered difference spectrum and lower immunochemical activity per unit phytochrome absorbance when measured by enzyme-linked immunosorbent assays (ELISA). Immunoblot analysis of the green-oat extracts indicated the presence of two different molecular species of phytochrome that could be separated by immunoprecipitation using

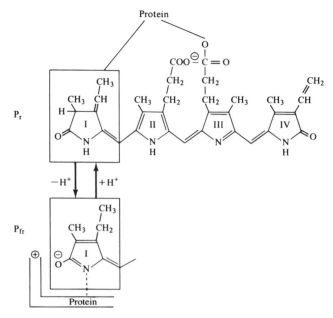

FIGURE 8.4. Proposed structure for phytochromobilin, the chromophore of phytochrome, its linkage to the protein, and phototransformation mechanism. [Redrawn, with permission, from Rüdiger, W. 1972. Chemistry of phytochrome chromophore, In: Mitrakos, K. and W. Shropshire. Jr., eds. *Phytochrome*. Pp. 129–141. Copyright by Academic Press, Inc. (London) Ltd.]

antibodies directed against etiolated-oat phytochrome. One species, etiolated-like in that it had a molecular mass of 124 kDa and was immunoprecipitable, represented about 30% of the total spectrally detectable phytochrome. The other species, representing the remaining spectral activity, is a distinct phytochrome in that it has a monomeric molecular mass of 118 kDa, and was not immunoprecipitated with the same antibody preparation. The abundance of this distinct green-tissue type phytochrome corresponded to only 1% of the spectral activity observed in etiolated tissue.

The abundance of the 124-kDa molecule increased at least 200-fold in etiolated *Avena* seedings over 72 hours, whereas in light-grown seedlings the level of this molecule is relatively constant. In contrast, the amount of the 118-kDa species increased only 2-fold in both dark- and light-grown seedlings over the same period of time. Thus it appeared that, whereas the abundance of 124-kDa phytochrome is regulated at the protein level by the well-documented, differential stability of the red- and far-red-absorbing forms *in vivo*, the 118-kDa molecule is present at a low constitutive (non-light-regulated) level, presumably reflecting no such difference in the stability of the two spectral forms.

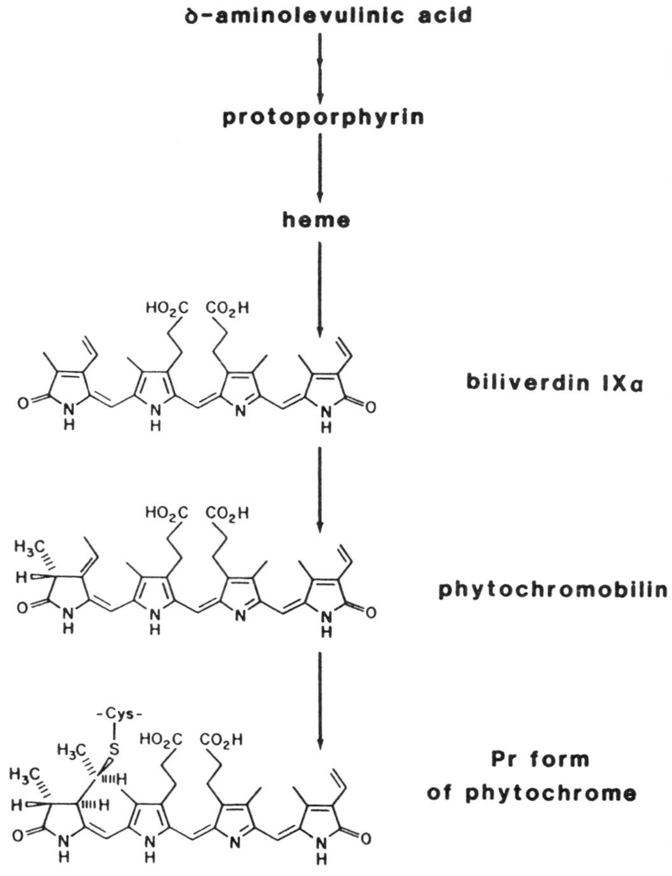

δ-aminolevulinic acid

protoporphyrin

heme

biliverdin IXα

phytochromobilin

**Pr form
of phytochrome**

FIGURE 8.5. Proposed biosynthetic pathway of the phytochrome chromophore, phytoch-romobilin. (Redrawn in slightly modified form, with permission, from Elich and Lagarias, 1987.)

The stable, "seed" phytochrome described by Hilton and Thomas (1985) appears to be comparable to the 118-kDa molecule (Tokuhisa and Quail, 1987), while the labile, "seedling" phytochrome is analogous to the 124-kDa molecule. Overall, the evidence seems to indicate that the total level of phytochrome appears to be modulated in transitions between light and dark by the levels of 124-kDa phytochrome above a stable background of the 118-kDa species (Hilton and Thomas, 1985; Hunt and Pratt, 1980b; Tokuhisa and Quail, 1987). There is some speculation that the 118-kDa type of phytochrome may be equivalent to the stable phytochrome that remains after the rapid degradation that occurs in etiolated tissue placed in the light. If so, it could represent a small, biologically active kind of phytochrome that has physiological significance.

Sequence Analysis of Phytochrome

Application of modern methods of plant molecular biology makes progress in the molecular analysis of phytochrome increasingly rapid and exciting. Peter H. Quail, formerly of the University of Wisconsin and now at the Plant Gene Expression Center at Albany, California, and his associates have isolated and characterized a gene encoding the phytochrome polypeptide in both *Cucurbita pepo* (zucchini) (Sharrock et al., 1986), a dicot, and *Avena sativa* (oat) (Hershey et al., 1987), a monocot.

The amino acid sequence of *Cucurbita* phytochrome was deduced from the nucleotide sequence of a cDNA clone that was initially identified by hybridization to an *Avena* phytochrome cDNA clone. Poly(A)$^+$ RNA was isolated from etiolated *Cucurbita* hypocotyl hooks and was enriched for phytochrome mRNA by size fractionation. Then cDNA copies of this RNA were prepared. Finally, these cDNA copies, annealed into pBR 322 and cloned in *Escherichia coli*, were colony-screened using a nick-translated 3.0-kb *Kpn*I–*Sac*I insert from an *Avena* phytochrome cDNA clone. The cDNA insert from the largest *Cucurbita* phytochrome cDNA clone isolated was sequenced and found to contain the entire amino acid coding region (Fig. 8.6). The *Cucurbita* phytochrome polypeptide is 1123 amino acids in length, corresponding to 125 kDa.

A short time later Quail and associates (Hershey et al., 1987) isolated and characterized a gene encoding the phytochrome polypeptide from *Avena*. Based on nucleotide sequence identity with previously sequenced cDNA clones, this gene is designated as type 3 (phy 3). The gene is about 5.9 kb long with six exons and five introns, one each of the latter in the 5′ and 3′ untranslated regions. The largest exon encodes the entire 74-kDa, chromophore-bearing, N-terminal domain of the photoreceptor postulated to be directly involved in its mechanism of action.

Overall the *Cucurbita* and *Avena* phytochrome sequences are 65% homologous at both the nucleotide and amino acid levels, but this sequence conservation is not evenly distributed. Most of the N-terminal two-thirds of the aligned polypeptide chains exhibits localized regions of high conservation, while the extreme N terminus and the C-terminal one-third are less homologous. Comparison of the predicted hydropathic properties of these polypeptides also indicates conservation of domains of phytochrome structure (Sharrock et al., 1986).

Interestingly, although dicot phytochromes, including those from *Cucurbita*, and monocot phytochromes, like that from *Avena*, are similar with respect to monomer M_r, subunit structure, spectral properties, and sensitivity to proteolysis, the two phytochromes show little immunological cross-reactivity. Amino acid compositions of *Cucurbita* and *Avena* phytochromes are presented in Table 8.2.

Cordonnier et al. (1986a) used a monoclonal antibody to pea phytochrome that was obtained from a mouse immunized with phytochrome purified from etiolated pea shoots to compare the antigenic properties of phytochrome from

```
AP3                                      -57      CAGGAGC ATA    GGG GTATA GA C  TTG GTG      AAT AC T  AG CAGGCG
CP    GCTCTAAATT CTCCTCCACC ATGCCTCTCA CCTTTCACTC TCAGCCCCTC TGATTCACCG GCGCCACTCT TGCCCCACCG GAAAATCTGT TCTTCTTTTG GTTGAGAAAC   -1

AP3             T     A   G       GC  TCC AG      TC   G AGG AAC   C CAG   C TCC CAA G A   GGG T A   A           A C C             C
CP    ATG TCT ACC TCT AGA CCT AGT CAG TCT TCT AGC AAC TCT GGG CGG TCC AGA CAT AGC ACT AGA ATT ATT GCT CAG ACA TCT GTT GAT GCG     90
CP    M   S   T   S   R   P   S   Q   S   S   S   N   S   G   R   S   R   H   S   T   R   I   I   A   Q   T   S   V   D   A        30
AP3       S               A   S               S   R   N       Q   S   S   Q   A       V   L           T   L

AP3   G G C C A T       A   A   A           T G C   C           C AG CTG   T GAA  C  CAG C G       G   CCA CCT TG  G  A
CP    AAC GTG CAA GCT GAT TTT GAG GAA TCT GGG AAT TCG TTT GAC TAC TCA AGT TCA GTG CGT GTC ACT AGT GAT GTT AGC GGA GAT CAA CAG    180
CP    N   V   Q   A   D   F   E   E   S   G   N   S   F   D   Y   S   S   S   V   R   V   T   S   D   V   S   G   D   Q   Q       60
AP3   E   L   N       E   Y           D                       K   L       E   A   Q   R       G   P   P   V

AP3   GGG C      G    G    G    G  C ... T   C  C TA G  C             A  G A C A          C           C T
CP    CCT AGG TCA GAC AAA GTT ACT ACA GCT TAT CTC CAT CAT ATT CAG AAA GGC AAA CTT ATT CAA CCA TTT GGT TGC TTG TTG GCC TTA GAT    270
CP    P   R   S   D   K   V   T   T   A   Y   L   H   H   I   Q   K   G   K   L   I   Q   P   F   G   C   L   L   A   L   D       90
AP3   G           E           I           Q                                           T

AP3         G  G GC     T   C  C      TC       G  C G A        C T A  C   C           G   A   C  TGT AT       CC
CP    GAC AAA ACA TTC AAG GTT ATT GCG TAT AGT GAA AAT GCC CCT GAA ATG TTG ACC ATG GTG AGC CAT GCT GTC CCA AGC ATG GGG GAT TAC   360
CP    D   K   T   F   K   V   I   A   Y   S   E   N   A   P   E   M   L   T   M   V   S   H   A   V   P   S   M   G   D   Y      120
AP3   E       S       N           F                                   T                               V   D       P

AP3     A AGG   G  G        C A        C T C     GT AC AA G     C A A          CAT    A C A  A       C T  T  G
CP    CCT GTT CTT GGC ATT GGC ACA GAT GTA AGA CGT ATT TTC ACC GCA CCT AGT GCT TCT GCA CTG TTG AAG GCC TTG GGC TTT GGA GAG GTT   450
CP    P   V   L   G   I   G   T   D   V   R   T   I   F   T   A   P   S   A   S   A   L   L   K   A   L   G   F   G   E   V      150
AP3   R                   N           S   L       S   D   Q   G       T           H                       A   D

AP3   T T T G  G              T G       A  A C G T          C          A  CA    T T T  G G G G A
CP    ACA CTT CTT AAT CCT ATC CTG GTG CAT TGC AAG ACT TCT GGA AAA CCC TTC TAT GCA ATT GTT CAT CGT GTT ACT GGA AGC TTA ATC ATT   540
CP    T   L   L   N   P   I   L   V   H   C   K   T   S   G   K   P   F   Y   A   I   V   H   R   V   T   G   S   L   I   I      180
AP3   S                           Q                               A               A           C   L   V   V

AP3                     A     ACA    TT  T CC      T       G   T G G C G  C       G  A  A T C AG
CP    GAC TTT GAG CCT GTG AAG CCT TAT GAA GGT CCA GTG ACT GCA GCT GGA GCT CTA CAA TCA TAT AAA CTT GCC GCC AAA GCG ATT CGA AGA   630
CP    D   F   E   P   V   K   P   Y   E   G   P   V   T   A   A   G   A   L   Q   S   Y   K   L   A   A   K   A   I   T   R      210
AP3                         T       F       A                                           S   K

AP3   A C     A        A   G          AG GT   A  C AT   T G       G AG     C       C T C  G        C A G
CP    TTG CAG TCT TTG CCT AGT GGA AGC ATG GCT AGG CTT TGT GAC ACA ATA GTT CAA GAA GTT TTT GAA CTA ACA GGT TAT GAT CGA GTT ATG   720
CP    L   Q   S   L   P   S   G   S   M   A   R   L   C   D   T   M   V   Q   E   V   F   E   L   T   G   Y   D   R   V   M      240
AP3   I               G                   E   V       N       V   K                       D

AP3         C  G T      A       C          T G A T C     A A          T  T       T  A C C     C
CP    GCT TAT AAA TTC CAT GAT GAT GAT CAT GGG GAA GTG ATC TCT GAA GTC GCA AAG CCC GGC CTT CAG CCA TAT CTT GGT TTG CAT TAT CCA   810
CP    A   Y   K   F   H   D   D   D   D   H   G   E   V   I   S   E   V   A   K   P   G   L   Q   P   Y   L   G   L   H   Y   P  270
AP3                       E                           F           I   T       E

AP3     C      C          A  CAG C  C T          G  C A A          TG       C       G  G TCC A        G  C
CP    GCA ACT GAC ATT CCT CAA GCT GCA CGT TTT TTG TTC ATG AAA AAT AAG GTC CGG ATG ATT GTT GAT TGT CGT GCA AAA CAT TTA AAA GTA   900
CP    A   T   D   I   P   Q   A   A   R   F   L   F   M   K   N   K   V   R   M   I   V   D   C   R   A   K   H   L   K   V      300
AP3                       L                                                   C               R   S   I

AP3   A T G G  C       GC  C C  CC        A T  GC C              G A CC  G  A         T  C C T             G
CP    CTC CAA GAT GAG AAA TTA CAG TTT GAT CTA ACT TTA TGT GGT TCA ACT TTA AGA GCT CCA CAC AGT TGC CAT TTA CAG TAT ATG GAA AAC   990
CP    L   Q   D   E   K   L   Q   F   D   L   T   L   C   G   S   T   L   R   A   P   H   S   C   H   L   Q   Y   M   E   N      330
AP3   I   E   A       A       P           I   S                   A

AP3             G  T  A C C  C         T                G AAT  A   G   T G         CT GAG TCT GAA   AA C A GCA
CP    ATG AAC TCT ATA GCC TCT TTG GTT ATG GCA GTT GTG GTT AAT GAA GGG GAT GAA GAA AAT GAA GGC ... ... ... CCT GCT TTG CAG CAG  1071
CP    M   N   S   I   A   S   L   V   M   A   V   V   V   N   E   G   D   E   E   N   E   G   .   .   .   P   A   L   Q   Q      357
AP3                                                       N   E   D   D       A   E   S   E   Q   P   A

AP3       G  A AG      A C         CCC T  T  C  C C  GAG C         A   C T       G GC T
CP    CAA AAG AGA AAG AGA TTA TGG GGC TTG GTA GTA TGT CAT AAT TCA AGT CCC AGA TTT GTT CCG TTT CCT CTT AGG TAT GCT TGT GAG TTC  1161
CP    Q   K   R   K   R   L   W   G   L   V   V   C   H   N   S   P   R   F   V   P   F   P   L   R   Y   A   C   E   F         387
AP3       K       K                       L                   H   E           Y

AP3   T    A G  G            G C          C   G G T A A G   G      G  T G CGT  G    C  A    AAG T   A       A G
CP    CTA GCT CAA GTA TTT GCT ATT CAT GTG AAC AAG GAA TTA GAG TTG GAA AAT CAA ATT ATA AAG AAG AAT ATT CTG CGT ACG CAG ACA CTC  1251
CP    L   A   Q   V   F   A   I   H   V   N   K   E   L   E   L   E   N   Q   I   I   E   K   N   I   L   R   T   Q   T   L      417
AP3                       V                   R       F           K       L   R               K   M       M

AP3   C C  CT   T       T G T C   A  A   C TCT     C C G AC     C    A  AG    CC    C T  C          C A        G      T
CP    TTG TGC GAC ATG CTA ATG CGT GCT ... CTA TTA GGT ATT GTG TCG AGG AGT CCT AAC ATA ATG GAT CTT GTC AAA TCT GAT GGG GCT     1338
CP    L   C   D   M   L   M   R   A   .   L   L   G   I   V   S   R   S   P   N   I   M   D   L   V   K   S   D   G   A        446
AP3   S                   F       E       S       T                           G   T                       C

AP3     T C T C G       GG  GG      G A        T C G C T AAT G T  A CG G  CT      A A AT   T   C  TC       A   A  T GT
CP    GCC TTG TTA TAT AAG AAA AAA ATT TGG CGA TTA GGA TTG ACA CCT AAT GAC TTC CAG TTG CTG GAC ATA GCT TCG TGG CTT TCC GAG TAT  1428
CP    A   L   L   Y   K   K   K   K   I   W   R   L   G   L   T   P   N   D   F   Q   L   L   D   I   A   S   W   L   S   E   Y  476
AP3               G   G       V               R   N   A       T   E   S       I   H           F               D   V

AP3     C G       C T C C            C C C C          T C . T  A      GC    C T          ATG A T       A
CP    CAT ATG GAT TCA ACG GGG TTG AGT ACT GAC GGT TTG TAT GAT GCA GGA TAC CCT GGA GCT ATT GCT TTA GGT GAT GAA GTG TGT GGG ATG  1518
CP    H   M   D   S   T   G   L   S   T   D   S   L   Y   D   A   G   Y   P   G   A   I   A   L   G   D   E   V   C   G   M     506
AP3   R                                           H                   A                   M   I

AP3           TG  CT A    C  AC TCC  G  T  T C          C AG  A  T A     GT  A C A          T            A   T         A   T
CP    GCA GCT GTG AGG ATA ACT AAT AAT GAC ATG ATT TTT TGG TTT CGA TCT CAC ACT GCT TCA GAG ATT CGA TGG GGT GGA GCA AAG CAT GAA  1608
CP    A   A   V   A   K       R   I   T   N   N   D   M   I   F   W   F   R   S   H   T   A   S   E   I   R   W   G   G   A   K   H   E  536
AP3   V   A   K           N   S   K       I   L                       A                   N   D

AP3   CA TCG G C  T        C AG          GG        C  T G TG           A               T      TG AG  C        T
CP    CAT GGT CAA AAG GAT GAT GAT GCA AGA AAA ATG CAT CCA AGA TCA TCT TTC AAG GCT TTC CTT GAA GTA GTC AAG ACA AGA AGT TTG CCC TGG  1698
CP    H   G   Q   K   D   D   D   A   R   K   M   H   P   R   S   S   F   K   A   F   L   E   V   V   K   T   R   S   L   P   W  566
AP3   P   S   D   M               S       R                   L                                   M   K
```

FIGURE 8.6. Nucleotide sequences of cDNAs and derived amino acid sequences of *Cucurbita* (zucchini) phytochrome and comparison with *Avena* type 3 phytochrome sequences. (With permission, from Sharrock et al., 1986.)

```
AP3  GT          A              T T T A G  .        A   G C  GGG   A C A  T     ...  CC AGC  AG CC   AG CGG GA  GCT
CP   AAG GAC TAT GAG ATG GAT GCA ATC CAC TCT TTA CAA CTT ATC CTT AGA AAT ACT TTT AAG GAT ACA GAT GCA ACT GAA ATA AAT AGA AAA   1788
CP   K   D   Y   E   M   D   A   I   H   S   L   Q   L   I   L   R   N   T   F   K   D   T   D   A   T   E   I   N   R   K       596
AP3  S                                                    G   L   N   .   A   S   K   P   K   R   E   A

AP3  AGT T AGT  AC CAG A        T              C   T       CTT GCT      C  CGG C  G C       A        T  T C    G  A
CP   TCA ATT CAA ACA ACA CTT GGT GAC CTA AAA ATT GAA GGG AGG CAA TTG GAA CTA TCA GAA AGT GAA ATG GTC CGA TTA ATC GAG ACA     1878
CP   S   I   Q   T   T   L   G   D   L   K   I   E   G   R   Q   E   L   E   S   V   T   S   E   M   V   R   L   I   E   T     626
AP3  L   D   N   Q   I                         L   D       L   A           Q   A                       M

AP3  A          T AC G A A      GGC A      A C GGC C                CAG     GCA  G  G        G  A AGA T
CP   GCT ACT GTG CCG ATT TTA GCT GTT GAT TTA GAT GGG TTA ATT AAT GGG TGG AAT ACA AAA ATT GCT GAA TTG ACT GGA CTG CCT GTG GAT   1968
CP   A   T   V   P   I   L   A   V   D   L   D   G   L   I   N   G   W   N   T   K   I   A   E   L   T   G   L   P   V   D       656
AP3                                G   N       V           Q       A               G       A   A   A

AP3  GT A  T A G  C A          C C T       G  C C         CC           CA  G    C  A ATC A  T C     G T A
CP   AAA GCT ATC GGC AAG CAT TTA CTT ACG TTA GTG GAG GAT TCA TCT GTA GAA GTT GTC AGG AAG ATG TTT CTG GCA TTG CAA GGA CAA       2058
CP   K   A   I   G   K   H   L   L   L   T   L   V   E   D   S   S   V   E   V   V   R   K   M   L   F   L   A   L   Q   G   Q   686
AP3  D                       R   I                           P           Q   R       Y               K

AP3          A G G A       G   T      G A       T  T CCGAG GG   T A    TCA GT T TG      G              C
CP   GAA GAG CAA AAC GTT CAA TTC GAG ATC AAG ACA CAC GGT TCT CAC ATT GAG GTT GGC TCC ATC AGC CTA GTT GTA AAT GCT TGT GCA AGT   2148
CP   E   E   Q   N   V   Q   F   E   I   K   T   H   G   S   H   I   E   V   G   S   I   S   L   V   V   N   A   C   A   S       716
AP3          K   E       R               V               P   K   R   D   D           P   V   I

AP3  C      C T A   T      T    GC          T C            G      TC T  T  G                              G GT
CP   AGG GAC TTG CGT GAA AAT GTC GGT GTG TTT TTT GCA CAA GAT ATC ACT GGT CAG AAG ATG GTA ATG GAC AAG TTC ACT CGA TTA           2238
CP   R   D   L   R   E   N   V   V   G   V   F   F   V   A   Q   D   I   T   G   Q   K   M   V   M   D   K   F   T   R   L       746
AP3          H   D   H                     C                       M       V   H   L                           V

AP3      G  T C         G  G  CAT C C G C       C C T     T        T GT C                       T  G
CP   GAA GGT GAT TAC AAA GCT ATT GTA CAA AAT CCC AAT TTG ATC CCT CCA ATA TTT GGA TCA GAT GAA TTT GGA TGG TGC TCA GAG TGG       2328
CP   E   G   D   Y   K   A   I   V   Q   N   P   N   L   I   P   P   I   F   G   S   D   E   F   G   W   C   S   E   W          776
AP3                    I   H                                                       A

AP3      G           AC  GTG          AAT A A  T         G C            C T  A G      AC AG AG   T G T
CP   AAT CCT GCA ATG GCG AAA CTA ACT GGG TGG TCA CGT GAA GAA GTA ATC GAT AAG ATG CTT TTG GGA GAG GTT TTT GGT GTT CAC AAA TCA   2418
CP   N   P   A   M   A   K   L   T   G   W   S   R   E   E   V   I   D   K   M   L   L   G   E   V   F   G   V   H   K   S       806
AP3  A           T               N       D       L                               D   S   S   N   A

AP3   CC  C C G      C AG   T A       A G      TTG  C T A C     G  A T A GCC  G G     A A A             C A
CP   TGT TGT CGT TTA AAG AAT CAA GAA TTT GTT AAT CTT GGG ATT GTC TTG AAC AAT GCC ATG TGT GGT CAA GAT CCT GAA AAG GCT TCC       2508
CP   C   C   R   L   K   N   Q   E   A   F   V   N   L   G   I   V   L   N   N   A   M   C   G   Q   D   P   E   K   A   S       836
AP3  S               R   D                       S       C   V   L   I       S           L   A       E   E   T       P

AP3       C        C ACA G  GT  A A        AT G         CA CA  C GA  AA GAA A   G G G    T CTC AC T  A GA
CP   TTT GGT TTC TTA GCT CGG AAC GGG ATG TAC GTG GAA TGT CTT CTA TGT GTC AAT AAG ATC TTG GAT AAA GAT GGC GCG GTT ACA GGG TTC   2598
CP   F   G   F   L   A   R   N   G   M   Y   V   E   C   L   L   C   V   N   K   I   L   D   K   D   G   A   V   T   G   F       866
AP3          F   D       S   K   I                   S   A       R   K   E   N   E   G       L   I   V

AP3   C T     AT T GG          T            C GGG G  AA GCC  CG           G T C A    A GCC AG
CP   TTT TGC TTT TTG CAG CTT CCT AGT CAT GAG TTG CAA CAA GCA CTA AAT ATC CAA CGC TTA TGT GAG CAA ACT GCA TTG AAG AGA TTG AGA   2688
CP   F   C   F   L   Q   L   P   S   H   E   L   Q   Q   A   L   N   I   Q   R   L   C   E   Q   T   A   L   K   R   L   R       896
AP3          I   H   V   A                     H           Q   V       Q   A   S                           K

AP3  T  C TCC       G  G CAT GC   C A C C      C A C G C   AC T A  A GC    A  A A   A  T      AAT A
CP   GCA TTA GGA TAC ATA AAA AGA CAA ATA CAA AAT CCT CTT TCT GGA ATA ATC TTT TCA CGA AGA TTA TTG CGC ACC GAG TTG GGA GTA       2778
CP   A   L   G   Y   I   K   R   Q   I   Q   N   P   L   S   G   I   I   F   S   R   R   L   L   E   R   T   E   L   G   V       926
AP3  F   S       M   R   H   A   N               M   L   Y               K   A       K   N       D       N   E

AP3       G  TG AG AG A T A  GT G A  AT AAT       C C C      A A A     A A  T CA C TG  T CAA      GC C CC  A
CP   GAA CAA AAA GAA CTT CTG CGT ACT AGC GGA CTC TGT CAA AAG CAG ATC TCC AAG GTT CTC GAT GAG TCA GAC ATC GAT AAA ATT ATT GAT   2868
CP   E   Q   K   E   L   L   R   T   S   G   L   C   Q   K   Q   I   S   K   V   L   D   E   S   D   I   D   K   I   I   D       956
AP3  M   K   Q   I   H   V   G   D   N               H   H           N       I   A   D   L           Q       S   T   E

AP3  AAA TCT A C GCT G        G G    CT A      CTG      A T G G G   CT GTG A         A  C A    CC TGC C G
CP   ...  GGG TTT ATT GAT CTA GAA ATG GAC TTT ACA TTG CAT GAA GTA TTG ATG GTA TCA ATT AGT CAA GTG ATG CTA AAG ATT AAA         2952
CP   .   .   G   F   I   D   L   E   M   D   F   T   L   H   E   V   L   M   V   S   I   S   Q   V   M   L   K   I   K           984
AP3  K   S   S   C   L               A           L           Q   D   V   A   A   V               L   I   T   C   Q

AP3        A  G       AGA  C TC  TGC A C CTG      G AGA TT       AAG C  T A G C           G   G T C A    G  A CC
CP   GGA AAG GGT ATC CAG ATA GTT AAT GAC CCA GAA GAG GCT ATG TCC GGA ACC TTA TAT GGA GAT AGT TTA AGG CTT CAA CAG GTC TTG       3042
CP   G   K   G   I   Q   I   V   N   E   T   P   E   E   A   M   S   E   T   L   Y   G   D   S   L   R   L   Q   Q   V   L       1014
AP3  R           S   C   N   L               R   F       K   Q   S   V                   G   V               I

AP3  T T        C T        T      G AG  TC T T     GTT        T TCT G T GAG         TT A G C          A  C AGC A    G A
CP   GCG GAC TTT CTA TTG ATA TCA GTT AGT TAT GCG CCT TCA GGG CAA CTA ATT TCA ACC GAT GTG ACC AAG AAT CAA TTA GGA AAG         3132
CP   A   D   F   L   L   I   S   V   S   Y   A   P   S   G   Q   L   T   I   S   T   D   V   T   K   N   Q   L   G   K         1044
AP3  S               F               K   F   S       V           S   V   E               S   K   L           I   G   E

AP3  AAC C T          TAT GCC T   ACT       C  AGC C CAA      TT       A G C    A CA GA  CCA  GCA C A           GAG  AG GA
CP   TCG GTC CAT CTG GTG CAT TTG GAG TTC ATA ACA TAT GCT GGA GGA GGT ATA CCG GGA TCG TTG CTG AAC GAG ATG TTT ... GGA AGC      3219
CP   S   V   H   L   V   H   L   E   F   R   I   T   Y   A   G   G   G   I   P   G   S   L   L   N   E   M   F   .   G   S      1073
AP3  N   L           I   D           L           H   Q       L       V       A   E           M   A   Q           E   E   D

AP3    A C A      G C A     A G         C   G C  C  AGT TC          C   C  C  C G C          T T T G C A
CP   GAG GAG GAC GCG TCC GAA GAG GGT TTC AGT CTG CTC ATC AGT AGA AAG CTG GTG AAG CTG ATG AAT GGA GAC GTA CGA TAT ATG AGG GAA   3309
CP   E   E   D   A   S   E   E   G   F   S   L   L   I   S   R   K   L   V   K   L   M   N   G   D   V   R   Y   M   R   E      1103
AP3  N   K   E   Q               L                   V           N       L   R               H   L

AP3      T  T GT   A C            C   C C  A           T C    CA CA G    T  GG CAA  GA TGA GCCAG TGGAAGT T  CA CTTA  G
CP   GCC GGG AAG TCG AGC TTC ATC ATA ACT ACT GTT GAA CTT GCT GCA GCT CAC AAG TCA AGG ACA ACG TAG GCCAAGACA ATTTGAAGCA TCAACCTTGT   3404
CP   A   G   K   S   S   F   I   I   T   V   E   L   A   A   A   H   K   S   R   T   T   *                                      1124
AP3  V           T       A               S       P   T   A   M   G   Q   *

AP3  TCATCAAATG TTC GT TGA A TCC AGTC ACGATAGCCG TGAGCATTGG TGAGTTGGTG GCATGTTCCT GGGGACGAGG ATGGAATGTG CTGCAGCCTG TAGTGTAGTC
CP   ATGAATTGCT GCATTCTATG TTGAAAAG

AP3  TTGCAGCTTG GTACTTCCGC TGTTGTTATG TTTCTGTCAT CCTACTGTGT AAGGTTCAAG TTTGAATTTA CCATGAATAA AGTGTGCAGA TGTACCTGCA CTTTGGTCTC

AP3  AAAAAAAAAA AAAAAAAAAA AAAAAAAAAA AAA
```

FIGURE 8.6. *Continued.*

TABLE 8.2. Phytochrome amino acid composition.[a]

Amino acid residue	Cucurbita (etiolated)		Avena (type 3)[b] (etiolated)	
Arg	54		53	
Lys	65		64	
Asn	40	} 103	45	} 113
Asp	63		68	
Gln	51	} 125	49	} 126
Glu	74		77	
His	29		34	
Pro	41		46	
Tyr	25		21	
Trp	10		10	
Ser	87		89	
Thr	58		43	
Gly	75		69	
Ala	76		96	
Met	34		34	
Cys	20		23	
Phe	45		44	
Leu	126		124	
Val	81		83	
Ile	69		56	
Total	1123		1128	

[a] Data of Sharrock et al. (1986) by permission.
[b] Data of Hershey et al. (1985) in Sharrock et al. (1986) by permission.

different sources. They found a domain common to phytochromes from a variety of etiolated and green angiosperms, as well as a moss and three algae, indicating that phytochrome has been highly conserved throughout evolution. The conserved domain is on the nonchromophore bearing, carboxyl half of phytochrome from etiolated oats.

Occurrence, Distribution, and Intracellular Localization

Phytochrome has been positively identified in numerous species of angiosperms, gymnosperms, green algae, liverworts, and plants of other phyletic groups. It seems safe to conclude, in fact, that the pigment occurs in all green plants. Quite probably also, it is physiologically active in all chlorophyllous plants.

The most intensive investigation of phytochrome distribution within seedlings was conducted by W. R. Briggs and H. W. Siegelman (1965). Using *in vivo* spectroscopy, they determined distribution in etiolated barley, corn, oat, pea, bean, and sunflower seedlings (Fig. 8.7). In general, the highest concentrations of phytochrome were found in tissues that were either meristematic or that had recently been meristematic. Regions with the highest levels of detectable phytochrome also had the highest phytochrome–protein ratios, suggesting that these regions might be active sites of phytochrome synthesis. The distribution of

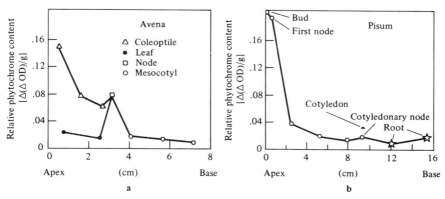

FIGURE 8.7. Distribution of phytochrome in dark-grown oat shoots, **a**, and dark-grown pea seedlings, **b**, per gram fresh weight. (Redrawn, with permission, from Briggs and Siegelman, 1965.)

phytochrome was compared with that of other physiological systems, and found generally to parallel auxin distribution, phototropic sensitivity, and red light sensitivity.

As regards intracellular localization of phytochrome, S. B. Hendricks and H. A. Borthwick first suggested, on the basis of some very rapid phytochrome-mediated responses that involve alterations in membrane permeability, that phytochrome might exist in association with membranes. Available evidence continues to support the notion that some but not all phytochrome is associated with some part of the membrane system of cells, including the plasmalemma and membranes of etioplasts and mature chloroplasts. There is considerable evidence also that P_{fr} becomes rapidly attached to some component of the membrane system after it is formed by red light.

In the tissues of most plants, P_{fr} appears to be labile, as previously noted (Fig. 8.3), undergoing a nonphotochemical transformation known as destruction. Destruction evidently occurs in both light and darkness, but it generally occurs more rapidly in light because of the constant generation of P_{fr} from P_r. Even though new phytochrome is synthesized, as P_r, as growth proceeds, it still is important to determine whether and how effective levels of phytochrome are maintained. Certainly a large portion of the phytochrome present in a dark-grown seedling is lost on the first exposure to light. D. T. Clarkson and W. S. Hillman (1967) reported on some experiments with pea seedlings that helped to clarify this point. They found that upon irradiation of decapitated, etiolated pea seedling shoots the concentration of total photoreversible phytochrome (P_r and P_{fr}) decreased to about 10% of the original level (Fig. 8.8). At that point a steady-state level was maintained in the light as synthesis of new P_r counteracted P_{fr} destruction. When the light was turned off, the concentration of phytochrome began to increase as the rate of synthesis of new P_r exceeded the rate of destruction of P_{fr}.

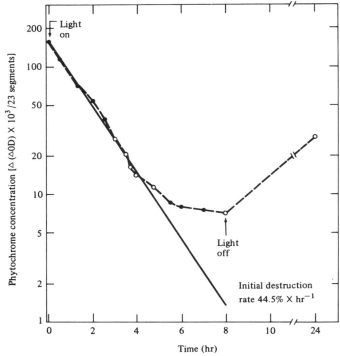

FIGURE 8.8. *In vivo* measurements of phytochrome concentration during the course of continuous illumination of decapitated, etiolated pea plants by red light, and after a subsequent dark period. (Redrawn, with permission, from Clarkson and Hillman, 1968.)

Induction–Reversion versus High Irradiance Responses

It is well to pause here to note that there are in fact two distinct types of phenomena that have been attributed to phytochrome—the so-called "induction-reversion" responses, with which this chapter is chiefly concerned, and the "high irradiance" responses. The former are the more familiar responses that are "light-triggered," which is to say that they are responses that are induced by a brief irradiation with relatively low intensity red light and reversible by immediate subsequent irradiation with far-red light. Light is viewed as merely "triggering" the response and as having no direct role in the inductive action of P_{fr}. Under conditions of nonsaturating light doses, these responses obey the law of reciprocity, according to which, for any given wavelength, intensity multiplied by time of irradiation is a constant. These are the classical phytochrome-mediated responses, and the photoreversibility of these responses is the generally accepted criterion, in fact, for establishing phytochrome involvement in a plant photoresponse.

High irradiance responses, in contrast, are "light-driven" photoresponses in the case of which continuous irradiation—with red, far-red, or blue light—at

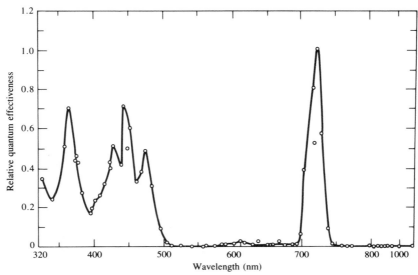

FIGURE 8.9. Action spectrum for inhibition of lettuce *(Lactuca sativa* L. cv. Grand Rapids) hypocotyl elongation, a "high irradiance" response. Continuous irradiation with high-quality monochromatic light was applied between 54 and 72 hours after planting. The increase in hypocotyl length during this period was measured. The far-red peak is exclusively due to phytochrome. How much, if any, the effect of blue and ultraviolet light can be attributed to phytochrome is unknown. (Redrawn, with permission, from Hartmann, 1967.)

relatively quite high intensity is necessary to sustain the response. Light is viewed as having a direct role in the inductive action of P_{fr}. They are effects that are observed only where a photostationary state of P_r:P_{fr} is established and maintained over relatively long periods. Reciprocity does not apply, and red, far-red photoreversibility is not always demonstrable. If irradiation ceases, the response simply reverts to the control condition, without the necessity of a final antagonistic irradiation. Of the high irradiance responses known, that of photoinhibition of lettuce hypocotyl elongation probably is the most studied. The action spectrum for this response shows a single sharp, symmetrical peak at approximately 720 nm (Fig. 8.9). Phytochrome may play a fundamentally different role in mediating the light effect in the high irradiance responses than in the "induction–reversion" responses to which most of this chapter is devoted. Moreover, where both blue and far-red action peaks are found, the blue peak probably is attributable to some pigment other than phytochrome.

Non-Phytochrome-Mediated Photoresponses to Blue Light

There is, in fact, another entirely separate photoreceptor system in lower and higher plants that many investigators consider to be equally as important as the phytochrome system. This "blue light photoreceptor" or "cryptochrome"

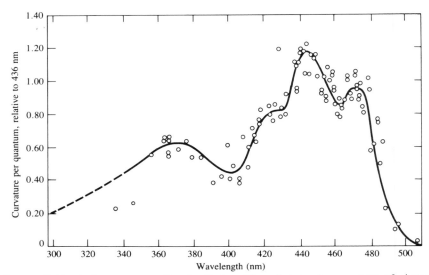

FIGURE 8.10. Action spectrum for the first-positive phototropic curvature of *Avena* coleoptiles, a non-phytochrome-mediated photoresponse to blue light. (Redrawn, with permission, from Curry, 1969.)

(Briggs, 1976) absorbs light in the long-wavelength ultraviolet and blue regions, and there is none of the photoreversibility that is characteristic of many phytochrome-mediated responses. Exact wavelengths vary somewhat from organism to organism, but the responses mediated by the blue light photoreceptor are always identical in the following features: a single broad peak near 370 nm, a shoulder near 420 nm, an action maximum between 445 and 460 nm, and another peak or shoulder between 470 and 490 nm (Fig. 8.10). Such action spectra are typical, for examples, for phototropism of *Avena* coleoptiles (first and second positive curvatures) and *Phycomyces* sporangiophores and for carotenoid synthesis in certain fungi. Indeed, it is likely that there is more than one blue light photoreceptor, based on the differences in the action spectra for different plant responses to blue light.

There is a substantial body of literature concerning the effects of blue light on biological processes in higher plants and fungi. Only recently has there been real progress toward the identification of the blue light photoreceptor associated with these effects. Recent studies have implicated a flavin-mediated reduction of a b-type cytochrome in the blue light photoreception process. The implication is based on the similarity of the action spectrum for the cytochrome reduction with that for many blue light responses. The actual photoreceptor supposedly involves a flavin moiety that specifically reduces the b-type cytochrome on excitation by blue light. There is some indication that both the flavin and the cytochrome could be associated with the same protein moiety.

Blue light-inducible reduction of a b-type cytochrome has been shown in membrane fractions from both *Neurospora* mycelium and etiolated corn coleoptiles (Brain et al., 1977). The photoreceptor evidently is associated specifically with the plasma membrane (Leong and Briggs, 1981).

One important photoresponse in higher plants that involves the blue light photoreceptor is light inhibition of stem elongation (see Cosgrove, 1981; Cosgrove and Green, 1981). Although phytochrome mediates a significant part of this response to light, there is evidence indicating the involvement of a blue light receptor, distinct from phytochrome, in the control of stem elongation. For example, blue irradiation rapidly inhibits the growth of etiolated cucumber seedlings, whereas red and far-red irradiations only slowly affect growth after a lag period of 60 minutes. There is also a difference in the timing of blue and red inhibition in light-grown cucumber seedlings. In contrast, no indication of involvement of a blue light photoreceptor was found in a study of the control of *Sinapis* hypocotyl growth.

Cosgrove (1981) reported that blue light rapidly suppressed stem elongation in etiolated seedlings of several species. In some species, the inhibition persisted only during the actual period of irradiation, after which time growth returned quickly to the higher dark rate, whereas, in other species, the light response had an additional long-term component that lasted for at least several hours in the dark. Cosgrove suggested that the long-term inhibition might be mediated by phytochrome, whereas the rapid short-term component is specific to a blue light photoreceptor. The rapid inhibition of growth observed in cucumber (*Cucumis sativus* L.) required high-energy blue irradiation, which was perceived directly by the growing region of the hypocotyl and inhibited all regions below the hook to the same extent. Investigation of the kinetics of the inhibition in cucumber and sunflower (*Helianthus annuus* L.) hypocotyls showed that, after a short lag period of 20 to 30 seconds in cucumber and 60 to 70 seconds in sunflower, the growth rate declined exponentially to a lower rate, with a half-time of 15 to 25 seconds in cucumber and 90 to 150 seconds in sunflower. Because of the rapid kinetics, Cosgrove concluded that the blue light photoreceptor cannot affect cell enlargement by altering the supply of growth hormone or the sensitivity to hormones, but probably operates more directly either on the biochemical process that loosens cell walls or on cell turgor.

Introduction to Mechanism of Phytochrome Action

It has been emphasized already that phytochrome is the photoreceptor for all of the red, far-red reversible photoreactions of plants. The many photoreactions or processes involving phytochrome are of three major types, namely (1) photoperiodic, photomorphogenetic responses, such as floral initiation in photoperiodically sensitive species, onset of winter bud dormancy, tuber formation, and many others; (2) nonphotoperiodic, phototomorphogenetic responses, such as greening of etiolated seedlings, germination of some kinds of seeds, elongation of fern rhizoids, and others; and (3) nonmorphogenetic photoresponses, such as synthesis of anthocyanins, induction of various chloroplastic enzymes, and others.

Some concepts of phytochrome action envisage a single mechanism of action of P_{fr} in all the many photoresponses for which phytochrome is the photoreceptor, and other concepts encompass two or more mechanisms of action. Whichever concept is favored, on the basis of information presently available, it is expedient to consider the action of phytochrome in the generally longer term and inductive photoperiodic reactions separately from the generally more rapid, noninductive photoreactions. This is the approach used in the following sections. We shall first consider the role of phytochrome in several of the noninductive photoresponses. Next, the evidence that P_{fr} stimulates GA biosynthesis and effux from plastids will be presented. Finally, we will turn to a discussion of the role of phytochrome in photoperiodism.

Phytochrome Action in Nonphotoperiodic Photoresponses

Effects on Membrane Permeability

An idea that originated with H. A. Borthwick and his group at Beltsville circa 1966, on the basis of some investigations of nyctinastic leaf movements, is that phytochrome (specifically the active form, P_{fr}) might act by regulating membrane permeability. Subsequent investigations by many scientists have substantiated repeatedly the fact that this is indeed a function of phytochrome in many photoresponses.

Fondeville et al. (1966) investigated "sleep movements" (nyctinasty) in the leaves of *Mimosa pudica*, a legume having doubly pinnately compound leaves. In this plant and other related legumes (e.g., *Albizzia julibrissin*, Fig. 8.11) the pinnules of the doubly compound leaves are open and separate in the light but "fold together" at night. The leaf movements result directly from differential turgor changes in structures called pulvini. In a doubly pinnately compound leaf, there is a primary pulvinus at the base of the rachis, a secondary pulvinus at the base of each rachilla, and a tertiary pulvinus (or pulvinule) at the base of each pinnule (Fig. 8.11). Fondeville et al. used excised pinnae, each consisting of approximately 12 paired pinnules, and observed opening and closing of the paired pinnules on the axis of the pinna.

Action spectra were determined for the potentiation of the closing movements and its reversal. If the plants were irradiated with red light immediately before a dark period, the leaves began to close within 5 minutes and were fully closed in 30 minutes (Fig. 8.12). But if the plants were irradiated with far-red light immediately prior to a dark period, the leaves remained open for several hours in darkness. The potentiated control of closing movements could be repeatedly established and reversed by alternating irradiations with red and far-red light, respectively.

Nyctinastic leaf movements are caused by relative changes in turgor pressure (and also cell size) on opposite sides of the pulvinus (or pulvinule). In darkness

FIGURE 8.11. *Albizzia julibrissin* leaves in normal daytime position (top) and in typical sleep (night) position (bottom).

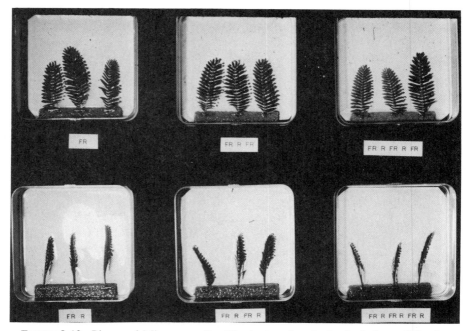

FIGURE 8.12. Pinnae of *Mimosa pudica* 30 minutes after placing in darkness following high-intensity light terminated with a succession of 2-minute exposures to red (r) and far-red (fr) irradiation. The pinnae remained open if exposure to far-red was last (top row) and closed if red irradiation was last (bottom row). (With permission, from Fondeville et al., 1966; photo courtesy of James D. Anderson, 1978.)

there is a movement of water out of the cells in the upper side of the pulvinus into cells in the lower side, causing a folding or closing movement of the leaflets. In the light, this process is reversed. The transfer of water is caused by an apparent alteration of membrane permeability and the transfer of K^+ and Cl^- ions (perhaps also other electrolytes). The cells comprising pulvini contain extraordinarily high concentrations of K^+, on the order of 0.5 M. Leaflet closure is associated with a loss of K^+ and Cl^- from the upper cells and a concomitant loss of water. The lower cells evidently do not gain the ions lost from the upper cells, but the lower cells, because of their negative water potential, do gain water lost from the upper cells. The role of phytochrome in all of this is either to cause a differential transient increase in the permeability of the plasma membranes of the upper cells, or perhaps to activate a membrane-bound ATPase.

An interesting and physiologically important effect of phytochrome on phototaxis of choroplasts in the filamentous green alga *Mougeotia* was reported by W. Haupt in 1963. Using microbeams of monochromatic light, Haupt observed that, when focused on or even very near the plate-like chloroplast in a single cell, the chloroplast would orient perpendicular to a microbeam of low-

FIGURE 8.13. Diagram of the alga *Mougeotia* (longitudinal section above, transverse section below), illustrating phototaxis of the plate-like chloroplast in red, **a**, and far-red, **b**, light. (Redrawn, with permission, from Haupt, 1963.)

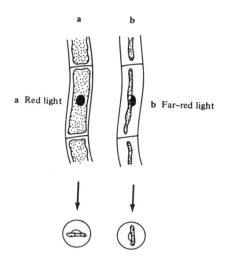

intensity red light and parallel to a beam of far-red light (Fig. 8.13). The effects of red and far-red light were fully reversible when alternating irradiations were used. Moreover, the change in plastid orientation was clearly evident in less than 10 minutes and was complete in 30 minutes. This phenomenon has yet to be elucidated, and there may be no effect of phytochrome on membrane permeability involved in this case. However, it is another example of a rapid kind of phytochrome-mediated response that evidently does not involve any effect of phytochrome on RNA or protein synthesis.

Further evidence for phytochrome being associated with cell membranes and having an effect on membrane properties has come from investigations of the so-called "Tanada effect." The empirical observations involved are as follows: Excised root tips from a plant such as mung bean (*Phaseolus aureus*) are suspended in solution (containing IAA, ATP, ascorbic acid, K^+, Mg^{2+}, Mn^{2+}, Ca^{2+}, and Cl^-) in a beaker in which is situated vertically a glass plate, the surface of which is made negative by washing with phosphate ions. When irradiated with red light, it is observed that the root tips adhere to the glass plate when the suspension is stirred. When irradiated with far-red light, the root tips release from the negatively charged surface (Fig. 8.14). By direct measurement it was confirmed that in red light the root tips developed a positive bioelectric potential at the tip, and in far-red light they developed a negative bioelectric potential. It was concluded that phototransformation of phytochrome changes the permeability characteristics of the cell membranes, with which the phytochrome probably is associated, resulting in a localized electrochemical gradient manifested as a bioelectric potential. Perhaps P_{fr} somehow affects a plasma membrane-bound ATPase, since K^+ and H^+ transport across the membrane is facilitated by such an enzyme.

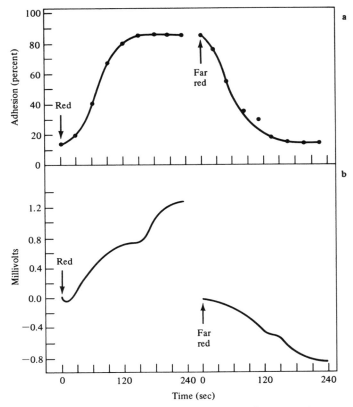

FIGURE 8.14. Kinetics of adhesion and release **a**, of mung bean root tips to a negatively charged glass plate and development of corresponding bioelectric potentials, **b**, (Redrawn, with permission, from Jaffe, M. J. 1968. Phytochrome-mediated bioelectric potentials in mung bean seedlings. *Science 162:* 1016–1018, 29 November 1968. Copyright 1968 by the American Association for the Advancement of Science.)

Differential Gene Activation

The responses described in the preceding section are all relatively rapid phenomena that evidently involve an effect of phytochrome on properties of cell membranes, with which the photoreceptor probably is intimately associated. There is, however, another quite different response that has been interpreted as indicating an effect of phytochrome in evoking *de novo* synthesis of particular enzymes. This is the so-called "differential gene activation hypothesis" regarding phytochrome action.

Illustrative of the kind of data on which the concept of gene activation by phytochrome is based are the extensive investigations conducted by Hans Mohr and his associates at the University of Freiburg on anthocyanin synthesis in the epidermis of the cotyledons of white mustard (*Sinapis alba*) seedlings.

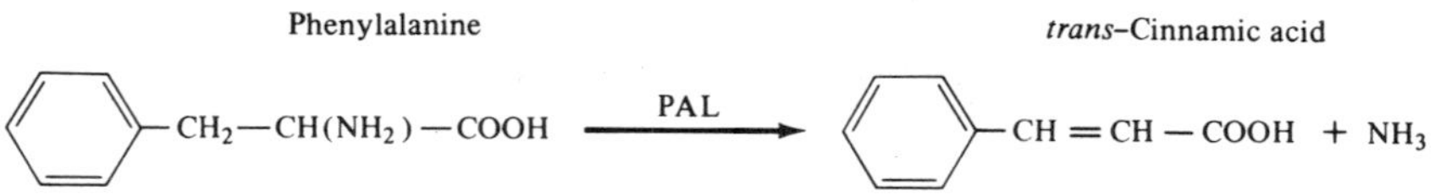

Phenylalanine   *trans*-Cinnamic acid

$$\langle \rangle - CH_2 - CH(NH_2) - COOH \xrightarrow{\text{PAL}} \langle \rangle - CH = CH - COOH + NH_3$$

FIGURE 8.15. The reaction catalyzed by phenylalanine ammonia lyase (PAL).

Anthocyanin synthesis is dependent on red light and specifically upon P_{fr}. Having noted early in their investigations that inhibitors of RNA (e.g., actinomycin) and protein (e.g., cycloheximide) synthesis block the effect of red light, Mohr and associates hypothesized circa 1965 that the action of P_{fr} might be to activate the one or more genes coding for one or more enzymes which catalyze anthocyanin biosynthesis. By this concept, P_{fr} would evoke the synthesis of one or more specific mRNAs and one or more specific gene products or enzymes.

Attention was focused early on one enzyme involved in anthocyanin production, namely phenylalanine ammonia lyase (PAL). PAL catalyzes the deamination of phenylalanine to form *trans*-cinnamic acid, which in turn is a precursor to the B ring of anthocyanins and other flavonoids (Fig. 8.15).

Their general procedure was to irradiate etiolated seedlings not with red light but with continuous far-red light, which maintains a low but nearly constant level of P_{fr} in the tissues over an extended period of time. It is important to understand that by this procedure, they were in fact investigating P_{fr}, even though far-red was the irradiation used. Characteristic of this system, PAL activity did increase up to approximately fourfold after the onset of far-red irradiation of etiolated seedlings, after an initial lag period of about 1.5 hours (Fig. 8.16). If, however, a seedling that had been preirradiated with 12 hours of

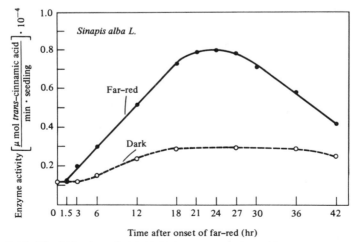

FIGURE 8.16. The induction of phenylalanine ammonia lyase (PAL) activity in the white mustard seedling by continuous standard far-red light. The primary (or initial) lag-phase of the response is 1.5 hours. (Redrawn, with permission, from Mohr et al., 1968.)

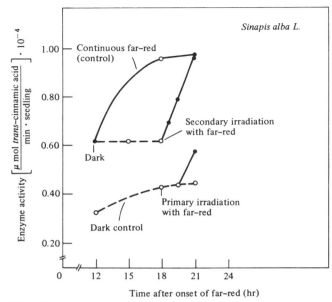

FIGURE 8.17. Initial and secondary lag-phases of far-red mediated increase of phenyla-lanine ammonia lyase (PAL) activity in the white mustard seedling. To determine a secondary lag-phase the seedlings were irradiated for 12 hours with standard far-red, placed in darkness for 6 hours, and then reirradiated with the same standard far-red. (Redrawn, with permission, from Mohr et al., 1968.)

far-red light was kept in darkness for a few hours and was reirradiated with far-red, the lag phase was eliminated. That is, PAL activity increased immediately after the onset of the second irradiation (Fig. 8.17). Mohr and the Freiburg group have interpreted their results as indicating that P_{fr} evokes *de novo* synthesis of PAL. Some evidence reported by Attridge and Johnson (1976) suggested, on the contrary, that synthesis of PAL was not involved but rather activation of pre-existing (but inactive) PAL in the tissues.

Evidence for stimulation of *de novo* synthesis of some other enzymes by P_{fr}—but at the translational, not the transcriptional level—also has been reported. The principal evidence is that red light (specifically P_{fr} thereby generated) causes conversion of monoribosomes to polyribosomes in certain tissues prior to photomorphogenetic responses, e.g., light-sensitive lettuce seeds, etiolated leaves, and pine seeds. There evidently is some sort of activating effect even on the monoribosomes, since monoribosomes isolated from corn leaves preirra-diated with red light were found to be more effective than monoribosomes from nonirradiated tissues in protein synthesis *in vitro*. In the case of etiolated bean leaves P_{fr} enhanced the conversion of monoribosomes to polyribosomes, from both cytoplasm and chloroplasts, even when mRNA synthesis was inhibited by cordycepin.

Thus as Lissemore et al. (1987) pointed out, despite much research, the mechanism by which P_{fr} generates morphogenic changes remains elusive. However, the idea that at least some of these responses are the result of phytochrome-mediated alterations in gene expression is supported by considerable direct evidence. Phytochrome has, in fact, been shown to regulate the steady-state transcript level of at least 20 different nuclear- and chloroplast-encoded genes (Lissemore et al., 1987 and references cited therein) as well as the transcription rate of the (nuclear) genes for the small subunit of ribulose bisphosphate carboxylase, chlorophyll a/b binding protein, protochlorophyllide reductase, phytochrome itself in *Avena*, and several unidentified gene products. This regulation has been shown to occur, at least partially, at the level of transcription by run-on transcription assays and by analysis in transgenic plant tissue of chimeric constructs involving upstream promoter regions of light-regulated genes.

A particularly interesting and important case is the regulation by phytochrome of the expression of its own genes, as elucidated by Peter H. Quail and his associates (Quail, 1984; Colbert et al., 1985; and Quail et al., 1986).

The intracellular concentration of phytochrome is known to be under both developmental and light control (Quail, 1984). The molecule is synthesized *de novo* in the P_r form, accumulating in dark-grown tissue until a plateau is reached. This plateau represents a steady-state balance between the synthesis and degradation of P_r. Transfer of tissue to the light results in a rapid decline in phytochrome levels, because the rate of degradation of P_{fr} is much higher than that of P_r. In plants transferred to continuous white light a plateau is reached at 1 to 3% of the phytochrome initially present, presumably representing a new steady-state balance between P_r synthesis and P_{fr} degradation. When light-treated tissue is returned to the dark, phytochrome reaccumulates in the P_r form by *de novo* synthesis.

Quail et al. focused on the negative feedback control that phytochrome exerts over its own genes (autoregulation) as an approach to understanding phytochrome-regulated gene expression generally. Pure phytochrome RNA sequence synthesized in an SP6-derived *in vitro* transcription system was used as a standard to quantify phytochrome mRNA abundance in *Avena* seedlings using a filter hybridization assay. In 4-day-old *Avena* seedlings phytochrome mRNA represents approximately 0.1% of the total poly(A)$^+$ RNA. Irradiation of such seedlings with a saturating red-light pulse or continuous white light induced a decline in this mRNA that is detectable within 30 minutes (even 15 minutes) after photoconversion and results in a 50% reduction by approximately 60 minutes and >90% reduction within 2 to 5 hours. The effect of the red-light pulse is reversed, approximately to the level of the far-red control, by an immediately subsequent far-red pulse. In seedlings maintained in extended darkness after the red-light pulse, the initial rapid decline in phytochrome mRNA level is followed by a slower reaccumulation, such that 50 to 60% of the initial abundance is reached by 48 hours. Less than 1% is sufficient to induce

60% of the maximum response, which is saturated at 20% P_{fr} or less. The data established that phytochrome-regulated changes in translatable phytochrome mRNA levels result from changes in the *physical abundance of mRNA* rather than from altered translatability (Colbert et al., 1985). Obviously, the rapidity of this autoregulatory control makes phytochrome itself an attractive system for investigating phytochrome-regulated gene expression.

Winslow R. Briggs and associates (Kaufman et al., 1985) at the Carnegie Institution of Washington at Stanford, California investigated phytochrome-regulated changes in transcript abundance for 11 different light-regulated mRNAs in developing pea (*Pisum sativum* L.) buds. Phytochrome regulation actually occurs over two fluence ranges of red light (R). The low fluence (LF) response has a threshold of approximately 10^1 μmol m^{-2} and is fully reversible by far-red light (FR). This is the common phytochrome response observed in most plants. The very low fluence (VLF) response has a threshold of approximately 10^{-3} μmol m^{-2}. It is not reversible by FR, and, in fact, is induced by most FR sources.

Fluence–response curves were measured for changes in transcript abundance in response to red light pulses in both the LF and VLF ranges. Most of the 11 transcripts showed only LF responses, with a threshold of approximately 10 μmol m^{-2}. One transcript showed a VLF response. They also used various fluences of red light as pretreatments before transferring seedlings to continuous white light. The threshold of these responses was about one order of magnitude greater than the threshold of the LF responses to red light alone.

Thus, by way of summary, it can be concluded that P_{fr} acts in some cases by evoking enzyme synthesis and in others by altering properties of membranes.

Effect on GA Biosynthesis and Efflux from Plastids

Compartmentation of GA biosynthesis in plastids and stimulation of GA biosynthesis by red light were discussed in Chapter 3. It is appropriate to consider now the extensive data showing an important interaction of the phytochrome system with the GA economy of plants.

Certain effects of phytochrome on GA metabolism in plastids have been studied intensively. Stoddart (1968) reported that extracts of chloroplast fractions of leaves of *Brassica oleracea* and *Hordeum vulgare* had biological activity resembling that of GAs. Subsequently he and others presented evidence that at least some GA biosynthesis is compartmentalized in chloroplasts and other plastids. There have been numerous reports since 1968 also indicating that light alters GA metabolism, including apparent effects on biosynthesis, interconversions, apparent release from membrane-bound forms, and phytochrome-mediated efflux from plastids. Involvement of plastids in the rapid effect of red light on GA metabolism (see Chapter 3) was based initially on the response of etioplasts *in vitro* and *in situ* to red and far-red light, suggesting the presence of phytochrome within or on the plastid envelope. Evans and Smith

(1976a,b) presented spectrophotometric evidence for the presence of phyto-chrome in the envelope membrane of barley etioplasts. Earlier, Cooke and Saunders (1975) had demonstrated a phytochrome-dependent increase in extractable GA-like activity in plastid preparations from etiolated wheat leaves, and it was reported a short time later that phytochrome and GA-like substances both were associated with etioplast envelopes (Cooke and Kendrick, 1976). Later GA_4 and GA_9 were shown to be tenaciously associated with chloroplast membranes, since the GAs were released by detergent treatment but not by sonication (Browning and Saunders, 1977).

Ellis and Hartley (1971) suggested that there must be a specific mechanism for transporting chloroplast proteins made on cytoplasmic ribosomes and other substances across the outer membrane of plastids. Some work suggests that the hypothetical transmembrane transport mechanism involves phytochrome. Many workers have reported a direct relation between irradiation with red light and the level of extractable GA-like activity or actual GAs from leaves of different species. This effect has been shown to be phytochrome dependent. Evans and Smith (1976a,b) demonstrated the presence of phytochrome within the etioplast and its relationship to the increase in extractable GA on illumination. They concluded that phytochrome would cause the transport of GA across the etioplast envelope. And, indeed, the transport of different compounds, including GAs, across plastid envelopes, has been reported by Wellburn and Hampp (1976) and the regulation of this process by phytochrome has been discussed (Hampp and Schmidt, 1977; Schmidt and Hampp, 1977). From the available data, we can conclude that some of the effects of red light (of P_{fr}) are specifically correlated with increases in free GAs available for growth regulation.

Phytochrome and Photoperiodic (Flowering) Responses

Introduction

Experiments were conducted in the 1940s on interrupting dark periods by flashes or longer exposures to light to ascertain effects on the flowering of photoperiodi-cally sensitive species. This "night break" or "night interruption" phenomenon is manifested as a nullification of the promotive effect of an otherwise inductive dark period for certain plants (the "short-day plants") and nullification of the inhibitory effect of the dark period for other plants (the "long-day plants"). By the early 1950s—soon after the action spectra for lettuce seed germination were worked out—it was observed that red light is most effective in the night interruption phenomenon, and that the effect of red can be nullified by immediate subsequent irradiation with far-red light. Indeed, it was the nature of the light sensitive seed germination response and of the night interruption phenomenon on which the theoretical existence of a pigment system such as phytochrome first was postulated. Promptly on the definitive discovery of phytochrome in 1959, it could be confidently concluded that phytochrome is, in fact, the photoreceptor for all photoperiodic responses, of which a very large number had been described.

In the remainder of this chapter attention will be directed to the role of phytochrome in photoperiodism, particularly reproductive photoperiodism, in angiosperms, and the interaction of phytochrome with endogenous circadian rhythms in photoperiodic timing. Discussion of evidence for flowering hormones and the effects of some of the known kinds of hormones on flowering will comprise the closing sections.

Photoperiodism

Photoperiodism is a response to the relative lengths of day and night, or more specifically, a response to the duration and timing of light and dark periods. This phenomenon is an integral part of the overall adaptation of many kinds of organisms—plant and animal—to their natural environments. Among the kinds of organisms that exhibit photoperiodism are at least some species among nonvascular and vascular cryptogams, gymnosperms, angiosperms, insects, fishes, reptiles, birds, and mammals.

In all but equatorial areas, of course, daylength changes at a constant rate with the change of the seasons. Photoperiod is therefore the "least variable variable in nature," as has often been noted. When this fact is appreciated, it does not seem at all surprising that photoperiodism has been of such prominent significance in the evolution of both plant and animal kingdoms.

Many processes and developmental events in both the vegetative and flowering stages of the life cycles of angiosperms are either controlled or greatly influenced by the duration and timing of daily light and dark periods. Among the many known examples of aspects of vegetative development that are affected by daylength, in at least some species, are (1) onset and breaking of bud dormancy in woody perennials, (2) leaf abscission in deciduous trees and shrubs, (3) bulb and tuber formation in herbaceous species, (4) seed germination, (5) cessation of cambial activity, and (6) development of frost resistance. Such purely vegetative responses frequently are termed "vegetative photoperiodism." More familiar to most students of biology, perhaps, is the effect of daylength on the flowering behavior of many species of angiosperms. It was primarily this phenomenon of "reproductive photoperiodism" in angiosperms that was described in the classical work of W. W. Garner and H. A. Allard in 1920. Garner and Allard discovered that angiosperms could be classified in three main groups, according to the ways in which flowering is affected by daylength. These groups are the short-day plants, long-day plants, and day-neutral plants. The short-day plants were said to be those that flower in response to daylengths shorter than some maximum. Long-day plants were identified as those that flower in response to daylengths longer than some critical duration. Day-neutral plants were regarded as those in which flowering is not specifically affected by daylength.

Actually, and as might have been expected, the diversity of response types is much greater than is reflected in the simple classification initially devised by Garner and Allard. For, example, a few species flower only on intermediate daylength, that is, when the days are neither too short nor too long. Other species are known that flower in response to a period of short days followed by a period of long days (short–long-day plants), and a few others flower in response to long days followed by short days (long–short-day plants). Winter annuals (e.g., winter varieties of wheat and rye) and biennials (e.g., carrot, cabbage, biennial henbane) require vernalization followed by long days to flower. "Vernalization" is an induction or acceleration of flowering by low temperature. Winter annuals normally are vernalized as seedlings, and biennials after the first season of growth.

Complicating the situation further, there are, in the categories of both short-day plants and long-day plants, species and varieties that exhibit more-or-less obligatory or qualitative responses and others that exhibit only facultative or quantitative responses. An obligate short-day plant typically has quite exacting requirements for photoperiodic floral induction and will not flower in their absence, whereas a facultative short-day plant does not exhibit an all-or-none response but flowers more readily or more vigorously if it experiences relatively short daylength.

In the preceding discussion emphasis has been placed on daylength. This is somewhat misleading, however, in that photoperiodism is, as stated earlier, a response to the duration and timing of light and dark periods. The single most important controlling factor is the length of the uninterrupted dark period. For all photoperiodic responses occurring in natural 24-hour light–dark cycles, it is possible to speak of both "critical daylength" and "critical nightlength." Critical daylength, in the case of a long-day plant, is the minimum daylength required for flowering; in the case of a short-day plant, it is the maximum daylength on which flowering can occur. The critical nightlength for a short-day plant is the minimum period of darkness required for flowering; for a long-day plant it is the maximum period of darkness on which flowering can occur. Thus for short-day plants the daylength must be shorter than the critical daylength and the dark period longer than the critical nightlength if flowering is to occur. For long-day plants, on the contrary, the daylength must be longer than the critical daylength and the dark period shorter than the critical nightlength for flowering to occur. The distinction between short-day plants and long-day plants has nothing to do with the absolute values of the critical daylength; the distinction is whether flowering is promoted by daylengths shorter or longer than the critical daylength. For example, the critical daylength for *Xanthium strumarium*, an absolute short-day plant, is about 15.5 hours, and for *Hyoscyamus niger*, an obligate long-day plant, about 11 hours. Both species flower well in response to 12- to 14-hour photoperiods in 24-hour light–dark cycles.

A general summary description of the requirements for photoperiodic induction of flowering in obligate short-day plants and long-day plants can be stated as follows:

Short-day plants require exposure to a certain minimum number of photoinductive cycles (which must be consecutive for maximum effectiveness) in each of which an uninterrupted dark period equals or exceeds a critical duration, after attaining "ripeness to flower."

Long-day plants require exposure to a certain minimum number of cycles (not necessarily consecutive) in each of which the photoperiod equals or exceeds some critical duration, after attaining "ripeness to flower;" they do not require a dark period.

Examples of obligate short-day plants are cocklebur (*Xanthium strumarium*), Japanese morning glory (*Pharbitis nil*), kalanchoë (*Kalanchoë blossfeldiana*), poinsettia (*Euphorbia pulcherrima*), Maryland Mammoth tobacco (*Nicotiana tabacum*), and the common weed *Chenopodium rubrum*. Among known obligate long-day plants are black henbane (*Hyoscyamus niger*), Darnel ryegrass (*Lolium temulentum*), sedum (*Sedum spectabile*), spinach (*Spinacia oleracea*), and some varieties of barley (*Hordeum vulgare*). Tomato (*Lycopersicum esculentum*) and corn (*Zea mays*) are familiar day-neutral plants.

The age or amount of development that a plant must attain before it is "ripe to flower" varies markedly with species. Some plants, such as *Pharbitis nil* and *Chenopodium rubrum*, can be induced to flower in the seedling stage. Others, such as *Xanthium strumarium* and *Hyoscyamus niger*, must be a few weeks of age, and some trees must be several years old before they can be induced to flower.

Species also differ markedly with respect to the number of photoinductive cycles required for flowering. Some very sensitive species, such as *Xanthium strumarium, Lolium temulentum*, and *Pharbitis nil*, will respond to a single photoinductive cycle. Others require several cycles and, for maximum effectiveness, in the case of short-day plants, the cycles must be consecutive. In those species that respond to a single photoinductive cycle, as in other species, the response (rate and vigor) increases as the number of cycles increases, up to some maximum number.

Previously reference was made to the night interruption phenomenon, which is manifested as a nullification of the promoting effect of an otherwise inductive dark period on the flowering of short-day plants and nullification of the inhibitory effect of a long dark period for long-day plants. Consider, for examples, certain hypothetical photoperiodic treatments of *Xanthium strumarium* (an obligate short-day plant with a critical daylength of 15.5 hours and critical nightlength, therefore, of 8.5 hours) and annual *Hyoscyamus niger* (an obligate long-day plant with a critical daylength of about 11 hours) (Table 8.3). A photoperiodic cycle consisting of a 16-hour day and an 8-hour night would not stimulate flowering in *X. strumarium* (because the dark period is too short) but would stimulate *H. niger* (because the critical daylength is equaled or exceeded). Conversely, a cycle consisting of an 8-hour day and a 16-hour night would stimulate flowering of *X. strumarium* (because the dark period exceeds the critical night) but would not evoke flowering of *H. niger* (because the dark period is too long). A photoperiodic cycle made up of a 12-hour day and a 12-hour night would stimulate flowering in both, chiefly because, for

TABLE 8.3. Expected flowering response of *Xanthium strumarium* (SDP) and *Hyoscyamus niger* (LDP) to various photoperiodic treatments.

	Expected flowering response	
Photoperiodic treatment	*X. strumarium*	*H. niger*
(1) 16-hour day, 8-hour night	−	+
(2) 8-hour day, 16-hour night	+	−
(3) 12-hour day, 12-hour night	+	+
(4) 10-hour day, 14-hour night— interrupted by light	−	+

X. strumarium, the dark period is longer than the critical nightlength of 8.5 hours, and, for *H. niger*, the photoperiod equals or exceeds the critical daylength. A 10-hour photoperiod and a 14-hour night, interrupted at the seventh hour with a flash of high-intensity white light, would not stimulate *X. strumarium* (because the interrupted dark period would be too brief) but would stimulate *H. niger* (because the interruption of the dark would nullify the otherwise inhibitory effect of a 14-hour night).

Determination of the action spectrum for the night interruption phenomenon would reveal that red light (in the region of 660 nm wavelength) is most effective, and that the effect of red light can be counteracted with immediate subsequent irradiation with far-red light (around 730 nm wavelength). All the criteria essential to prove that phytochrome is the photoreceptor for the night interruption phenomenon would be fully met.

As the transformation reactions for phytochrome were discovered, there quickly developed the hypothesis that "pigment shift" represents the basic timing mechanism in photoperiodic timing. "Pigment shift" is the process whereby, during darkness, all the phytochrome comes to be present as P_r. Component reactions of "pigment shift" in all plants are destruction of P_{fr} and synthesis of new phytocrome as P_r. In some dicots at least, a third reaction is reversion of P_{fr} to P_r in darkness. By this hypothesis the critical nightlength would simply be the time required for all the phytochrome to come to be present as P_r in darkness. However, it turned out that this hypothesis is inadequate by itself. In the few cases where "pigment shift" has been measured the process appears to be complete in only 2–4 hours or less. Moreover, it has become clear that endogenous circadian rhythms are very much involved in photoperiodic timing.

Endogenous Circadian Rhythms

A fundamentally important question regarding photoperiodism is: How do plants measure time in darkness? We have seen that the process of "pigment shift" involving phytochrome cannot, by itself, account for measurement of dark periods. What is involved evidently is a complex interaction of the phytochrome system with endogenous circadian rhythms. Endogenous rhythms are characteristic of all living things except perhaps for bacteria. An endogenous rhythm was discussed in Chapter 1 in relation to bud dormancy in a tropical tree. In that

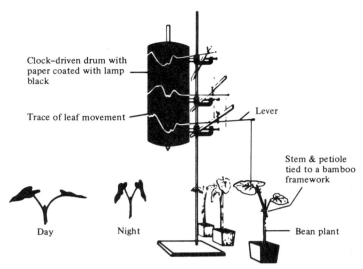

Clock–driven drum with
paper coated with lamp
black

Trace of leaf movement

Lever

Stem & petiole
tied to a bamboo
framework

Day   Night

Bean plant

FIGURE 8.18. Circadian movements of a leaf of a bean plant *(Phaseolus vitellinus)* show a rising in the day and a falling at night. (From *Plant Physiology,* 3rd ed., by Frank B. Salisbury and Cleon W. Ross. © 1985 by Wadsworth Publishing Company, Inc., Belmont, California 94002. Reprinted by permission of the publisher.)

example the rhythm had a period of approximately 26 days. It is to be remembered that there are many endogenous rhythms known for plants, and there is variation among all the known rhythms with respect to period length. However, the most familiar type of rhythms, are the endogenous "circadian" rhythms or "biological clocks" that have a period of approximately 24 hours ("circa" = about; "diem" = day) (Fig 8.18; cf. Figs. 1.11 and 1.12).

Under normal circumstances, of course, plants grow in an environment in which a number of parameters (e.g., light intensity and temperature) fluctuate rhythmically with a period of 24 hours. It is not surprising, therefore, that plants have adapted their physiological activity to cycles of approximately 24 hours. Rhythmic behavior in plants has been known for centuries, yet the mechanisms underlying endogenous rhythms, circadian and otherwise, remain largely a mystery. This is extremely challenging, because of the very great biological significance of "biological clocks." Presently we understand only some of the more salient features of endogenous circadian rhythms such as (1) the natural free-running period is not precisely 24 hours but usually is somewhere between 21 and 28 hours; (2) the rhythms are not merely the consequences of fluctuations in physical environmental conditions, although it often is possible to entrain circadian rhythms to periods of other than about 24 hours by manipulation of physical factors; (3) however, even an entrained system normally returns to its natural value immediately if the organism is returned to the natural environment; and (4) endogenous circadian rhythms are only slightly affected by temperature over a wide range.

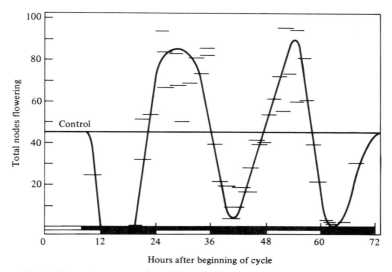

FIGURE 8.19. Flowering response of Biloxi soybean (a short-day plant) to 4-hour light interruptions given at various times during a 72-hour cycle. Lines indicate the 4-hour interruption periods. The control plants were exposed to seven cycles, each cycle consisting of an 8-hour photoperiod followed by a 64-hour dark period. For the various treatments the 64-hour dark period was interrupted at various times by illumination with high-intensity light of 4 hours duration. Flowering of the controls with no light interruptions for 72-hour-long cycles is represented by the horizontal line across the graph. The upper bar on the bottom of the graph indicates the 8 hours of light and the 64 hours of darkness given in each of the seven cycles to which the plants were exposed. The lower bar indicates the daily cycle of light and darkness under normal conditions. (Redrawn, with permission, from: Hamner, K. 1963. Endogenous rhythms in controlled environments. In: L. T. Evans, ed. *Environmental Control of Plant Growth*. Pp.215–232. Copyright by Academic Press, Inc., New York.)

Experiments conducted with photoperiodic flowering responses have indicated rather vividly the role of endogenous circadian rhythms in photoperiodic reactions, as Bünning suggested is the case in the early 1930s. A classical experiment with the Biloxi variety of soybean (*Glycine max*), a SDP, (Fig 8.19) is illustrative of this very crucial point. K. C. Hamner exposed Biloxi soybeans to seven cycles of 8 hours of light followed by 64 hours of darkness and interrupted the long dark period with light at various times. The plants had been previously grown ("entrained") to 12-hour photoperiods alternating with 12-hour dark periods. The striking result was that the night interruption phenomenon (manifested as inhibition of flowering) showed three peaks, approximately 24 hours apart and corresponding to the times when plants normally would have been in darkness. Obviously, from all that is known about the process, "pigment shift" was complete during the early hours of the 64-hour dark period. Moreover, we can confidently assume that night interruption at any time caused a certain amount of P_r to be converted to P_{fr}. Yet the consequence of P_{fr} presence depended clearly on when interruption of the dark period occurred.

Or, to put it another way, the susceptibility of the flowering process, specifically the dark processes leading to flowering, to inhibition by P_{fr} varied with the timing of the interruption of the dark period.

In summary, it appears that endogenous circadian rhythms play the basic and fundamental role in timing of light and dark periods in angiosperms generally. In photoperiodic responses, the phytochrome system seems to be superimposed as a modifying influence on the basic timing mechanism. The nature of the control of timing in photoperiodic phenomena seems to reside in a complex interaction between the state of the phytochrome and the endogenous circadian rhythm.

One important final point remains to be clarified regarding the role of phytochrome in the flowering of photoperiodically sensitive species. One might be led to conclude, erroneously, that P_{fr} is only inhibitory to flowering of short-day plants but that it is essential to flowering of long-day plants. As a matter of fact, the evidence is clear that P_{fr} is essential to the flowering of both. The difference evidently is only a quantitative one. Short-day plants depend on P_{fr} action occurring during a photoperiod that must not be too long (longer than the critical daylength), and they must be free of the presence of P_{fr} during a certain period of darkness. Long-day plants, in contrast, either have a greater need for P_{fr} action or are less sensitive to P_{fr} action, since they flower best when P_{fr} is continually present (i.e., in constant light).

Hormonal Regulation of Flowering

Within less than two decades after the discovery of photoperiodism, the hypothesis developed that one or more specific flowering hormones are responsible for floral initiation. Chailakhyan (1936) coined the word "florigen" for the hypothetical flowering hormone. A short time later, Melchers (1939) suggested the term "vernalin" for the hypothetical stimulus thought to develop during vernalization of cold-requiring plants. Many years later Chailakhyan proposed the term "anthesins" for hypothetical flowering hormones (Chailakhyan, 1970) and gave emphasis to the role GAs play in the flowering of many plants.

Early experiments, conducted in the 1930s and 1940s, with classical obligate short-day plants such as cocklebur (*Xanthium strumarium*), which responds to a single photoinductive cycle (Fig. 8.20), yielded results that provided a very logical basis for the florigen (anthesin) concept. It was observed that, in the case of cocklebur but not all short-day plants, induction of a single leaf was sufficient to cause the plant to flower systemically (Fig. 8.21). Furthermore, a single photoinduced leaf or whole plant could be grafted onto another plant kept on noninductive long days and the latter would flower (Fig. 8.22). These and countless related experiments involving a large number of photoperiodically sensitive species suggested strongly that the leaf is the organ that perceives the photoperiodic stimulus, that phytochrome is the photoreceptor, and that a substance or substances is formed in the leaves of short-day plants that is translocated to vegetative meristems and causes their morphogenetic transforma-

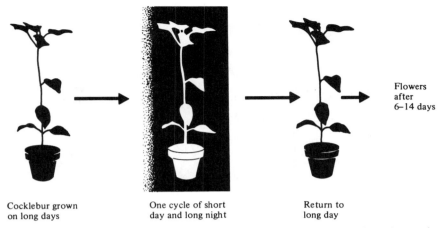

Cocklebur grown
on long days

One cycle of short
day and long night

Return to
long day

Flowers
after
6–14 days

FIGURE 8.20. Photoperiodic induction in the cocklebur by a single short-day cycle.
(Redrawn, with permission from Bonner and Galston, 1952.)

tion into floral meristems. Furthermore, from numerous grafting experiments, it
appeared that there was only one "florigen," or if two or more substances, they
were physiologically equivalent among many species. The various photoperiodic
response types would differ not in their requirement for florigen but in the
environmental requirements for florigen production. For example, "ripe-to-

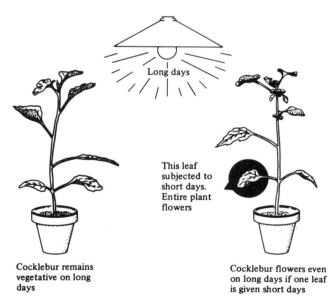

Long days

This leaf
subjected to
short days.
Entire plant
flowers

Cocklebur remains
vegetative on long
days

Cocklebur flowers even
on long days if one leaf
is given short days

FIGURE 8.21. Photoinduction of a single leaf causes cocklebur to flower systemically.
(Redrawn, with permission, from Bonner and Galston, 1952.)

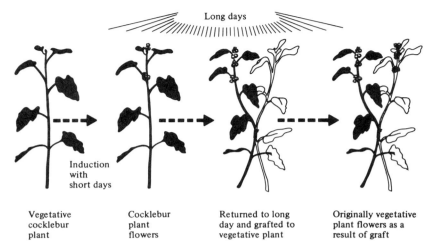

| Vegetative cocklebur plant | Cocklebur plant flowers | Returned to long day and grafted to vegetative plant | Originally vegetative plant flowers as a result of graft |

FIGURE 8.22. The flowering stimulus can be transmitted from plant to plant across a graft union. (Redrawn, with permission, from Bonner and Galston, 1952.)

flower'' vegetative scions of *Sedum spectabile* (a long-day plant) could be grafted onto stocks of *Kalanchoë blossfeldiana* (a short-day plant) and the grafted plants maintained under short-day conditions and the *Sedum spectabile* would flower (Fig. 8.23). The reciprocal graft under long days worked equally well. A logical explanation of these results is that a florigen was formed in the leaves of the stock under photoperiodic conditions inductive for it that passed the graft union and caused flowering of the scions under noninductive conditions. By 1963, approximately 25 such graft unions had proved successful. The 25 cases involved species from long-day, short-day and day-neutral types.

With such evidence compatible with a florigen concept, a logical next step would be to attempt to isolate and chemically characterize the flowering stimulus or stimuli. In fact, there have been many such efforts. Generally the attempts have ended in failure; occasionally there have been modest successes. One group of investigations, which looked particularly promising, was conducted by Richard G. Lincoln and associates at Long Beach State College (1961, 1966) and by K. C. Hamner and associates (1970) at the University of California at Los Angeles. In this work, extracts were prepared from the leaves and buds of flowering cocklebur (*Xanthium strumarium*) that stimulated flowering in the short-day plant *Lemna perpusilla* under noninductive conditions and had slight flower-inducing activity also toward vegetative cocklebur plants maintained under noninductive conditions if supplemented with GA (Tables 8.4 and 8.5). ''Florigenic acid'' is the name given to the yet unidentified active substance or substances in the extracts. Active extracts have been prepared from day-neutral sunflower plants and from a fungus (*Calonectria*) but not from vegetative cocklebur plants.

Thus, as Frank B. Salisbury and Cleon W. Ross (1985) have noted, ''There is much circumstantial evidence that flower initiation is controlled by hormones:

FIGURE 8.23. Transmission of flowering stimulus from the short-day plant *Kalanchoë blossfeldiana* (lower part) to the long-day plant *Sedum spectabile* (upper part) under short-day conditions. The axillary shoots of *Kalanchoë* flowered under the influence of short days. The *Sedum* scion was induced to flower also under short-day conditions by the *Kalanchoë* stock. The photograph was taken 97 days after grafting (By permission, from Zeevaart, 1958; see also Zeevaart, 1962.)

one or more positively acting florigens and one or more negatively acting inhibitors. These substances remain to be identified.'' That certain of the known hormones are involved in regulation of flowering is definite as indicated by numerous investigations conducted with GAs since circa 1957 and with ABA since circa 1967.

TABLE 8.4. Effects of an extract from *Xanthium strumarium* on flowering of cocklebur.[a]

Treatment, extract or chemical	Flowering stage				
	0	1	2–5	6–10	>10
Flowering plants	91	9	0	0	0
Flowering plants + GA	11	13	44	21	11
Vegetative plants	97	3	0	0	0
Vegetative plants + GA	84	16	0	0	0
Gibberellic acid	88	12	0	0	0
Water	100	0	0	0	0

[a] Extract activity evaluated by the *Xanthium* biossay. One hundred plants were used in each of the six treatments. Ten plants were used to evaluate each of the 20 extracts from flowering or vegetative plants. The figures represent the total number of plants in each stage of flowering. Plants in stage 2 or above are considered to be flowering. GA, gibberellic acid. (With permission from Hodson, H. K. and K. C. Hamner. 1970. Floral inducing extract from *Xanthium. Science* **167**: 384–385, 23 January 1970. Copyright 1970 by the American Association for the Advancement of Science.)

TABLE 8.5. Effects of an extract from *Xanthium strumarium* on flowering of *Lemna perpusilla*.[a]

Treatment	Plants flowering (%)									
Flowering plants	45	31	54	37	58	46	45	51	39	60
Flowering plants + GA	0	0	0	0	0	0	0	0	0	0
Vegetative plants	0	0	9	0	7	0	0	0	2	0
Vegetative plants + GA	0	0	0	0	0	0	0	0	0	0
Gibberellic acid	0	0	0	0	0	0	0	0	0	0
Water	0	0	5	0	11	3	0	0	1	5

[a] Extract activity evaluated by the *Lemna* bioassay. A total of 30,000 plants were used to evaluate the activity of the 20 extracts. The extracts were tested with and without the addition of gibberellic acid (GA). Each figure represents the average response of five lots of 100 plants each. (With permission from Hodson, H. K. and K. C. Hamner, 1970. Floral inducing extract from *Xanthium*. *Science* **167**: 384–385, 23 January 1970. Copyright 1970 by the American Association for the Advancement of Science.)

Anton Lang first reported in 1957 that exogenous GA could cause flowering in numerous species of long-day and vernalization-requiring plants under noninductive environmental conditions (Fig. 8.24), but GA is not the long sought after florigen. One reason is that GAs do not cause flowering of short-day plants under noninductive conditions, or even in all long-day plants. Moreover, there is now good evidence that flowering and flower-bearing stem elongation (bolting) are separate processes in plants such as *Silene armeria* (Cleland and Zeevaart, 1970), with GA promoting stem elongation only. The latter conclusion was based on the findings that (1) treatment of *Silene armeria* with Amo-1618, an inhibitor of GA biosysthesis, inhibited stem elongation but not flowering under long-day conditions; and (2) with GA treatment on short days, stem elongation sometimes occurred in the absence of flowering. Thus it seems clear that although flowering and stem elongation are closely related in rosette-type long-day plants under normal conditions, the two processes can be experimentally separated and thus represent separate developmental processes. While GA is not "florigen," it conceivably might be "vernalin."

Jan A. D. Zeevaart of the MSU-DOE Plant Research Laboratory at East Lansing, Michigan and his associates have elucidated in remarkable detail the GA metabolism that evidently is causally involved in the photoperiodic control of stem growth in spinach (*Spinacia oleracea*), a long-day rosette plant. Very early in their research it was determined that photoperiodic control of stem growth in spinach, as in other long-day, rosette plants like *Silene armeria* and *Agrostemma githago*, is mediated by GA (Zeevaart, 1971). They identified six endogenous C-13 hydroxylated GAs in spinach shoots: GA_{17}, GA_{19}, GA_{20}, GA_{29}, GA_{44}, and GA_{53}. Five of these (except GA_{17}) were shown to be metabolically related as illustrated in Fig. 8.25 (Metzger and Zeevaart, 1980a; Gilmour et al., 1986, 1987). In spinach, long-day treatment of plants did not increase the total GA level, but levels of individual GAs did vary (Fig. 8.26), indicating a profound change in GA metabolism. Thus GA_{19} showed a 5-fold decline, while the levels of GA_{20} and GA_{29} increased dramatically during the same period. The levels of GA_{17} and GA_{44} did not change significantly with long-day treatment. GA_{53} occurred in quantities too small to be measured.

FIGURE 8.24. Carrot (*Daucus carota*) plants showing the application of GA as a substitute for a cold requirement for flowering. Treatments: Left, no GA, no cold treatment. Center, 10 μg GA per day, no cold treatment. Right, no GA but 8 weeks cold treatment. The plants were grown in long days. (From Lang, 1957, with permission; photo courtesy of A. Lang, 1978.)

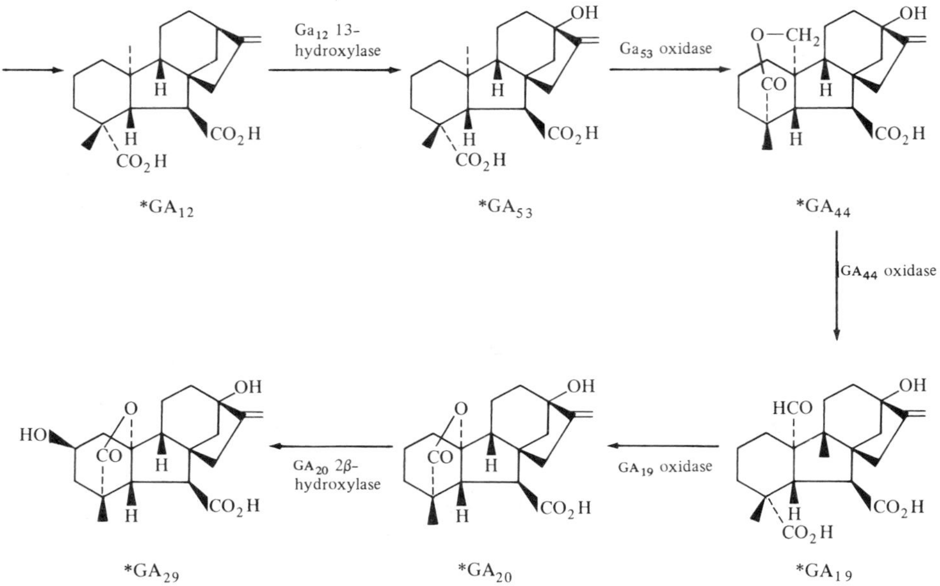

FIGURE 8.25. GA conversions in cell-free systems from spinach leaves grown in long days, except for the step $GA_{20} \rightarrow GA_{29}$ that was not observed. GAs marked with an asterisk are endogenous to spinach. GA_{12} has not been shown to be endogenous to spinach. (Redrawn in slightly modified form, with permission, from Gilmour et al., 1987).

Gilmour et al. (1986) investigated further the metabolic relations among the endogenous GAs by using cell-free enzyme extracts prepared from spinach leaves and by identifying the products by gas chromatography-mass spectrometry. The pathway shown in Fig. 8.25 was found to operate in the cell-free enzyme extracts, except that the conversion of GA_{20} to GA_{29} was not observed. Extracts from plants given long days (LD) or short days (SD) were examined, and enzymic activities were measured as a function of exposure to LD, as well as to darkness following 8 LD. The results indicate that the activities of the enzymes oxidizing GA_{53} and GA_{19} are increased in LD and decreased in SD or darkness, but that the enzyme activity oxidizing GA_{44} remained high irrespective of light or dark treatment. This photoperiodic control of enzyme activity was shown not to be due to the presence of an inhibitor in plants grown in SD. These observations provided an explanation for the higher GA_{20} content of spinach plants in LD than in SD. It is thought to be the increased endogenous GA_{20} that brings about stem growth in LD (Metzger and Zeevaart, 1980b). More recently Gilmour et al. (1987) reported on the extraction and partial purification of four enzyme activities catalyzing the following oxidative steps in the GA pathway in spinach: $GA_{12} \rightarrow GA_{53} \rightarrow GA_{44} \rightarrow GA_{19} \rightarrow GA_{20}$.

During the early investigations of the effects of exogenous ABA on plants, it

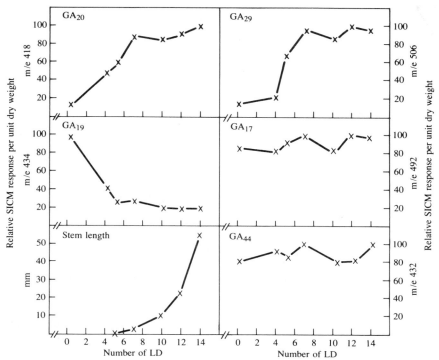

FIGURE 8.26. Changes in the relative levels of five GAs in spinach as measured by GLC-SICM, and stem length as affected by different durations of LD treatment. The highest concentration (SICM response/unit dry weight) of each GA was arbitrarily assigned a value of 100 and the other concentrations were expressed in proportion to this value. The SICM response of the molecular ion was used in all calculations except for GA_{19}, in which case the base peak was used. (Redrawn in slightly modified form, with permission, from Metzger and Zeevaart, 1980b.)

was reported that this growth-inhibiting hormone could cause flowering of certain short-day plants (e.g., *Chenopodium rubrum, Pharbitis nil,* and some others) under long-day conditions. But ABA is not a florigen because there is no effect of exogenous ABA on flowering of some short-day plants or on long-day plants. It may be one of the postulated flowering inhibitors, which have long been thought to be produced in the leaves of long-day plants subjected to noninductive short-day conditions.

Another interesting report of the effect of a growth regulator on flowering was published by Charles F. Cleland in 1974. He showed that the naturally occurring regulator salicylic acid could cause the long-day plant *Lemna gibba* L. strain G3 (a duckweed) to flower under short-day conditions. Salicylic acid was effective on both a marginal photoperiodic regimen of 10 L:14 D and under a strict short-day regimen of 9 L:15 D. The main effect of the salicylic acid was on flower initiation, although it also had a slight stimulating effect on flower development. Analyzing the flower-inducing effect further, Cleland and Tanaka (1979) investigated the effect of daylength on the capacity of salicylic acid to induce flowering in *Lemna gibba* G3. They noted that when the plant is grown on control medium the critical daylength is just under 10 hours. With 3.2 μM salicylic acid added to the medium, substantial flower promotion was obtained on 9-, 10-, and 11-hour daylengths. On an 8-hour daylength salicyclic acid caused only a very small flowering response, and with daylengths less than 8 hours flowering was never obtained. Thus, salicyclic acid treatment caused a shift in the critical daylength of about 2 hours, from just less than 10 hours to just under 8 hours.

Salicylic acid was also observed to cause flower promotion in the short-day plant *Lemna paucicostata* Helgelm., strain 6746. The shift in critical daylength for this plant was only about 1 hour, and was extended rather than shortened. Thus, in control medium the critical daylength for this short-day plant was about 14 hours. When the plants were given 3.2 μM salicylic acid, substantial flower promotion was obtained on daylengths of 13 and 14 hours, and the critical daylength was close to 15 hours.

Recently there have been many investigations of flower bud formation in thin-layer explants of day-neutral and photoperiodically sensitive species, e.g., tobacco (*Nicotiana tabacum*) and other *Nicotiana* species. Thin-layer explants, usually consisting of epidermis and underlying parenchyma, are taken generally from pedicels, but sometimes from inflorescence branches or the calyx, and are cultured on a basal agar medium containing glucose and one or more hormones. Both auxin and cytokinin are required for flower bud formation. Buds typically develop between 8 and 14 days of culture. Under certain conditions, as many as 20 to 30 buds can be formed on one explant, with the number being strongly dependent on the cytokinin concentration. Such experimental systems should prove useful in helping to elucidate the hormonal regulation of flowering and floral development. For example, one promising line of investigation is to attempt to isolate and study specific genes and gene products associated with the flowering process. The thin-layer explant systems afford exciting opportunities for studying developmental morphology at the molecular level.

Calcium Messenger System in Plants

Obviously, plants receive many diverse signals from hormones and physical environmental factors throughout their life cycles. How the cells sense these signals and go on to manifest responses has long been a subject of great interest to plant physiologists. The biochemical basis for the transduction of extracellular signals into intracellular events has been studied extensively in animals. In animal systems a signal received on the surface of an individual cell is transmitted into the metabolic machinery of that cell by two major signal pathways: (1) one employing cyclic adenosine monophosphate (cAMP) as a "second messenger;" and (2) another employing a combination of "second messengers" that include Ca^{2+} ions, inositol 1,4,5-trisphosphate, and diacyl glycerol.

In higher plants it appears that only one of these signal pathways is operative. While cAMP evidently occurs in higher plants, there is no evidence that it plays a role as a "second messenger" in them. However, recent investigations, since circa 1979, from many laboratories indicate that the other signal pathway, involving calcium, is operative in higher plants. In stimulus–response coupling involving various physiological processes elicited by extracellular signals such as light, gravity, and hormones, calcium has been implicated. The further discovery of various components of the Ca^{2+} messenger system of animal systems—such as calmodulin and calmodulin-dependent enzymes—thus has led plant physiologists to regard Ca^{2+} as a "second messenger" in plants (Poovaiah and Reddy, 1987).

If Ca^{2+} does act as a "second messenger" or mediator in translating a stimulus into a response, it must meet the following criteria: (1) cytoplasmic levels of Ca^{2+} must change upon impingement of the signal, and this change must precede the physiological response; (2) it must be possible to evoke the physiological response by inducing a change in the cytoplasmic Ca^{2+} in the absence of the external signal; further, blocking the change in concentration of cytoplasmic Ca^{2+} in the presence of the stimulus must prevent the response; (3) cells must have a mechanism for sensing the changes in concentration of cytosolic Ca^{2+} and translating them into the physiological response; and (4) blocking the operation of the Ca^{2+}-sensing system must block the physiological response to the stimulus.

There are many physiological processes in plants in which Ca^{2+} evidently acts as a "second messenger" and that meet some or all of the above criteria (Poovaiah and Reddy, 1987). A few of the best known cases in higher plants will be described briefly.

One such response is the regulation of gravitropism by light in the hybrid corn (*Zea mays*) variety Merit. Although most roots show a positive gravitropic response when kept in total darkness, roots of the Merit corn become sensitive to gravity only after exposure of the root cap to white (specifically red) light. This red light induction of positive gravitropism (specifically orthogravitropism) is a phytochrome-mediated response that saturates at very low fluences. In the absence of red light treatment the roots show diagravitropic curvature (at right angles to the line of gravity).

Root tips contain high amounts of Ca^{2+}. A Carl Leopold and associates (Perdue et al., 1988), using Merit corn, have shown that (1) calcium chelating agents inhibit the light response; (2) calcium channel blockers (verapamil, lanthanum) also inhibit the light response; and (3) a calcium ionophore (A 23187) can substitute for light. Treatments like cold shock or heat that trigger Ca^{2+} influx into the cytosol can also substitute for red light. Agents like serotonin, 2,4-D, or deoxycholate that are antagonists of the phosphatidylinositol second messenger system can each partially substitute for the red light, and Li^+ can inhibit the light effect. Calmodulin antagonists inhibit the gravitropic response. The experiments by Perdue et al. suggested to them that the induction of positive gravitropism by red light involves a rise in cytoplasmic Ca^{2+} concentration. There are data from other investigations that show that calcium movement within the root tip is associated with corn root gravitropism. The variety Golden Cross Bantam X70 that, like Merit, requires red light for positive gravitropism, also requires red light for the characteristic movement of Ca^{2+} across the root tip after gravistimulation. Thus there is considerable evidence for a potential role of Ca^{2+} as a "second messenger" in the photocontrol of gravitropism. Interestingly, Ca^{2+} appears to be involved in stimulus–response coupling in some other phytochrome-mediated phenomena (Hepler and Wayne, 1985; Poovaiah and Reddy, 1987).

Research during the last decade has indicated that Ca^{2+} has significant modifying effects on the functions of each of the major kinds of plant hormones, in some cases amplifying the hormonal response and in other cases suppressing it (Poovaiah and Reddy, 1987). For example, an involvement of Ca^{2+} in auxin action has been recognized for many years. Auxin-induced elongation in pea epicotyl sections is blocked by treating the segments with Ca^{2+} chelators (EGTA and CTC) and channel blockers (La^{3+} and verapamil). Auxin delays abscission in bean explants, and this effect can be prevented by EGTA or La^{3+}. In those cases it is believed that auxin lowers cytosolic Ca^{2+}.

Drugs that inhibit calmodulin-dependent processes in plants have been reported to inhibit the cytokinin-induced synthesis of betacyanin in *Amaranthus tricolor* seedlings and cytokinin-promoted growth of soybean callus tissue. Examples of such calmodulin-binding drugs are trifluoperazine, a phenothiazine tranquilizer, and tetracaine, a local anesthetic (Elliott, 1983). They also inhibit auxin-promoted growth of wheat coleoptile segments and the induction by GA of α-amylase in barley aleurone layers (Elliott et al., 1983).

Russell L. Jones and associates at the University of California at Berkeley have reported extensively on the regulation of synthesis of α-amylase in barley aleurone cells by GA and Ca^{2+}. The presence of Ca^{2+} in the incubation medium is required for maximum production of α-amylase. Most studies on the mechanism of induction of α-amylase synthesis and secretion by GA in the cereal aleurone have included Ca^{2+} in the incubation medium. They have shown that Ca^{2+} is specifically required for the synthesis and secretion of certain isoenzymes of α-amylase. Recently they have shown that Ca^{2+} acts as a "second messenger" in this system. There is also an interaction of Ca^{2+} with cytokinin in such responses as leaf senescence. Cytokinin is ineffective in delaying senescence if excised leaf discs of *Xanthium* are depleted of Ca^{2+}. The cytokinin effect is restored by the addition of Ca^{2+} (Poovaiah and Reddy, 1987). Finally, a synergistic interaction between Ca^{2+} and ABA in maintaining stomatal closure has been reported by Hetherington et al. (1986). Their data suggest that ABA increases the permeability of guard cells to Ca^{2+}, and that the Ca^{2+} acts as a "second messenger" in the hormonal regulation of ion fluxes that determine guard cell turgor.

In most cases where Ca^{2+} acts as the "second messenger" of a stimulus, the chemical signal that initiates the response is not free Ca^{2+} itself but the complex between Ca^{2+} and any one of a number of proteins of a class called Ca^{2+}-binding proteins. Calmodulin is the best known and more generally distributed Ca^{2+}-binding protein, which mediates Ca^{2+} messages in most eukaryotic cells, and has been reported to occur in plants since circa 1978. It has been characterized extensively (Poovaiah and Reddy, 1987). The Ca^{2+}-calmodulin complex can act basically in two ways: (1) directly on an effector system; or (2) on a regulatory system, usually a protein kinase that through phosphorylation promotes or inhibits the activity of other enzymes. These two modes of action permit both fast and slow responses to be mediated by Ca^{2+} and calmodulin. In plants enzymes that have been shown to be regulated by Ca^{2+} and calmodulin include NAD kinase, Ca^{2+}-transporting ATPase, H^+-transporting ATPase, quinate:NAD^+ oxidoreductase, and protein kinases and others (Poovaiah and Reddy, 1987). Proteins undergo many types of posttranslational modifications. Protein phosphorylation, catalyzed by protein kinases, is one such process and is one of the major regulatory mechanisms by which cellular metabolic activities are controlled. Phosphorylation of proteins alters their activity and properties.

Although it is widely accepted that external signals such as light, gravity, and hormones alter cytosolic calcium, it is not clear how these changes occur. In animal cells, turnover of phosphoinositide plays a key role in this process. In these systems, external signals stimulate the hydrolysis of phosphatidylinositol 4,5-bisphosphate, resulting in the production of inositol 1,4,5-trisphosphate and diacylglycerol. Inositol 1,4,5-trisphosphate increases cytosolic calcium by releasing calcium from nonmitochrondrial stores in the cell. Whether such a pathway exists in plants currently is being investigated, but some positive evidence has already been reported (Poovaiah and Reddy, 1987).

Therefore, as B. W. Poovaiah at Washington State University and others have emphasized, it is an attractive working hypothesis that Ca^{2+} acts as a "second messenger" coupling external stimuli to physiological responses in plants. In responsive tissues, a given signal may cause an increase in the concentration of cytoplasmic Ca^{2+}, perhaps via an intervention of inositol trisphosphate. This increased concentration could then activate Ca^{2+}-calmodulin-dependent enzymes, which causes changes in the phosphorylation status of proteins in the cell. Additionally, diacylglycerol produced by the hydrolysis of inositol phospholipid turnover could activate protein kinase C leading to changes in protein phosphorylation. Such changes in protein phosphorylation then could modulate various biochemical processes resulting in a physiological response to a given signal (Poovaiah, 1988, personal communication). As A. Carl Leopold recently suggested, the involvement of common "second messengers" could help account for the extensive overlap in regulatory functions of the various plant hormones.

Thus, as the final section of the last chapter of this discussion of plant hormones ends, it is clear that much remains to be discovered about the hormonal regulation of flowering as well as many other aspects of hormonal regulation of plant growth and development. Undoubtedly, fascinating new discoveries lie ahead.

References

Attridge T. H. and C. B. Johnson. 1976. Photocontrol of enyzme levels. In: H. Smith, ed. *Light and Plant Development*. Butterworth & Company, Limited, London. Pp. 185–192.

Batschauer, A. and K. Apel. 1984. An inverse control by phytochrome of the expression of two nuclear genes in barley (*Hordeum vulgare* L.). *Eur. J. Biochem.* **132**: 593–597.

Ben-Tal, Y. and C. F. Cleland. 1982. Uptake and metabolism of [^{14}C] salicylic acid in *Lemna gibba* G3. *Plant Physiol.* **70**: 291–296.

Bevington, J. M. and M. C. Hoyle. 1981. Phytochrome action during prechilling induced germination of *Betula papyrifera* Marsh. *Plant Physiol.* **67**: 705–710.

Biro, R. L., S. Daye, B. S. Serlin, M. E. Terry, N. Datta, S. K. Sopory, and S. J. Roux. 1984. Characterization of oat calmodulin and radioimmunoassay of its subcellular distribution. *Plant Physiol.* **75**: 382–386.

Boeshore, M. L. and L. H. Pratt. 1981. Characterization of a molecular modification of phytochrome that is associated with its conversion to the far-red-absorbing form. *Plant Physiol.* **68**: 789–797.

Bolton, G. W. and P. H. Quail. 1982. Cell-free synthesis of phytochrome apoprotein. *Planta* **155**: 212–217.

Bonner. J. and A. W. Galston. 1952 *Principles of Plant Physiology*. W. H. Freeman and Company, San Francisco.

Borthwick, H. 1972. History of phytochrome. In: K. Mitrakos and W. Shropshire, Jr., eds. *Phytochrome*. Academic Press. New York. Pp. 3–23.

Brain, R. D., J. A. Freeberg, C. V. Weiss, and W. R. Briggs. 1977. Blue light-induced absorbance changes in membrane fractions from corn and *Neurospora. Plant Physiol.* **59**: 948–952.

Briggs, W. R. 1976. The nature of the blue light photoreceptor in higher plants and fungi. In: H. Smith, ed. *Light and Plant Development*. Butterworth & Company, Limited, London. Pp. 7–18.

Briggs, W. R. and M. Iino. 1983. Blue-light-absorbing photoreceptors in plants. *Phil. Trans. R. Soc. London B Biol. Sci.* **303**: 347–359.

Briggs, W. R., E. Mösinger, and E. Schäfer. 1988. Phytochrome regulation of greening in barley—effects on chlorophyll accumulation. *Plant Physiol.* **86**: 435–440.

Briggs, W. R. and H. V. Rice. 1972. Phytochrome: chemical and physical properties and mechanism of action. *Annu. Rev. Plant Physiol.* **23**: 293–334.

Briggs, W. R. and H. W. Siegelman. 1965. Distribution of phytochrome in etiolated seedlings. *Plant Physiol.* **40**: 934–941.

Browning, G. and P. F. Saunders. 1977. Membrane localised gibberellins A_9 and A_4 in wheat chloroplasts. *Nature (London)* **265**: 375–377.

Bünning, E. 1973. *The Physiological Clock*. 3rd ed. Academic Press, New York.

Bünning, E. 1977. Fifty years of research in the wake of Wilheim Pfeffer. *Annu. Rev. Plant Physiol.* **28**: 1–22.

Butler, W. L., K. H. Norris, H. W. Seigelman, and S. B. Hendricks. 1959. Detection, assay, and preliminary purification of the pigment controlling photoresponsive development of plants. *Proc. Natl. Acad. Sci. U.S.A.* **45**: 1703–1708.

Campbell, B. R. and B. A. Bonner. 1986. Evidence for phytochrome regulation of gibberellin A_{20} 3β-hydroxylation in shoots of dwarf (le le) *Pisum sativum* L. *Plant Physiol.* **82**: 909–915.

Chailakhyan, M. Kh. 1968. Internal factors of plant flowering. *Annu. Rev. Plant Physiol.* **19**: 1–36.

Chailakhyan, M. Kh. 1970. Flowering and photoperiodism of plants. *Plant Sci. Bull.* **3**(3): 1–7.

Chailakhyan, M. Kh. 1975. Forty years of research on the hormonal basis of plant development—some personal reflections. *Bot. Rev.* **41**: 1–29.

Clarkson, D. T. and W. S. Hillman. 1967. Apparent phytochrome synthesis in *Pisum* tissue. *Nature (London)* **213**: 468–470.

Clarkson, D. T. and W. S. Hillman. 1968. Stable concentrations of phytochrome in *Pisum* under continuous illumination with red light. *Plant Physiol.* **43**: 88–92.

Cleland, C. F. 1974. The influence of salicylic acid on flowering and growth in the long-day plant *Lemna gibba* G3. In: Bieleski, R. L., A. R. Ferguson, and M. M. Cresswell, eds. *Mechanisms of Regulation of Plant Growth*. Bulletin 12, Royal Society of New Zealand, Wellington. Pp. 553–557.

Cleland, C. F. and A. Ajami. 1974. Identification of the flower-inducing factor isolated from honeydew as being salicylic acid. *Plant Physiol.* **54**: 904–906.

Cleland, C. F. and O. Tanaka. 1979. Effect of day length on the ability of salicylic acid to induce flowering in the long-day plant *Lemna gibba* G3 and the short-day plant *Lemna paucicostata* 6746. *Plant Physiol.* **64**: 421–424.

Cleland, C. F. and J. A. D. Zeevaart. 1970. Gibberellins in relation to flowering and stem elongation in the long day plant *Silene armeria*. *Plant Physiol.* **46**: 392–400.

Colbert, J. T., H. P. Hershey, and P. H. Quail. 1983. Autoregulatory control of translatable phyochrome mRNA levels. *Proc. Natl. Acad. Sci. U.S.A.* **80**: 2248–2252.

Colbert, J. T., H. P. Hershey, and P. H. Quail. 1985. Phytochrome regulation of phytochrome mRNA abundance. *Plant Mol. Biol.* **5**: 91–101.

Cooke, R. J. and R. E. Kendrick. 1976. Phytochrome controlled gibberellin metabolism in etioplast envelopes. *Planta* **131**: 303–307.

Cooke, R. J. and P. F. Saunders. 1975. Phytochrome mediated changes in extractable gibberellin activity in a cell-free system from etiolated wheat leaves. *Planta* **123**: 299–302.

Cordonnier, M.-M., H. Greppin, and L. H. Pratt. 1984. Characterization by enzyme-linked immunosorbent assay of monoclonal antibodies to *Pisum* and *Avena* phytochrome. *Plant Physiol.* **74**: 123–127.

Cordonnier, M.-M., H. Greppin, and L. H. Pratt. 1986a. Monoclonal antibodies with differing affinities to the red-absorbing and far-red absorbing forms of phytochrome. *Biochemistry* **24**: 3246–3253.

Cordonnier, M.-M., H. Greppin, and L. H. Pratt. 1986b. Identification of a highly conserved domain on phytochrome from angiosperms to algae. *Plant Physiol.* **80**: 982–987.

Cordonnier, M.-M. and L. H. Pratt. 1982a. Immunopurification and initial characterization of dicotyledonous phytochrome. *Plant Physiol.* **69**: 360–365.

Cordonnier, M.-M. and L. H. Pratt. 1982b. Comparative phytochrome immunochemistry as assayed by antisera against both monocotyledonous and dicotyledonous phytochrome. *Plant Physiol.* **70**: 912–916.

Cosgrove, D. J. 1981. Rapid suppression of growth by blue light. Occurrence, time course, and general characteristics. *Plant Physiol.* **67**: 584–590.

Cosgrove, D. J. and P. B. Green. 1981. Rapid suppression of growth by blue light. Biophysical mechansim of action. *Plant Physiol.* **68**: 1447–1453.

Cumming, B. G. and E. Wagner. 1968. Rhythmic processes in plants. *Annu. Rev. Plant Physiol.* **19**: 381–416.

Curry, G. M. 1969. Phototropism. In: Wilkins, M. B., ed. *The Physiology of Plant Growth and Development.* McGraw-Hill Publishing Company (UK) Limited, Berkshire. Pp. 245–273.

Daniels, S. M. and P. H. Quail. 1984. Monoclonal antibodies to three separate domains on 124-kilodalton phyochrome from *Avena*. *Plant Physiol.* **76**: 622–626.

Downs, R. J. and H. Hellmers. 1975. *Environment and the Experimental Control of Plant Growth.* Academic Press, New York.

Downs, R. J. and J. F. Thomas. 1982. Phytochrome regulation of flowering in the long-day plant *Hyoscyamus niger*. *Plant Physiol.* **70**: 898–900.

Elich, T. D. and J. C. Lagarias. 1987. Phytochrome chromophore biosynthesis. Both 5-aminolevulinic acid and biliverdin overcome inhibition by gabaculine in etiolated *Avena sativa* L. seedlings. *Plant Physiol.* **84**: 304–310.

Elliott, D. C. 1983. Inhibition of cytokinin-regulated responses by calmodulin-binding compounds. *Plant Physiol.* **72**: 215–218.

Elliot, D. C., S. M. Batchelor, R. A. Cassar, and N. G. Marinos. 1983. Calmodulin-binding drugs affect responses to cytokinin, auxin, and gibberellic acid. *Plant Physiol.* **72**: 219–224.

Ellis, R. J. and M. R. Hartley. 1971. Sites of synthesis of chloroplast proteins. *Nature (London) New Biol.* **233**: 193–196.

Evans, A. and H. Smith. 1976a. Spectrophotometric evidence for the presence of phytochrome in the envelope membranes of barley etioplasts. *Nature (London)* **259**: 323–325.

Evans, A. and H. Smith. 1976b. Localization of phytochrome in etioplasts and its regulation *in vitro* of gibberellin levels. *Proc. Natl. Acad. Sci. U.S.A.* **73**: 138–142.

Evans, L. T., 1975. *Daylength and the Flowering of Plants.* W. A. Benjamin, Inc., Menlo Park, California.

Feldman, L. J. and W. R. Briggs. 1987. Light-regulated gravitropism in seedling roots of maize. *Plant Physiol.* **83**: 241–243.

Flint, L. H. and E. D. McAlister. 1935. Wave lengths of radiation in the visible spectrum inhibiting the germination of light-sensitive lettuce seed. *Smithsonian Inst. Misc. Coll.* **94**(5): 1–11.

Flint, L. H. and E. D. McAlister. 1937. Wavelengths of radiation in the visible spectrum promoting the germination of light-sensitive lettuce seed. *Smithsonian Inst. Misc. Coll.* **96**: 1–8.

Fondeville, J. C., H. A. Borthwick, and S. B. Hendricks. 1966. Leaflet movement of *Mimosa pudica* L. indicative of phytochrome action. *Planta* **69**: 357–364.

Galston, A. W. 1974. Plant photobiology in the last half-century. *Plant Physiol.* **54**: 427–436.

Galston, A. W. and R. L. Satter. 1976. Light, clocks, and ion flux: an analysis of leaf movement. In: H. Smith, ed. *Light and Plant Development.* Butterworth & Company, Limited, London. Pp. 159–184.

Gardner, G. and H. L. Gorton. 1985. Inhibition of phytochrome synthesis by gabaculine. *Plant Physiol.* **77**: 540–543.

Gardner, G., H. L. Gorton, and S. A. Brown. 1988. Inhibition of phytochrome synthesis by the transaminase inhibitor, 4-amino-5-fluoropentanoic acid. *Plant Physiol.* **87**: 8–10.

Garner, W. W. and H. A. Allard. 1920. Effect of the relative length of day and night and other factors of the environment on growth and reproduction in plants. *J. Agr. Res.* **18**: 553–606.

Garner, W. W. and H. A. Allard. 1923. Further studies in photoperiodism, the response of the plant to the relative length of day and night. *J. Agr. Res.* **23**: 871–920.

Gilmour, S. J., A. B. Bleecker, and J. A. D. Zeevaart. 1987. Partial purification of gibberellin oxidases from spinach leaves. *Plant Physiol.* **85**: 87–90.

Gilmour, S. J., J. A. D. Zeevaart, L. Schwenen, and J. E. Graebe. 1986. Gibberellin metabolism in cell-free extracts from spinach leaves in relation to photoperiod. *Plant Physiol.* **82**: 190–195.

Hamner, K. 1963. Endogenous rhythms in controlled environments. In: L. T. Evans, ed. *Environmental Control of Plant Growth.* Academic Press, New York. Pp. 215–232.

Hampp, R. and H. W. Schmidt. 1977. Regulation of membrane properties of mitochondria and plastids during chloroplast development. I. The action of phytochrome in situ. *Z. Pflanzenphysiol.* **82**: 68–77.

Hartman, K. M. 1967. Ein Wirkungsspektrum der Photomorphogenese unter Hochenergiebedingungen un seine Interpretation auf der Basis des Phytochroms (Hypokotylwachstums hemmung bei *Lactuca sativa* L.). *Z. Naturforschg.* **22b**: 1172–1175.

Haupt, W. 1963. Photoreceptorprobleme der Chloroplastenbewegung. *Ber. Deutsch. Bot. Gesellschaft* **76**: 313–322.

Haupt, W. 1973. Role of light in chloroplast movement. *BioScience* **23**: 289–296.

Hepler, P. K. and R. O. Wayne. 1985. Calcium and plant development. *Annu. Rev. Plant Physiol.* **36**: 397–439.

Hershey, H. P., R. F. Barker, K. B. Idler, M. G. Murray, and P. H. Quail. 1987. Nucleotide sequence and characterization of a gene encoding the phytochrome polypetide from *Avena. Gene* **61**: 339–348.

Hetherington, A. M., D. L. R. DeSilva, R. C. Cox, and T. A. Mansfield. 1986. Abscisic acid, calcium ions and stomatal function. In: Trewavas, A. J., ed. *Molecular and Cellular Aspects of Calcium in Plant Development.* Plenum Press, New York.

Hillman, W. S. 1962. *The Physiology of Flowering.* Holt, Rinehart & Winston, New York.

Hillman, W. S. 1976. Biological rhythms and physiological timing. *Annu. Rev. Plant Physiol.* **27**: 159–179.

Hilton, J. R. and B. Thomas. 1985. A comparison of seed and seedling phytochrome in *Avena sativa* L. using monoclonal antibodies. *J. Exp. Bot.* **36**: 1937–1946.

Hodson, H. K. and K. C. Hammer. 1970. Floral inducing extract from *Xanthium. Science* **167**: 384–385.

Hunt, R. E. and L. H. Pratt. 1979a. Phytochrome radioimmunoassay. *Plant Physiol.* **64**: 327–331.

Hunt, R. E. and L. H. Pratt. 1979b. Phytochrome immunoaffinity purification. *Plant Physiol.* **64**: 332–336.

Hunt, R. E. and L. H. Pratt. 1980a. Partial characterization of undegraded oat phytochrome. *Biochemistry* **19**: 390–394.

Hunt, R. E. and L. H. Pratt. 1980b. Radioimmunoassay of phytochrome content in green, light-grown oats. *Plant Cell Environ.* **3**: 91–95.

Jaffe, M. J. 1968. Phytochrome-mediated bioelectric potentials in mung bean seedlings. *Science* **162**: 1016–1017.

Jaffe, M. J. and A. W. Galston. 1967. Phytochrome control of rapid nyctinastic movements and membrane permeability in *Albizzia julibrissin. Planta* **77**: 135–141.

Jones, A. M., C. D. Allen, G. Gardner, and P. H. Quail. 1986. Synthesis of phytochrome apoprotein and chromophore are not coupled obligatorily. *Plant Physiol.* **81**: 1014–1016.

Jones, A. M. and P. H. Quail. 1986. Quaternary structure of 124-kilodalton phytochrome from *Avena sativa* L. *Biochemistry* **25**: 2987–2995.

Jones, A. M., R. D. Vierstra, S. M. Daniels, and P. H. Quail. 1985. The role of separate molecular domains in the structure of phytochrome from etiolated *Avena sativa* L. *Planta* **164**: 501–506.

Kaufman, L. S., W. R. Briggs, and W. F. Thompson. 1985. Phytochrome control of specific mRNA levels in developing pea buds. The presence of both very low fluence and low fluence responses. *Plant Physiol.* **78**: 388–393.

Kaufman, L. S., L. L. Roberts, W. R. Briggs, and W. F. Thompson. 1986. Phytochrome control of specific mRNA levels in developing pea buds. Kinetics of acccumulation, reciprocity, and escape kinetics of the low fluence reponse. *Plant Physiol.* **81**: 1033–1038.

King, R. W., R. P. Pharis, and L. N. Mander. 1987. Gibberellins in relation to growth and flowering in *Pharbitis nil* Chois. *Plant Physiol.* **84**: 1126–1131.

Lagarias, J. C. 1985. Progress in the molecular analysis of phytochrome. *Photochem. Photobiol.* **42**: 811–820.

Lagarias, J. C. and H. Rappaport. 1980. Chromopeptides from phytochrome. The structure and linkage of the Pr form of the phytochrome chromophore. *J. Am. Chem. Soc.* **102**: 4821–4828.

Lang, A. 1957. The effect of gibberellin upon flower formation. *Proc. Natl. Acad. Sci. U.S.A.* **43**: 709–717.

Leong, T.-Y. and W. R. Briggs. 1981. Partial purification and characterization of a blue light-sensitive cytochrome-flavin complex from corn membranes. *Plant Physiol.* **67**: 1042–1046.

Lhoste, J. M. 1972. Some physical aspects of the phytochrome phototransformation. In: K. Mitrakos and W. Shropshire, Jr., eds. *Phytochrome.* Academic Press, New York. Pp. 47–74.

Lincoln, R. G., A. Cunningham, B. H. Carpenter, J. Alexander, and D. L. Mayfield. 1966. Florigenic acid from fungal culture. *Plant Physiol.* **41**: 1079–1080.

Lincoln, R. G., D. L. Mayfield, and A. Cunningham. 1961. Preparation of a floral initiating extract from Xanthium. *Science* **133**: 756.

Lissemore, J. L., J. T. Colbert, and P. H. Quail. 1987. Cloning of cDNA for phytochrome from etiolated *Cucurbita* and coordinate photoregulation of the abundance of two distinct phytochrome transcripts. *Plant Mol. Biol.* **8**: 485–496.

Mancinelli, A. L. 1986. Comparison of spectral properties of phytochrome from different preparations. *Plant Physiol.* **82**: 956–961.

Mancinelli, A. L. 1988. Phytochrome photoconversion *in vivo. Plant Physiol.* **86**: 749–753.

Mandoli, D. F. and W. R. Briggs. 1981. Phytochrome control of two low-irradiance responses in etiolated oat seedlings. *Plant Physiol.* **67**: 733–739.

Marmé, D. 1977. Phytochrome: membranes as possible sites of primary action. *Annu. Rev. Plant Physiol.* **27**: 173–198.

Metzger, J. D. and J. A. D. Zeevaart. 1980a. Identification of six endogenous gibberellins in spinach shoots. *Plant Physiol.* **65**: 623–626.

Metzger, J. D. and J. A. D. Zeevaart. 1980b. Effect of photoperiod on the levels of endogenous gibberellins in spinach as measured by combined gas chromatography-selected ion current monitoring. *Plant Physiol.* **66**: 844–846.

Metzger, J. D. and J. A. D. Zeevaart. 1982. Photoperiodic control of gibberellin metabolism in spinach. *Plant Physiol.* **69**: 287–291.

Mitrakos, K. and W. Shropshire. Jr., eds. 1972. *Phytochrome.* Academic Press. New York.

Mohr, H., C. Huault, H. Lange, L. Lohmann, I. Rissland, and M. Weidner. 1968. Lag-phases in phytochrome-mediated enzyme synthesis (PAL). *Planta* **83**: 267–275.

Morgan, P. W. and J. R. Quinby. 1987. Genetic regulation of development of *Sorghum bicolor*. IV. GA$_3$ hastens floral differentiation but not floral development under nonfavorable photoperiods. *Plant Physiol.* **85**: 615–620.

Mösinger, E., A. Batschauer, K. Apel, E. Schäfer, and W. R. Briggs. 1988. Phytochrome regulation of greening in barley. Effects on mRNA abundance and on transcriptional activity of isolated nuclei. *Plant Physiol.* **86**: 706–710.

Parks, B. M., A. M. Jones, P. Adamse, M. Koornneef, R. E. Kendrick, and P. H. Quail. 1987. The *aurea* mutant of tomato is deficient in spectrophotometrically and immunochemically detectable phytochrome. *Plant Mol. Biol.* **9**: 97–107.

Perdue, D. O., A. K. LaFavre, and A. C. Leopold. 1988. Calcium in the regulation of gravitropism by light. *Plant Physiol.* **86**: 1276–1280.

Pharis, R. P., L. T. Evans, R. W. King, and L. N. Mander. 1987. Gibberellins, endogenous and applied, in relation to flower induction in the long-day plant *Lolium temulentum*. *Plant Physiol.* **84**: 1132–1138.

Pharis, R. P. and C. G. Kuo. 1977. Physiology of gibberellins in conifers. *Can. J. Forest Res.* **7**: 299–325.

Poovaiah, B. W. and A. S. N. Reddy. 1987. Calcium messenger system in plants. *CRC Crit. Rev. Plant Sci.* **6**: 47–103.

Pratt, L. H. 1976. Re-examination of photochemcial properties and absorption characteristics of phytochrome using high-molecular-weight preparations. In: H. Smith, ed. *Light and Plant Development.* Butterworth & Company, Limited, London. Pp. 19–30.

Pratt, L. H. 1979. Phytochrome: Functions and properties. *Photochem. Photobiol. Rev.* **4**: 59–124.

Pratt, L. H. 1982. Phytochrome: The protein moiety. *Annu. Rev. Plant Physiol.* **33**: 557–582.

Pratt, L. H. and R. A. Coleman. 1974. Phytochrome distribution in etiolated grass seedlings as assayed by an indirect antibody-labelling method. *Am. J. Bot.* **61**: 195–202.

Pratt, L. H., J. E. Wampler, and E. S. Rich, Jr. 1984–85. An automated dual-wavelength spectrophotometer optimized for phytochrome assay. *Anal. Instrument.* **13**: 269–287.

Quail, P. H. 1976. Phytochrome. In: Bonner, J. and J. E. Varner, eds. *Plant Biochemistry.* 3rd ed. Academic Press, New York. Pp. 683–711.

Quail, P. H. 1980. Phytochrome: the first five minutes from P$_{fr}$ formation. In: DeGreef, J., ed. *Photoreceptors and Plant Development.* Proceedings of the Annual

European Symposium on Plant Photomorphogenesis held at Antwerpen, Belgium, July 22–28, 1979. Antwerpen University Press, Antwerpen.

Quail, P. H. 1984. Phytochrome: A regulatory photoreceptor that controls the expression of its own gene. *Trends Biochem. Sci.* **9**: 450–453.

Quail, P. H., J. T. Colbert, H. P. Hershey, and R. D. Vierstra. 1983. Phytochrome: Molecular properties and biogenesis. *Phil. Trans. R. Soc. London B* **303**: 387–402.

Quail, P. H., J. T. Colbert, N. K. Peters, A. H. Christensen, R. A. Sharrock, and J. L. Lissemore. 1986. Phytochrome and the regulation of the expression of its genes. *Phil. Trans. R. Soc. London B* **314**: 469–480.

Quail, P. H., E. Schäfer, and D. Marmé. 1973. *De novo* synthesis of phytochrome in pumpkin hooks. *Plant Physiol.* **52**: 124–127.

Raghothama, K. G., Y. Mizrahi, and B. W. Poovaiah. 1985. Effect of calmodulin antagonists on auxin-induced elongation. *Plant Physiol.* **79**: 28–33.

Rüdiger, W. 1972. Chemistry of phytochrome chromophore. In: Mitrakos, K. amd W. Shropshire, Jr., eds. *Phytochrome*. Academic Press, New York. Pp. 129–141.

Salisbury, F. B. 1963. *The Flowering Process*. Pergamon Press, New York.

Salisbury, F. B. 1965. The initiation of flowering. *Endeavour* **24**: 74–80.

Salisbury, F. B. 1971. *The Biology of Flowering*. Natural History Press, Garden City, New York.

Salisbury, F. B. and C. W. Ross. 1985. *Plant Physiology*. 3rd ed. Wadsworth Publishing Company, Belmont, California.

Satter, R. L. and A. W. Galston. 1976. The physiological functions of phytochrome. In: L. W. Goodwin, ed. *Chemistry and Biochemistry of Plant Pigments*. Vol. 1. Academic Press, New York. Pp. 680–735.

Satter, R. L., S. E. Guggino, T. A. Lonergan, and A. W. Galston. 1981. The effects of blue and far red light on rhythmic leaflet movements in *Samanea* and *Albizzia*. *Plant Physiol.* **67**: 965–968.

Satter, R. L., P. Marinoff, and A. W. Galston. 1972. Phytochrome-controlled nyctinasty in *Albizzia julibrissin*. IV. Auxin effects on leaflet movement and K flux. *Plant Physiol.* **50**: 235–241.

Schaer, J. A., D. F. Mandoli, and W. R. Briggs. 1983. Phytochrome-mediated cellular photomorphogensis. *Plant Physiol.* **72**: 706–712.

Schmidt, H. W. and R. Hampp. 1977. Regulation of membrane properties of mitochondria and plastids during chloroplast development. II. The action of phytochrome in a cell-free system. *Z. Pflanzenphysiol.* **82**: 428–434.

Schopfer, P. 1977. Phytochrome control of enzymes. *Annu. Rev. Plant Physiol.* **28**: 223–252.

Sharrock, R. A., J. L. Lissemore, and P. H. Quail. 1986. Nucleotide and amino acid sequence of a *Cucurbita* phytochrome cDNA clone: Identification of conserved features by comparison with *Avena* phytochrome. *Gene* **47**: 287–295.

Shimazaki, Y., M.-M. Cordonnier, and L. H. Pratt. 1986. Identification with monoclonal antibodies of a second antigenic domain on *Avena* phytochrome that changes upon its photoconversion. *Plant Physiol.* **82**: 109–113.

Shinkle, J. R. 1986. Photobiology of phytochrome-mediated growth responses in sections of stem tissue from etiolated oats and corn. *Plant Physiol.* **81**: 533–537.

Sieglman, H. W. and W. L. Butler. 1965. Properties of phytochrome. *Annu. Rev. Plant Physiol.* **16**: 383–392.

Siegelman, H. W. and E. M. Firer. 1964. Purification of phytochrome from oat seedlings. *Biochemistry* **3**: 418–423.

Silverthorne, J. and E. M. Tobin. 1984. Demonstrator of transcriptional regulation of specific genes by phytochrome action. *Proc. Natl. Acad. Sci. U.S.A.* **81**: 1112–1116.

Smith, H. ed. 1976. *Light and Plant Development*. Butterworth & Company, Limited, London.

Sponga, F., G. F. Deitzer, and A. L. Mancinelli. 1986. Cryptochrome, phytochrome, and the photoregulation of anthocyanin production under blue light. *Plant Physiol.* **82**: 952–955.

Stiekema, W. J., C. F. Wimpee, J. Silverthorne, and E. M. Tobin. 1983. Phytochrome control of the expression of two nuclear genes encoding chloroplast proteins in *Lemna gibba* L. G-3. *Plant Physiol.* **72**: 717–724.

Stoddart, J. L. 1968. The association of gibberellin-like activity with the chloroplast fraction of leaf homogenates. *Planta* **81**: 106–112.

Stoddart, J. L. 1969. Incorporation of kaurenoic acid into gibberellins by chloroplast preparations of *Brassica oleracea*. *Phytochemistry* **8**: 831–837.

Suzuki, Y., Y. Soejima, and T. Matsui. 1980. Influence of afterripening on phytochrome control of seed germination in two varieties of lettuce (*Lactuca sativa* L.). *Plant Physiol.* **66**: 1200–1201.

Sweeney, B. M. 1969. *Rhythmic Phenomena in Plants*. Academic Press, New York.

Tanada, T. 1968. A rapid photoreversible response of barley root tips in the presence of 3-indoleacetic acid. *Proc. Natl. Acad. Sci. U.S.A.* **59**: 376–379.

Tanada, T. 1982. Effect of far-red and green irradiation on the nyctinastic closure of *Albizzia* leaflets. *Plant Physiol.* **70**: 901–904.

Thien, W. and P. Schopfer. 1982. Control by phytochrome of cytoplasmic precursor rRNA synthesis in the cotyledons of mustard seedlings. *Plant Physiol.* **69**: 1156–1160.

Tobin, E. M. and J. Silverthorne. 1985. Light regulation of gene expression in higher plants. *Annu. Rev. Plant Physiol.* **36**: 569–593.

Tokuhisa, J. G., S. M. Daniels, and P. H. Quail. 1985. Phytochrome in green tissue: Spectral and immunochemical evidence for two distinct molecular species of phytochrome in light-grown *Avena sativa* L. *Planta* **64**: 321–332.

Tokuhisa, J. G. and P. H. Quail. 1987. The levels of two distinct species of phytochrome are regulated differently during germination in *Avena sativa* L. *Planta* **172**: 371–377.

Travis, R. L., J. L. Key, and C. W. Ross. 1974. Activation of 80S maize ribosomes by red light treatment of dark-gown seedlings. *Plant Physiol.* **53**: 28–31.

Veluthambi, K. and B. W. Poovaiah. 1984. Calcium- and calmodulin-regulated phosphorylation of soluble and membrane proteins from corn coleoptiles. *Plant Physiol.* **76**: 359–365.

Verbelen, J.-P., L. H. Pratt, W. L. Butler, and K. Tokuyasu. 1982. Localization of phytochrome in oats by electron microscropy. *Plant Physiol.* **70**: 867–871.

Vierstra, R. D., M.-M. Cordonnier, L. H. Pratt, and P. H. Quail. 1984. Native phytochrome: Immunoblot analysis of relative molecular mass and in-vitro proteolytic degradation for several plant species. *Planta* **160**: 521–528.

Vierstra, R. D. and P. H. Quail. 1982a. Proteolysis alters the spectral properties of 124-kilodalton phytochrome from *Avena*. *Planta* **156**: 158–165.

Vierstra, R. D. and P. H. Quail. 1982b. Purification and initial characterization of 124-kilodalton phytochrome from *Avena*. *Biochemistry* **22**: 2498–2505.

Vierstra, R. D. and P. H. Quail. 1983a. Purification and initial characterization of 124–

kilodalton phytochrome from *Avena. Biochemistry* **22**: 2498–2505.

Vierstra, R. D. and P. H. Quail. 1983b. Photochemistry of 124-kilodalton *Avena* phytochrome *in vitro. Plant Physiol.* **72**: 264–267.

Vierstra, R. D. and P. H. Quail. 1985. Spectral characterization and proteolytic mapping of native 120–kilodalton phytochrome from *Cucurbita pepo* L. *Plant Physiol.* **77**: 990–998.

Vince-Prue, D. 1975. *Photoperiodism in Plants.* McGraw-Hill Book Company, New York.

Wayne, R. and P. K. Hepler. 1984. The role of calcium ions in phytochrome-mediated germination of spores of *Onoclea sensibilis* L. *Planta* **160**: 12–20.

Wellburn, A. R. and R. Hampp. 1976. Fluxes of gibberellic and abscisic acids, together with that of adenosine 3',5'-cycle phosphate across plastid envelopes during development. *Planta* **131**: 95–96.

Wong, Y.-S., H.-C. Cheng, D. A. Walsh, and J. C. Lagarias. 1986. Phosphorylation of *Avena* phytochrome in vitro as a probe of light-induced conformational changes. *J. Biol. Chem.* **261**: 12089–12097.

Zeevaart, J. A. D. 1958. Flower formation as studied by grafting. *Mededel. Landbouwhogeschool Wageningen* **58**: 1–88.

Zeevaart, J. A. D. 1962. Physiology of flowering. *Science* **137**: 723–731.

Zeevaart, J. A. D. 1971. Effects of photoperiod on growth rate and endogenous gibberellins in the long-day rosette plant spinach. *Plant Physiol.* **47**: 821–827.

Zeevaart, J. A. D. 1976. Physiology of flower formation. *Annu. Rev. Plant Physiol.* **27**: 321–348.

Zhu, Y. S., S. D. Kung, and L. Bogorad. 1985. Phytochrome control of levels of mRNA complementary to plastid and nuclear genes of maize. *Plant Physiol.* **79**: 371–376.

Index